THE MOLECULAR BASIS OF CELLULAR DEFENCE MECHANISMS

The Ciba Foundation is an international scientific and educational charity (Registered Charity No. 313574). It was established in 1947 by the Swiss chemical and pharmaceutical company of CIBA Limited — now Ciba-Geigy Limited. The Foundation operates independently in London under English trust law.

The Ciba Foundation exists to promote international cooperation in biological, medical and chemical research. It organizes about eight international multidisciplinary symposia each year on topics that seem ready for discussion by a small group of research workers. The papers and discussions are published in the Ciba Foundation symposium series. The Foundation also holds many shorter meetings (not published), organized by the Foundation itself or by outside scientific organizations. The staff always welcome suggestions for future meetings.

The Foundation's house at 41 Portland Place, London W1N 4BN, provides facilities for meetings of all kinds. Its Media Resource Service supplies information to journalists on all scientific and technological topics. The library, open five days a week to any graduate in science or medicine, also provides information on scientific meetings throughout the world and answers general enquiries on biomedical and chemical subjects. Scientists from any part of the world may stay in the house during working visits to London.

Ciba Foundation Symposium 204

THE MOLECULAR BASIS OF CELLULAR DEFENCE MECHANISMS

1997

JOHN WILEY & SONS

Chichester · Weinheim · New York · Brisbane · Toronto · Singapore

National 01243 779777
International (+44) 1243 779777
e-mail (for orders and customer service enquiries): cs-books@wiley.co.uk
Visit our Home Page on http://www.wiley.co.uk
or http://www.wiley.com

Other Wiley Editorial Offices

John Wiley & Sons, Inc., 605 Third Avenue,
New York, NY 10158-0012, USA

VCH Verlagsgesellschaft mbH Pappalallee 3,
D-69469 Weinheim, Germany

Jacaranda Wiley Ltd, 33 Park Road, Milton,
Queensland 4064, Australia

John Wiley & Sons (Canada) Ltd, 22 Worcester Road,
Rexdale, Ontario M9W 1L1, Canada

John Wiley & Sons (Asia) Pte Ltd, 2 Clementi Loop #02-01,
Jin Xing Distripark, Singapore 0512

Ciba Foundation Symposium 204
ix+250 pages, 23 figures, 10 tables

Library of Congress Cataloging-in-Publication Data

The molecular basis of cellular defence mechanisms.
p. cm. — (Ciba Foundation symposium ; 204)
Edited by Gregory R. Bock and Jamie A. Goode.
Proceedings of a symposium held Mar. 19–21, 1996 at the Walter and Eliza Hall Institute of Medical Research, Melbourne.
Includes bibliographical references and indexes.
ISBN 0-471-96567-7 (alk. paper)
1. Molecular immunology — Congresses. 2. Cellular immunity — Congresses. 3. Granulocyte-macrophage colony-stimulating factor — Congresses. I. Bock, Gregory. II. Goode, Jamie. III. Series.
QR185.6.M6525 1997
616.07'9 — dc20 96-43728
CIP

British Library Cataloguing in Publication Data

A catalogue record for this book is available from the British Library

ISBN 0 471 96567 7

Typeset in 10/12pt Garamond by Dobbie Typesetting Limited, Tavistock, Devon.
Printed and bound in Great Britain by Biddles Ltd, Guildford.
This book is printed on acid-free paper responsibly manufactured from sustainable forestation, for which at least two trees are planted for each one used for paper production.

Donald Metcalf, Jacques Miller and Sir Gustav Nossal.

Contents

Symposium on The molecular basis of cellular defence mechanisms, held at The Walter and Eliza Hall Institute of Medical Research, Melbourne 19–21 March 1996

Editors: Gregory R. Bock (Organizer) and Jamie A. Goode

This symposium was held in the year in which Sir Gustav Nossal, Donald Metcalf and Jacques Miller retired, to honour their contributions to Ciba Foundation symposia over many years

Participants

A. Burgess The Ludwig Institute for Cancer Research, Melbourne Tumor Biology Branch, Post Office, Royal Melbourne Hospital, Victoria 3050, Australia

F. R. Carbone Department of Pathology and Immunology, Monash Medical School, Alfred Hospital, Prahran, Victoria 3181, Australia

S. Cory The Walter and Eliza Hall Institute of Medical Research, Post Office, Royal Melbourne Hospital, Victoria 3050, Australia

M. M. Davis Howard Hughes Medical Institute, Beckman Center, Stanford University School of Medicine, Stanford, CA 94305–5428, USA

T. M. Dexter Paterson Institute for Cancer Research, Christie Hospital NHS Trust, Wilmslow Road, Manchester M20 4BX, UK

C. C. Goodnow Howard Hughes Medical Institute and Department of Microbiology and Immunology, Stanford University School of Medicine, Stanford, CA 94305–5428, USA

P. Hodgkin Centenary Institute of Cancer Medicine and Cell Biology, Locked Bag No.6, Newtown, NSW 2042, Australia

J. Kirberg *(Ciba Foundation Bursar)* The Netherlands Cancer Institute, Division of Molecular Genetics (H4), Plesmanlaan 121, NL-1066 CX Amsterdam, The Netherlands

G. J. Lieschke Whitehead Institute for Biomedical Research, 9 Cambridge Center, Cambridge, MA 02142–1479, USA

D. Mathis Institut de Génétique et de Biologie Moléculaire et Cellulaire, 1 rue Laurent Fries, 67404 Illkirch Cedex, C U de Strasbourg, France

F. Melchers Basel Institute for Immunology, Grenzacherstrasse 487, Postfach, CH-4005 Basel, Switzerland

D. Metcalf The Walter and Eliza Hall Institute of Medical Research, Post Office, Royal Melbourne Hospital, Victoria 3050, Australia

J. F. A. P. Miller The Walter and Eliza Hall Institute of Medical Research, Post Office, Royal Melbourne Hospital, Victoria 3050, Australia

G. Morstyn Amgen Inc., 1840 Dehavilland Drive, Thousand Oaks, CA 91320–1789, USA

T. R. Mosmann Department of Medical Microbiology and Immunology, University of Alberta, Edmonton, Canada T6G 2H7

N. A. Nicola The Walter and Eliza Hall Institute of Medical Research, Post Office, Royal Melbourne Hospital, Victoria 3050, Australia

G. J. V. Nossal *(Chairman)* The Walter and Eliza Hall Institute of Medical Research, Post Office, Royal Melbourne Hospital, Victoria 3050, Australia

W. E. Paul Laboratory of Immunology, National Institute of Allergy and Infectious Diseases, National Institutes of Health, Bethesda, MD 20892, USA

K. Rajewsky Institute for Genetics, University of Cologne, Weyertal 121, D-50931 Köln, Germany

K. Shortman The Walter and Eliza Hall Institute of Medical Research, Post Office, Royal Melbourne Hospital, Victoria 3050, Australia

A. Strasser The Walter and Eliza Hall Institute of Medical Research, Post Office, Royal Melbourne Hospital, Victoria 3050, Australia

D. Tarlinton The Walter and Eliza Hall Institute of Medical Research, Post Office, Royal Melbourne Hospital, Victoria 3050, Australia

H. von Boehmer Institut Necker, INSERM 373, rue de Vaugirard, Paris Cedex 75015, France

D. Williams Herman B Wells Center for Pediatric Research, Department of Pediatric Hematology/Oncology, Riley Hospital for Children, Howard Hughes Medical Institute, Indiana University School of Medicine, 702 Barnhill Drive, Indianapolis, IN 46202–5225, USA

R. M. Zinkernagel Institute for Experimental Immunology, University Hospital of Zurich, Schmelzstrasse 12, CH-8091, Zurich, Switzerland

Chairman's introduction

Sir Gustav Nossal

The Walter and Eliza Hall Institute of Medical Research, Post Office, Royal Melbourne Hospital, Victoria 3050, Australia

It is daunting for me to recognize that 30 years ago, about 200 m from here, I attended my first Ciba Foundation Symposium, at the time of Macfarlane Burnet's retirement. I don't know where those three decades have gone! Don Metcalf and Jacques Miller also took part in that classic meeting on the thymus, just as its biology was beginning to be unravelled.

To all of you, the finest colleagues that we have around the world, my heartfelt thanks for coming to this meeting to mark our triple retirement. Remarkably, all the scientists who were approached almost immediately accepted the invitation to attend. Of course, there were others we might also have liked to invite, but you are the scientists that we three long-time colleagues most wanted to have here for this occasion, and I thank you all for sacrificing the time from your busy lives. I am sure it is going to be a superb meeting.

What about three retirees? There is a nice story here, because we do go back a very long way. Jacques and I go back to 1941 — we attended the same primary school, and Jacques was a year behind me. Subsequently, we went to the same secondary school and attended Sydney University together. In our fourth year there we met Don Metcalf, who was two years ahead of me. The circumstances in which we met him were also noteworthy. We had all taken a year-off from our medical course to study for a new degree, B.Sc. (Med), with a crusty, eccentric and very gifted virologist, Pat de Burgh, who was our first scientific mentor. During that period, Don and I actually visited the Hall Institute with de Burgh. To a significant degree, that week spent in Melbourne as young medical students set the pattern of our lives. It made a big impression on us to come from the very unprofessional (at that time) medical research scene at Sydney University to one of the few places in Australia where work of true excellence was going on. We have been 'labourers in the vineyard' in our own different ways ever since. Don was here at the Hall Institute when I arrived in 1957, and Jacques was later invited by me, as one of the first acts of my directorship, to return from London and he arrived in 1966 to set up a new unit.

By the way things have evolved, we have been engaged in three related but separate fields of endeavour. Jacques has studied everything to do with the thymus, which was also Don's first love. Don has since 1965 devoted his attention chiefly to haemopoiesis, and very particularly to the generation of the non-specific defence cells (macrophages and granulocytes), but more generally to the rules governing the molecular control of haemopoiesis. I have studied antibody formation and thus B cells. In the last 10 years, Jacques and I have both been interested in the opposite side of immune activation, namely immunological tolerance, as a major thread in our work.

In a way, our lives have mirrored perhaps the most important change that has occurred in biology in the last 20 years or so — congruence. Many significant discoveries have united and helped to interlock these three fields of endeavour: for instance, the T cells collaborating with the B cells, discovered severally by Mitchison, Rajewsky, and Miller and Mitchell, each contributing uniquely to the solution of a major puzzle; in haemopoiesis, the recognition that the stem cell is a progenitor for all the white cells, including the lymphocytes; also, the unfolding recognition down the decades that haematology and immunology are united through the great influence of growth factors and cytokines, and their receptors, on cellular division and differentiation. More recently, the discovery that GM-CSF and the Flk ligand are powerful stimuli for dendritic cells, and therefore are strongly involved in the induction of immune responses, provides yet another link. Of course, I need hardly add that signalling pathways and the general concept of receptor–ligand interaction and downstream events unite virtually the whole of biology. When the story of this remarkable quarter-century is finally written, the gene cloning revolution will come into real focus, not because it spawned the biotechnology industry, but because of the way it unified the biosciences. Since Paul Berg, Herb Boyer and Sidney Cohen, we have all spoken the same language. In the context of this meeting, these T cells, B cells, stem cells and macrophages are all talking to each other in strange ways, every day of the week at the Hall Institute.

As you look at the publication lists of the three retirees, you won't see many co-publications. However, this hides the fact that each day we are in intellectual communication through the wide network of the Hall Institute. This daily discourse has been extremely important in our lives. When Suzanne Cory addressed the staff on the announcement of her appointment, she made reference to something that has been fairly special at the Hall Institute. This is the way that the cellular and whole animal tradition handed down from Burnet resonates with and is supported by the molecular biological tradition, so that it is now completely natural to look at bodily systems with state-of-the-art technology in both of these areas. I hope that this comes through in some of the papers at this symposium.

I would like to thank the organizers for their hard work in bringing us here together and, once again, my warm thanks to all of you for coming. Let's get to town with the real business.

How do stem cells decide what to do?

Michael A. Cross, Clare M. Heyworth and T. Michael Dexter

Paterson Institute for Cancer Research, Christie Hopsital NHS Trust, Wilmslow Road, Manchester M20 9BX, UK

Abstract. The continuous replenishment of mature blood cells from multipotent stem cells proceeds under the influence of haemopoietic growth factors which clearly regulate both cell survival and proliferation. The extent to which these factors might influence lineage choice is still unclear, however, and it seems likely that resolution of this issue will require direct analysis of multipotent cells undergoing commitment rather than determination of their productivity in colony assays. Chromatin analysis of a multipotent progenitor cell line indicates that many of the genes relevant to alternative lineage fates are maintained in an accessible (primed) state prior to lineage commitment. Furthermore, multipotent cells have been found to co-express a number of lineage-restricted genes, suggesting that commitment proceeds as the consolidation of an existing programme. There are indications that the patterns of gene expression in multipotent progenitors change over time, raising the possibility of temporal priming towards different lineages. In multipotential cell lines, exogenous growth factors are necessary for survival, but not for lineage commitment, implying a largely supportive role in early progenitors. In contrast, recent work on primary bipotent granulocyte/macrophage progenitors does demonstrate an inductive role for growth factors in these more lineage-restricted cells.

1997 The molecular basis of cellular defence mechanisms. Wiley, Chichester (Ciba Foundation Symposium 204) p 3–18

Around 2×10^{11} erythrocytes and 10^{10} white blood cells are required each day in order to maintain human adult haemopoiesis. They are produced by the differentiation and extensive proliferation of progenitors derived from a relatively small number of pluripotent stem cells in the bone marrow. In order to prevent long-term depletion of these stem cells and thus ensure an adequate supply of mature blood cells throughout life, commitment to differentiation must be balanced by self-renewal within the stem cell pool. This choice between self-renewal and commitment is just one of a number of decisions which punctuate blood cell production. Other options include survival versus apoptosis, proliferation versus quiescence and periodic choice between alternative developmental lineages. It is the combination of these decisions which ultimately governs the numbers of mature cells emerging from each lineage, and tailors production to the changing demands imposed by stresses such as infection or injury.

Haemopoiesis takes place in the bone marrow within a complex regulatory milieu offering interactions with stromal cells, extracellular matrix, and a rich variety of both stimulatory and inhibitory growth factors. In bone marrow cultures designed to emulate these conditions *in vitro*, long-term maintenance of haemopoiesis does not require the addition of exogenous growth factors. It is, however, entirely dependent on physical interaction between haemopoietic progenitors and the stromal cells which synthesize, localize and present factors in organized microenvironments allowing close-contact signalling. Our knowledge of the functional arrangement of haemopoietic microenvironments is currently very limited. However, it seems likely that the presentation of stimuli within an ordered niche serves both to protect and to contain the stem cells, and that co-localization of multiple stimuli facilitates cooperative interactions between them (Dexter et al 1990). In this respect it is important to note that there may well be overlap between the roles of cell adhesion molecules and growth factor–receptor interactions: the integrins which mediate progenitor–stromal cell interactions may modify growth factor responses by communicating with common signalling pathways; while the recognition of the membrane-bound form of stem cell factor (SCF; also known as Steel factor) by its receptors on progenitors has been found to potentiate a degree of cell adhesion (Kodama et al 1994).

Fortunately, the requirement for stromal cells to support haemopoietic cell differentiation can be circumvented *in vitro* by the addition of soluble growth factors, a large number of which have now been identified, purified and characterized in terms of their effects on haemopoiesis both *in vivo* and *in vitro*. Analysis of haemopoietic cell production *in vitro*, for instance, has demonstrated that a number of growth factors used individually can facilitate the development of specific subsets of lineages from committed (developmentally restricted) progenitor cells, while combinations of factors are generally required to support the multilineage development of more primitive progenitors (Metcalf 1993). The precise nature of the molecular events which underlie the ability of such growth factors to promote haemopoiesis have, however, been very difficult to identify. This is partly because of the difficulties of purifying homogenous populations of progenitor cells of a defined type, and partly because the relevant biological readout systems are based on the number and types of mature blood cells produced from progenitor cells over a number of days. These 'productivity' assays provide a measure of the developmental *consequences* of growth factor treatments, but not of the decisive events themselves.

Only relatively recently, for instance, have we come to appreciate that haemopoietic growth factors do not merely support proliferation and differentiation of the progenitor cells, but that they are also able to suppress apoptosis (Williams et al 1990). As a consequence of this, the idea has emerged that mature cell production may be regulated not only by the modulation of proliferation of progenitor cells, but also by regulating their life and death. In considering the relative contributions of survival and proliferation, it is worth pointing out that they are separable processes. It is obvious that many cells survive for long periods without proliferating, but perhaps more surprising is the observation that cell cycling can be promoted in the absence of an adequate survival stimulus, leading to apoptosis (Evan et al 1992). For this reason it would be unwise to assume that all factors which promote proliferation are necessarily good survival stimuli, and *vice versa*. SCF, for example, is a particularly effective survival factor both for primitive stem cells (Keller et al 1995) and for more committed progenitors (Heyworth et al 1992). However, SCF alone has only a limited proliferative effect and it is only when SCF is added to progenitor cells in the presence of other growth factors (e.g. granulocyte colony-stimulating factor or interleukin 3) that a substantial stimulation of haemopoiesis is seen. Since blood cell production, by necessity, requires not only proliferation but also survival of the developing cells, it is easy to see how complementation between factors acting predominantly on either process may lead to synergistic effects on productivity.

On this basis, it is now generally accepted that the numbers of blood cells produced *in vivo* are regulated largely by the action of haemopoietic growth factors with effects on the survival and proliferation of specific lineages. However, the obvious responses of lineage-committed cells to specific haemopoietic growth factors raises the question of the status of the multipotential progenitors, and the role, if any, of haemopoietic growth factors in lineage commitment. The most obvious question concerns the presence of receptors: if multipotent progenitors possess functional receptors for the haemopoietic growth factors that are known to regulate the subsequent development of individual lineages, then we must consider the possibility that these factors may play a role also in the commitment decisions that are taken by the multipotent cells. In this scenario, the subsequent lineage-specific expression of receptors would be due largely to receptor loss during differentiation along inappropriate lineages. An alternative possibility is that the multipotent cells are restricted in their expression of receptors, and that these are acquired as a result of a commitment decision in which they play no part (Fig. 1). It has been remarkably difficult to obtain an unequivocal answer to this (relatively) straightforward question of receptor expression since the studies are complicated not only by the low numbers of receptors which may be involved, but also by the problems of obtaining pure populations of multipotent progenitors (Testa et al 1993), and the uncertainty (referred to above) in extrapolating short-term effects from long-term cell productivity assays. Hence, whereas combinations of growth factors certainly enhance the productivity of multipotent progenitor cells and have been shown in some cases to be required simultaneously in order to exert their effect (Heyworth et al 1992) this can only be interpreted as being consistent with, rather than direct evidence for, the presence of responsive receptor combinations on the multipotent cell surface (the receptor loss model of Fig. 1).

Results and discussion

It is clear that we need to identify more immediate assays of growth factor responsiveness, and to analyse the multipotent progenitors themselves rather than their progeny, if these questions are to be addressed satisfactorily. To approach this,

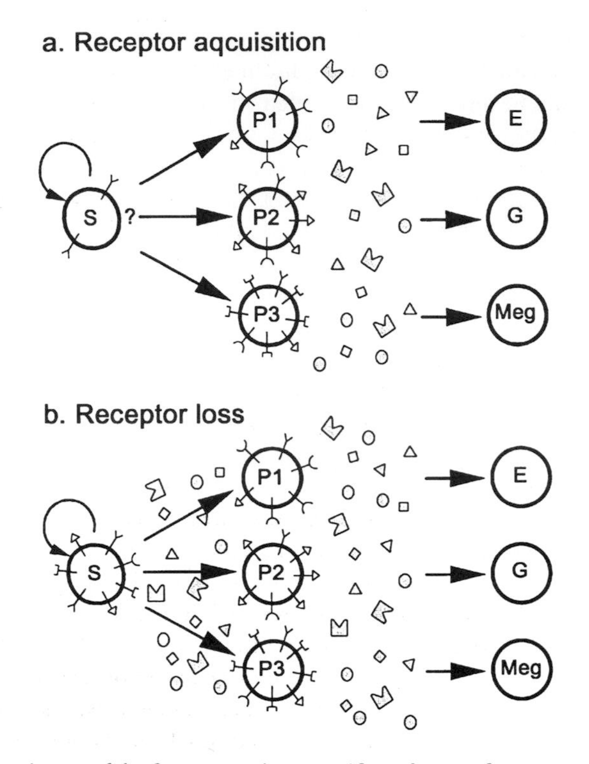

FIG. 1. Alternative models for generating specific subsets of receptors on committed progenitors P1–3. (a) Receptor acquisition: the stem cell (S) displays receptors for no (or very few) factors relevant specifically to the development of restricted lineages. Expression of these receptors occurs only as a consequence of commitment to differentiation. (b) Receptor loss: the stem cell displays a mixture of receptor types. Inappropriate receptors are then down-regulated to produce the combinations characteristic of each progenitor.

we have exploited a number of continuously growing multipotent progenitor cell lines, exemplified by the murine multimyeloid progenitor FDCP-mix A4 (factor-dependent cell Paterson, mixed potential, isolate A4; Spooncer et al 1986). FDCP-mix cells self-renew when cultured in high concentrations of interleukin 3 (IL-3), but on lowering the IL-3 concentration, will undergo differentiation along alternative lineages depending on the other haemopoietic growth factors present (Heyworth et al 1990, 1995). The availability of large numbers of these cells has facilitated biochemical and molecular studies which would be impractical with primary cells. DNase I hypersensitivity studies have, for instance, demonstrated that the genes for the macrophage-specific lysozyme M and the erythroid β-globin both exist in an accessible, 'primed' state in the multipotent FDCP-mix population (Jimenez et al 1992, Mollers et al 1992). These observations suggest that at least at the chromatin level, multipotent progenitors keep their developmental options open by poising or priming a number of alternative programmes.

Consistent with the poised state of lineage-restricted genes in multipotent progenitors, undifferentiated populations of FDCP-mix co-express relatively high levels of mRNA for the lineage-restricted transcription factors PU.1, SCL and GATA-2. The subsequent lineage restriction of PU.1 (to macrophages) and SCL (to erythroid cells) mRNAs is due predominantly to gene inactivation during differentiation along inappropriate lineages (Cross et al 1994). The functional significance of this observation has since become apparent from the multi-lineage haemopoietic deficiencies in mice lacking the *PU.1*, *SCL* or *GATA-2* genes, indicating a requirement for lineage-restricted transcription factors in multipotent progenitors (Robb et al 1995, Scott et al 1994, Tsai et al 1994).

While these molecular studies have provided some valuable information about multipotent progenitors, in common with similar studies of primary progenitors, interpretation is limited by heterogeneity within the population. For example, although the data support the hypothesis that there is co-expression of lineage-restricted transcription factors and simultaneous priming of lineage-restricted genes in populations highly enriched for stem cells, firm conclusions can only be drawn from studies at the single cell level. However, even these have not provided the clear answers which might have been anticipated. By combining single cell PCR analysis with determination of developmental potential of sibling cells from 4–8 cell colonies grown *in vitro* from primary progenitors, Brady et al (1995) have mapped the expression of a number of genes within the murine haemopoietic hierarchy, and have provided evidence for quite marked heterogeneity in gene expression even between single multipotent progenitors with identical colony forming abilities. This suggests either that there are distinct subpopulations of progenitors with similar or identical multipotentialities but widely differing patterns of gene expression, or that the levels of mRNA of many of the genes vary over time.

An adaptation of this technique designed to detect very low levels of specific mRNAs has recently revealed that significant and comparable proportions of both FDCP-mix and primary $CD34^+$ cells co-express myeloperoxidase (granulocyte) and β-globin (erythroid) mRNAs. This same study demonstrated co-expression of mRNAs for the receptors for IL-3, SCF, granulocyte colony-stimulating factor (G-CSF), macrophage (M)-CSF, granulocyte/macrophage (GM)-CSF and erythropoietin, although a high degree of heterogeneity is evident even at this level of sensitivity. Of course, the presence of mRNA does not necessarily indicate the presence of functional protein, but it is interesting that there is a remarkable correlation between the proportion of FDCP-mix cells expressing mRNAs for the erythropoietin, G-CSF and SCF receptors, and the short-term survival response of these cells to the respective factor. In this case, since there is no significant degree of proliferation or differentiation in response to any of these factors alone, the use of short-term survival as the readout in these experiments has proven to be far more informative than colony assay in demonstrating the presence of a responsive receptor.

The accessibility (in regions of open chromatin) of genes that are relevant to a number of alternative lineages and the co-expression of genes for 'lineage-restricted' transcription factors, growth factor receptors and maturation markers suggest that multipotent cells are in a primed state from which lineage commitment proceeds as the consolidation of an existing programme with concurrent silencing of the alternatives. In other words, lineage commitment involves a resolution of complexity to relative simplicity. The indications that there may well be temporal changes in gene expression are intriguing in this respect, since they raise the possibility that single progenitors may be primed towards different lineages at different times. It is of course possible that temporal variations reflect the concomitant and intermittent expression of genes appropriate to each lineage potential of a single progenitor, and that shifts in expression pattern are relatively small and non-coordinated (Fig. 2, bottom). Alternatively, one could imagine each cell progressing through states of concerted priming due to the biased expression of genes relevant to individual lineages. The coincidence of mRNAs appropriate to different lineages would then be due either to overlap between primed states or to the persistence of long-lived mRNAs. An extreme version of this model in which priming follows a set, sequential pattern is shown in Fig. 2 (top). A degree of concerted priming and coordination (though not necessarily that required for sequential priming) would indeed be consistent with evidence emerging from a number of sources, including studies of *Drosophila* development (Jiang & Levine 1993) and mammalian myogenesis (Wentraub 1993), which suggest that lineage commitment involves the attainment of threshold levels of decisive transcription factors. Beyond this level, positive feedback is expected to reinforce identity, while below it negative feedback would keep the options open.

Since transcription factors in haemopoietic cells probably respond to signalling from haemopoietic growth factor receptors (the expression of which in turn depends on the activity of transcription factors) there is ample scope for feedback at all levels. In addition to the probable cross- and auto-regulation of transcription factors, there are examples both of receptor transmodulation (Walker et al 1985) and of positive feedback between the stimulation and transcription of a single receptor type (Heberlein et al 1992), each of which could act to reinforce the response of cells to a specific signal.

Unfortunately, the concept of temporal priming does little to resolve the questions surrounding the involvement of growth factors in commitment, other than to define a specific area of uncertainty: if commitment occurs at the attainment of threshold levels of receptors, signalling molecules or transcription factors, then are these parameters affected at all below the critical threshold by signalling from the haemopoietic growth factor receptors, or do growth factors act simply to support the survival and proliferation of cells which have progressed beyond the commitment point?

It should be noted that lineage commitment *in vitro* can be demonstrated in the absence of individual lineage-restricted growth factors. Mice lacking the erythropoietin receptor gene (Wu et al 1995) or polycythaemic mice making no detectable erythropoietin (von Wangenheim et al 1977) both generate erythroid-committed precursors which can be rescued by replacing the appropriate stimulus.

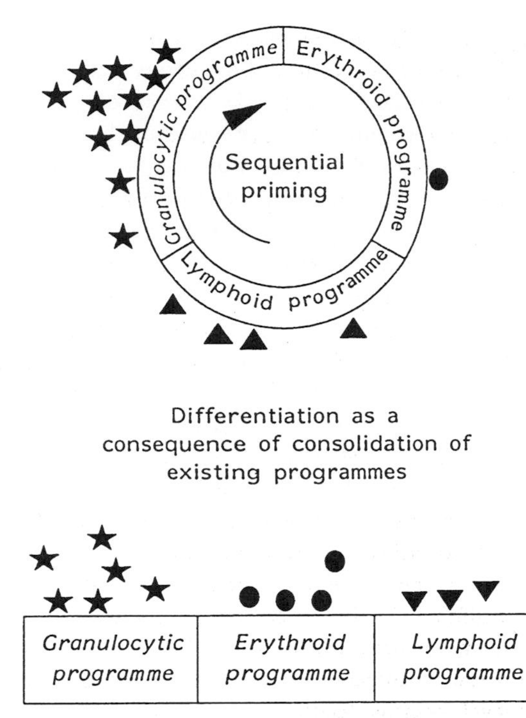

FIG. 2. Single multipotent progenitors co-express mRNAs relevant to alternative lineage fates: ★, granulocytic; ●, erythroid; ▼, lymphoid. In the models shown, temporal priming occurs as a bias in mRNA representation resulting either from the simultaneous but random transcription of genes from all lineage programmes (bottom), or the concerted transcription of genes from a single programme. In the extreme case, temporal shifts in transcription might follow a determined sequence (top).

Perhaps more compelling, however, are data showing that multipotent cells which have been denied the option of apoptosis (by forcing expression of *bcl-2* or of the *bcl-2*-related *A1* gene) can still undergo multilineage commitment and differentiation when deprived of exogenous growth factors (Fairbairn et al 1993, Lin et al 1996, Rodel & Link 1996). Hence there is convincing evidence that lineage choice in multipotent progenitors can be made without reference to externally derived factors. This strengthens the contention that the decisive 'thresholds' (that lead to commitment) in these cases at least, are likely to involve transcription factors.

This is not to say that lineage choice must be purely intrinsic under all circumstances. On the contrary, recent studies of primary murine granulocyte/macrophage colony-forming cells (GM-CFCs) suggest strongly that the differential commitment of these cells can indeed be influenced by the growth factors to which they are exposed. M-CSF and SCF, which signal through related receptor tyrosine kinases, result in the alternative development of GM-CFCs into macrophages or granulocytes, respectively. Although plating efficiencies of $<50\%$ can not rule out recruitment of independent populations, [^{3}H]thymidine suicide assays show that the two factors do elicit proliferative responses from a common target population. Stimulation with mixtures of M-CSF and SCF showed that macrophage induction is the dominant signal, and a search for biochemical signalling events associated with M-CSF treatment revealed a marked, chronic translocation of protein kinase C (PKC) α to the nucleus. This contrasts with a decrease in PKCβ2 levels seen in response to SCF (Whetton et al 1994). The link between PKCα translocation and macrophage commitment has been confirmed in three ways. Firstly, inclusion of the PKC activators TPA (12-*O*-tetradecanoylphorbol 13-acetate) or bryostatin in combination with SCF resulted in the production of macrophages instead of the granulocyte development normally associated with this factor (Heyworth et al 1993), while a PKC inhibitor (calphostin) converted the M-CSF (macrophage) response to one of granulopoiesis (Whetton et al 1994). Secondly, pursuing the observation that IL-4 can elicit a short term survival and proliferation response from GM-CFC, we found that this factor also causes both PKCα translocation and macrophage development in the presence of G-CSF (Nicholls et al 1995). Finally, transduction of GM-CFC with a retrovirus expressing activated PKCα results in a strong bias towards macrophage and away from granulocyte differentiation under conditions which support both lineages (G. Wark, unpublished results). These experiments show that lineage choice in bipotent granulocyte/macrophage progenitors can indeed be influenced by growth factors which signal through PKCα (Fig. 3).

In the presence of GM-CSF, GM-CFCs produce a combination of granulocyte, macrophage and mixed granulocyte/macrophage colonies in a concentration dependent manner, macrophage differentiation being favoured at high factor concentrations, and granulocyte at low. This is somewhat surprising in the light of the reported bias towards granulocyte differentiation at high GM-CSF concentrations in a similar system (Metcalf 1980). Rather than calling either observation into question, it is likely that this apparent inconsistency simply draws attention to the complexity and versatility of growth factor responses (Metcalf 1993), and our ignorance of the ways in which the responses to specific growth factors may depend very strongly on less well-defined signals. As an example of this, we have recently found that the tendency of high concentrations of GM-CSF to favour the formation of macrophage colonies from GM-CFCs is dependent on the concentration of glucose in the growth medium. Simplistically, this would be consistent with macrophage commitment requiring relatively high energy levels, perhaps for correspondingly high levels of phosphorylation in the relevant signalling pathways. A link between GM-CSF signalling and glucose transport has indeed recently been established with the report that GM-CSF regulates glucose transport using signals transduced through the GM-CSF-specific α-subunit of the heterodimeric receptor (Ding et al 1994). The bipotent GM-CFCs provide the best evidence to date that growth factor signalling can influence lineage choice in haemopoietic progenitors. It is intriguing

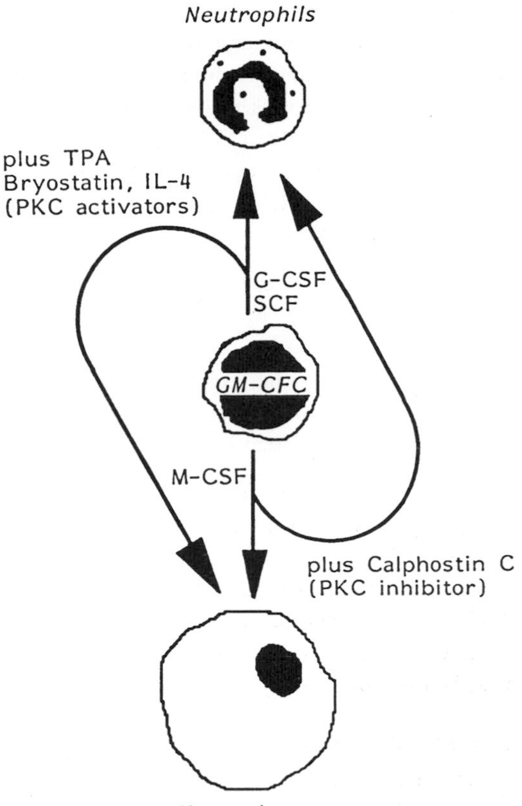

FIG. 3. Lineage choice in bipotent granulocyte/macrophage progenitors is influenced by lineage-restricted growth factors, and by the activation state of protein kinase C. GM-CFC, granulocyte/macrophage colony-forming cell; G-CSF, granulocyte colony-stimulating factor; SCF, stem cell factor; M-CSF, macrophage colony-stimulating factor; IL-4, interleukin 4; PKC, protein kinase C; TPA, 12-*O*-tetradecanoylphorbol 13-acetate.

that the clearest indication of a similar role for transcription factors is the ability of Egr-1 to promote macrophage and block granulocyte differentiation in a similar system (Nguyen et al 1993). Egr-1 is induced in early responses to mitogenic stimuli in a number of cell types, and in response to GM-CSF in a human myeloid leukaemia cell line (Wong & Sakamoto 1995). It is possible that the recent emergence of data supporting an inductive role for haemopoietic growth factors in granulocyte/macrophage lineage choice is simply a consequence of the relative ease of studying bipotent rather than multipotent progenitors. It should be borne in mind, however, that whereas coercion of relatively abundant bipotent progenitors into alternative fates by extrinsic signals is unlikely to have far-reaching adverse effects on the alternative lineage, similar induction of the relatively small number of very early, multipotent progenitors (into erythroid and away from myeloid lineages, for instance) would run an increased risk of depleting input into other lineages. Perhaps the input of extrinsic factors into lineage commitment decisions really does increase as lineage potential becomes progressively more restricted.

Summary

Multipotent progenitor cells appear to be primed for commitment by the activation of genes relevant to alternative lineage programmes. This implies that lineage commitment proceeds as the consolidation of an existing programme rather than initiation of a new one, and also that the large scale inactivation of inappropriate genes accompanies the up-regulation of those appropriate to the lineage concerned.

There is strong evidence that multipotent cells can undergo commitment without reference to lineage-specific, inductive factors. However, the demonstration of growth factor-dependent lineage choice in bipotent granulocyte/macrophage progenitors suggests that inductive processes do indeed play a role in haemopoiesis, and that their influence may be stronger in downstream progenitors.

Acknowledgements

We wish to thank numerous members of the Department of Experimental Haematology, PICR, as well as our colleagues Dr A. D. Whetton (LRF Unit, UMIST), and Dr Tariq Enver and M. Hu (Institute of Cancer Research, London) for their important contributions. This work was supported by the Cancer Research Campaign of Great Britain and the Leukaemia Research Fund. T. M. D. is a Gibb Research Fellow of the Cancer Research Campaign.

References

Brady G, Billia F, Knox J et al 1995 Analysis of gene expression in a complex differentiation hierarchy by global amplification of cDNA from single cells. Curr Biol 5:909–922

Cross MA, Heyworth CM, Murrell AM, Bockamp EO, Dexter TM, Green AR 1994 Expression of lineage restricted transcription factors precedes lineage specific differentiation in a multipotent haemopoietic progenitor cell line. Oncogene 9:3013–3016

Dexter TM, Coutinho LH, Spooncer E et al 1990 Stromal cells in haemopoiesis. In: Molecular control of haemopoiesis. Wiley, Chichester (Ciba Found Symp 148) p 76–95

Ding DX, Rivas CI, Heaney ML, Raines MA, Vera JC, Golde DW 1994 The alpha subunit of the human granulocyte-macrophage colony-stimulating factor receptor signals for glucose transport via a phosphorylation-independent pathway. Proc Natl Acad Sci USA 91:2537–2541

Evan GI, Wyllie AH, Gilbert CS et al 1992 Induction of apoptosis in fibroblasts by c-myc protein. Cell 69:119–128

Fairbairn LJ, Cowling GJ, Reipert BM, Dexter TM 1993 Suppression of apoptosis allows differentiation and development of a multipotent hemopoietic cell line in the absence of added growth factors. Cell 74:823–832

Ford AM, Bennett CA, Healy LE, Navarro E, Spooncer E, Greaves MF 1992 Immunoglobulin heavy-chain and CD_3 delta-chain gene enhancers are DNase I-hypersensitive in hemopoietic progenitor cells. Proc Natl Acad Sci USA 89:3424–3428

Heberlein C, Fischer KD, Stoffel M et al 1992 The gene for erythropoietin receptor is expressed in multipotential hematopoietic and embryonal stem cells: evidence for differentiation stage-specific regulation. Mol Cell Biol 12:1815–1826

Heyworth CM, Dexter TM, Kan O, Whetton AD 1990 The role of hemopoietic growth factors in self-renewal and differentiation of IL-3-dependent multipotential stem cells. Growth Factors 2:197–211

Heyworth CM, Whetton AD, Nicholls S, Zsebo K, Dexter TM 1992 Stem cell factor directly stimulates the development of enriched granulocyte–macrophage colony-forming cells and promotes the effects of other colony-stimulating factors. Blood 80:2230–2236

Heyworth CM, Dexter TM, Nicholls SE, Whetton AD 1993 Protein kinase C activators can interact synergistically with granulocyte colony-stimulating factor or interleukin-6 to stimulate colony formation from enriched granulocyte-macrophage colony-forming cells. Blood 81:894–900

Heyworth CM, Alauldin M, Cross MA, Fairbairn LJ, Dexter TM, Whetton AD 1995 Erythroid development of the FDCP-Mix A4 multipotent cell line is governed by the relative concentrations of erythropoietin and interleukin 3. Br J Haematol 91:15–22

Jiang J, Levine M 1993 Binding affinities and cooperative interactions with bHLH activators delimit threshold responses to the dorsal gradient morphogen. Cell 72:741–752

Keller JR, Ortiz M, Ruscetti FW 1995 Steel factor (c-kit ligand) promotes the survival of hematopoietic stem/progenitor cells in the absence of cell division. Blood 86:1757–1764

Kodama H, Nose M, Niida S, Nishikawa S, Nishikawa S 1994 Involvement of the c-kit receptor in the adhesion of hematopoietic stem cells to stromal cells. Exp Hematol 22:979–984

Lin EY, Orlofsky A, Wang H, Reed JC, Prystowsky MB 1996 A1, a *Bcl-2* family member, prolongs cell survival and permits myeloid differentiation. Blood 87:983–992

Metcalf D 1980 Clonal analysis of proliferation and differentiation of paired daughter cells: action of granulocyte–macrophage colony-stimulating factor on granulocyte–macrophage precursors. Proc Natl Acad Sci USA 77:5327–5330

Metcalf D 1993 Hematopoietic regulators: redundancy or subtlety? Blood 82:3515–3523

Mollers B, Klages S, Wedel A et al 1992 The mouse M-lysozyme gene domain: identification of myeloid and differentiation specific DNase I-hypersensitive sites and of a 3′-cis acting regulatory element. Nucleic Acids Res 20:1917–1924

Nguyen HQ, Hoffman-Liebermann B, Liebermann DA 1993 The zinc finger transcription factor Egr-1 is essential for and restricts differentiation along the macrophage lineage. Cell 72:197–209

Nicholls SE, Heyworth CM, Dexter TM, Lord JM, Johnson GD, Whetton AD 1995 IL-4 promotes macrophage development by rapidly stimulating lineage restriction of bipotent granulocyte–macrophage colony-forming cells. J Immunol 155:845–853

Robb L, Lyons I, Li R et al 1995 Absence of yolk sac hematopoiesis from mice with a targeted disruption of the *Scl* gene. Proc Natl Acad Sci USA 92:7075–7079

Rodel JE, Link DC 1996 Suppression of apoptosis during cytokine deprivation of 32D cells is not sufficient to induce complete granulocytic differentiation. Blood 87:858–864

Scott EW, Simon MC, Anastasi J, Singh H 1994 Requirement of transcription factor PU.1 in the development of multiple hematopoietic lineages. Science 265:1573–1577

Spooncer E, Heyworth CM, Dunn A, Dexter TM 1986 Self-renewal and differentiation of interleukin-3-dependent multipotent stem cells are modulated by stromal cells and serum factors. Differentiation 31:111–118

Testa U, Pelosi E, Gabbianelli M et al 1993 Cascade transactivation of growth factor receptors in early human hematopoiesis. Blood 81:1442–1456

Tsai FY, Keller G, Kuo FC et al 1994 An early haematopoietic defect in mice lacking the transcription factor GATA-2. Nature 371:221–226

von Wangenheim HR, Schofield R, Kyffin S, Klein B 1977 Studies on erythroid-committed precursor cells in the polycythaemic mouse. Biomedicine 27:337–340

Walker F, Nicola NA, Metcalf D, Burgess AW 1985 Hierarchical down-modulation of hemopoietic growth factor receptors. Cell 43:269–276

Wentraub H 1993 The myoD family and myogenesis, networks and thresholds. Cell 75:1241–1244

Whetton AD, Heyworth CM, Nicholls SE et al 1994 Cytokine-mediated protein kinase C activation is a signal for lineage determination in bipotential granulocyte macrophage colony-forming cells. J Cell Biol 125:651–659

Williams GT, Smith CA, Spooncer E, Dexter TM, Taylor DR 1990 Haemopoietic colony stimulating factors promote cell survival by suppressing apoptosis. Nature 343:76–79

Wong A, Sakamoto KM 1995 Granulocyte–macrophage colony-stimulating factor induces the transcriptional activation of egr-1 through a protein kinase A-independent signaling pathway. J Biol Chem 270:30271–30273

Wu H, Lui X, Jaenisch R, Lodish HF 1995 Generation of committed erythroid BFU-e and CFU-e does not require erythropoietin or the erythropoietin receptor. Cell 83:59–67

DISCUSSION

Nossal: Before we get overwhelmed by this binary choice between macrophages and granulocytes that will engage a lot of discussion, what is your precise definition of the term 'stem cell'? I want to ask you very precisely whether or not you are talking about the Gerry Spangrude type of stem cell, i.e. a cell which is not only CD34$^+$, lineage-marker negative, but which is also non-cycling G0, small and low in mitochondria. That is quite critical to the conclusions you have drawn. If instead you are talking about a cell that is already in cycle, the implications of it possessing mRNAs for all these cytokines would be quite different.

Dexter: I agree that we have to be careful about how we define stem cells (see Lord & Dexter 1995), but perhaps the precise nature of the cells we are using is not so critical for this discussion. Let's take it back a stage: we know that we are dealing with multipotential cells, that these cells express genes that are characteristic of multiple lineages and that during differentiation there is down-regulation of genes that would be inappropriately expressed depending on the cell lineage. For example, there is the down-regulation of β-globin message when the neutrophil lineage is chosen and of myeloperoxidase when the erythroid lineage is chosen. The important point, then, is that the parent cells were multipotent, and whether or not they are the 'true' stem cells is not really central to the argument. Having said this, it is clear that the freshly isolated CD34$^+$ cells that we looked at did contain stem cells, as measured by their ability to regenerate haemopoesis *in vivo* and give rise to lymphoid and myeloid populations. We don't know if *all* the cells in this population are stem cells; we simply haven't got the ability to measure this. However, within this primitive cell population there must be at least some cells, perhaps a majority, co-expressing receptors for multiple growth factors and also genes that we normally associate with lineage commitment.

Shortman: You have taken these stem cells out of the animal and found that some of them spontaneously differentiate. Can you exclude the possibility that within the animal these cells had already had their key exposure to cytokines, and that is why they moved on?

Dexter: We have never shown that freshly isolated stem cells 'spontaneously' differentiate. The experiments you are referring to were performed with growth factor-dependent multipotent haemopoietic stem cell lines that were transfected with the *bcl-2* gene. To my knowledge no one has got normal freshly isolated stem cells to survive long enough *in vitro*, in the absence of growth factors, in order to see whether they can also undergo differentiation and development.

Cory: Was the portfolio of transcription in the stem cells determined on cells as they came out of the animal?

Dexter: Yes. We have examined the patterns of gene expression both in cultured FDCP mix and in stem cells taken directly from animals. The concordance of the data is remarkable: no matter where we collect the stem cells from, the patterns are surprisingly similar.

Cory: Have you compared those portfolios of transcription from stem cells straight out of the animal with those of cells that have been held in SCF for 24 h, for example?

Dexter: Not yet, but this is well worth doing. The idea is to examine patterns of gene expression and to see how these change if multipotent cells are held for varying lengths of time in a 'survival' stimulus, such as SCF. This is fairly high on our list of priorities but, as you can appreciate, these are very time consuming experiments.

Strasser: I would like to remark on your point that giving a survival stimulus might be sufficient to induce full differentiation, as you did by putting a Bcl-2 retrovirus into FDCP-mix cells. This might be sufficient in some systems but definitely not in all. You alluded to a paper by Rodel & Link (1996). One thing the authors noticed was that by putting Bcl-2 into 32D cells they only got partial differentiation after cytokine deprivation. In our own experiments with lymphoid cells in transgenic mice, there are certainly some events in maturation that never occur in the absence of the real physiological stimulus, even with the addition of a very strong survival stimulus such as Bcl-2. For example, you can make double-positive thymocytes survive *in vivo* and *in vitro* for a long time, but if their T cell receptor is not engaged they will not become mature T cells (Strasser et al 1994).

Dexter: I certainly wouldn't rule out a role for growth factors in development, but our data, and perhaps also your data, suggest that they have a greater influence on downstream, more committed cells. Indeed, in the papers you referred to (Lin et al 1996, Rodel & Link 1996) the authors reported that the cells didn't acquire a fully mature phenotype. My understanding is that they had lost their self-renewal potential, differentiated, but did not produce mature cells, i.e. they had gone most of the way there, and they simply hadn't got some of the functional genes activated. Maybe, then, one of the important functions of the growth factors is to allow expression of a fully mature phenotype. Indeed, our data indicate that growth factors may be a lot more important for lineage commitment and development of the more mature cells than the earlier cells. Perhaps differentiation of stem cells is regulated differently than it is in the more developmentally restricted cells.

Burgess: Do you believe that a self-renewing stem cell contains β-globin mRNA?

Dexter: Yes, I believe that it can — which is not the same as saying that *all* stem cells *must* express this RNA.

Burgess: You didn't mention self-renewal at all in your paper, as though none of these factors influence the decision not to commit. Are there any factors that prevent the differentiation programme going forward?

Dexter: That's what we've all been searching for. The 'holy grail' of haemopoiesis research is to take a population of putative stem cells out of the body, put them in the right combination of adhesion molecules, cytokines and matrix molecules, and then watch them self renew. But I've not yet seen a paper with data that would convince me that this has been achieved. It really needs to be done with single cells, and the read-out assay system has to be robust. The one example where there is evidence for self renewal *in vitro* is based upon the ability of retrovirally marked cells isolated from long-term bone marrow cultures to regenerate haemopoiesis *in vivo*. To my knowledge that is the only good evidence that stem cell self-renewal has occurred *in vitro*. The problem here, of course, is to understand exactly what it is in the environment of the bone marrow that influences self-renewal decisions.

Nicola: In the first half of your paper you said that cell survival alone would allow differentiation of these cells; later on, you presented evidence that growth factors can influence those differentiation choices. In the initial experiments with the FDCP-mix cells expressing Bcl-2, have you looked at whether PKCα or β is translocated to the nucleus?

Dexter: We have not done those experiments.

Metcalf: I noticed in your experiments where you mixed IL-4 with various CSFs that the clonogenicity was only about 20–30%. Formally, since we all live with the nightmare of heterogeneity in such progenitor cells, you could be stimulating different populations with your different combinations. A solution to this problem is to work with paired daughter cells, and culture one daughter with one combination and one with another. IL-4 plus G-CSF seems a good combination to try: have you tried that one?

Dexter: No. We really should do this. However, we are well aware of the importance of the question that you have raised and have performed a variety of experiments that convince us that we are not recruiting different sub-populations of cells (Nicholls et al 1995).

Metcalf: The problem is that it is not possible to culture one and the same cell in two different ways. The only way I can think of to get around this is to use daughter cell pairs.

Dexter: We're beginning to do that with IL-4 in combination with other growth factors, because of the 'commitment' included by this cytokine. What has really motivated us here is the short time (4 h) for a commitment decision to be reached. This gives us an opportunity to examine events at the molecular as well as the biological level.

Goodnow: What do you predict would be the phenotype of a person that couldn't self-renew their stem cells?

Dexter: I think they would die *in utero*, i.e. that it would be a dominant lethal condition.

Paul: When you mentioned IL-4 my ears pricked up. We have become interested in dissecting the receptor and looking at the domains that are involved in transducing IL-4-mediated growth and differentiation. They can be distinguished. I wonder whether your IL-4 effect could be replaced by either insulin or insulin-like growth factor 1 (IGF-1), because these stimulants share one of the two principal signalling pathways induced by IL-4.

Dexter: Some of these experiments are being done at the moment, and may also be relevant to the effects of M-CSF (which also requires a serum component, perhaps IGF-1, to exert its effects). Presumably you are arguing that these experiments might help us to decide which signal is really responsible for the commitment decision as opposed to the proliferative response.

Paul: There are two approaches that can be used. One is the replacement of the IL-4 signal with that of IGF-1. The other is the use of mice that are deficient in a principal signal transducer. Now that STAT-6 knockouts are available, that question could be asked very directly.

If the alternative differentiation outcomes are achievable with a cell line, one has the option of putting engineered mutated receptors in to determine which receptor domains are required.

Dexter: Thank you, those are useful suggestions.

Zinkernagel: Is there any evidence from clinical observations that could relate to your observation of high or low glucose on granulocytes versus macrophage pathways? For example, can this be related to a patient's lack of resistance to local infections?

Dexter: I doubt whether you would see it, because the range of glucose concentrations we had to use to pick up this difference between macrophages versus neutrophils was much greater than one would ever find in a patient.

References

Lin EY, Orlofsky A, Wang H, Reed JC, Prystowsky MB 1996 A1, a *Bcl-2* family member, prolongs cell survival and permits myeloid differentiation. Blood 87:983–992

Lord BI, Dexter TM 1995 Which are the hematopoietic stem cells? [or: Don't debunk the history!] Exp Cell Hematology 23:1237–1241

Nicholls SE, Heyworth CM, Dexter TM, Lord JM, Johnson GD, Whetton AD 1995 IL-4 promotes macrophage development by rapidly stimulating lineage restriction of bipotent granulocyte–macrophage colony forming cells. J Immunol 155:845–853

Rodel JE, Link DC 1996 Suppression of apoptosis during cytokine deprivation of 32D cells is not sufficient to induce complete granulocyte differentation. Blood 87:858–864

Strasser A, Harris AW, von Boehmer H, Cory S 1994 Positive and negative selection of T cells in T cell receptor transgenic mice expressing a Bcl-2 transgene. Proc Natl Acad Sci USA 91:1376–1380

The structural basis of the biological actions of the GM-CSF receptor

Nicos A. Nicola, Alison Smith, Lorraine Robb, Donald Metcalf and C. Glenn Begley

The Walter and Eliza Hall Institute of Medical Research, Post Office, Royal Melbourne Hospital, Victoria 3050, Australia

Abstract. The receptor for granulocyte/macrophage colony-stimulating factor (GM-CSF) consists of a ligand-specific low-affinity binding chain (GM-CSFRα) and a second chain that is required for high-affinity binding and signal transduction. This second chain is shared by the ligand-specific α-chains for the interleukin 3 (IL-3) and IL-5 receptors and is therefore called β common (β_c). In mice but not humans the IL-3 receptor can also use a closely related but IL-3-specific β-chain (β_{IL-3}). In order to define the contributions of each chain to receptor signalling we generated mice in which either β_c or β_{IL-3} expression was deleted. β_{IL-3} null mice were phenotypically normal but displayed a decreased responsiveness to IL-3 *in vitro*. β_c null mice, on the other hand, were unresponsive to GM-CSF or IL-5 but still responded to IL-3. These data demonstrated that GM-CSF and IL-5 receptors can use only one β-chain for signalling (β_c) while IL-3 can effectively use either β-chain. The hierarchical basis of receptor transmodulation was shown to result from this differential usage of β-chains. To define the regions required for different types of cell signalling, we constructed human β_c mutants with successive cytoplasmic truncations. By the use of appropriate biological read-out systems we found that the cytoplasmic region of the receptor has a modular design with distinct domains required for cell proliferation, cell survival, differentiation and growth suppression. Appropriate targeting of these domains and the signalling pathways they initiate may provide highly specific cell therapies in the future.

1997 The molecular basis of cellular defence mechanisms. Wiley, Chichester (Ciba Foundation Symposium 204) p 19–32

Granulocyte/macrophage colony-stimulating factor (GM-CSF) was named for its ability to stimulate the formation, in semi-solid cultures, of colonies of granulocytes and macrophages from individual precursor cells in the bone marrow (Metcalf & Nicola 1995). It has since been shown to enhance the survival of haemopoietic progenitor cells; stimulate the proliferation of neutrophilic and eosinophilic granulocyte, macrophage and megakaryocytic progenitor cells; induce differentiation in some myeloid leukaemic cell lines; and enhance the functional capacities of mature neutrophils, eosinophils and macrophages (Metcalf & Nicola 1995). As a result of these actions it has found clinical utility in stimulating bone marrow recovery, particularly after cancer chemotherapy (Grant & Heel 1992). Strains of mice have been generated recently which lack functional expression of the GM-CSF gene. These mice are nearly normal haematologically but they display an increased susceptibility to certain infections and lung pathology (alveolar proteinosis) suggestive of abnormal macrophage function (Stanley et al 1994, Dranoff et al 1994). This may suggest that the primary role of GM-CSF is in maintaining end-cell function and responding to haemopoietic stress rather than in steady-state haemopoiesis.

Some of the biological actions of GM-CSF are shared by two structurally related cytokines, interleukin 3 (IL-3; also known as multi-CSF) and IL-5. In the case of IL-3 almost all of these activities are shared and IL-3 has additional actions on mast and erythroid cells, while for IL-5 only the actions on eosinophils are shared.

Whereas GM-CSF, IL-3 and IL-5 show little amino acid sequence identity, they have been shown directly or by prediction to adopt very similar 3D folds (short chain four-α-helical bundles) (Nicola 1994). Perhaps more importantly from a functional point of view, they bind to and activate homologous receptor complexes in which one chain is identical in all three complexes. Each ligand binds with high specificity but low affinity to a private receptor α-chain. Each of these three α-chains has a very similar overall structure and, in particular, contains a 200 amino acid domain (the haemopoietin receptor domain) in the extracellular region, which is the defining feature of a very large family of type I cytokine receptors that includes receptors for interleukins 2–7, 9, 11–13, leukaemia inhibitory factor (LIF), ciliary neurotrophic factor (CNTF), OSM, cardiotrophin 1, GM-CSF, G-CSF, erythropoietin, thrombopoietin, leptin, growth hormone and prolactin (Hilton 1994). This domain is characterized by the relative disposition of four cysteine residues and the sequence element WSXWS.

The α-chain receptors for GM-CSF, IL-3 and IL-5 contain a very short cytoplasmic tail and, by themselves, appear to be unable to transmit most of the biological signals mediated by these cytokines. A possible exception may be the stimulation of enhanced glucose transport in some cell types (Golde et al 1994). In order to form a high-affinity receptor capable of signal transduction the ligated α-chain receptors must interact with a second receptor subunit termed the β-chain. In humans, a single β-chain (β common or β_c) is used by all three α-chain complexes to form a high-affinity signalling complex, but in the mouse all three α-chain complexes can use the β_c equivalent while the IL-3–IL-3 receptor α-chain complex can also use an alternate IL-3-specific β-chain ($\beta_{IL\text{-}3}$) (Fig. 1). $\beta_{IL\text{-}3}$ and β_c are highly homologous in their structures and are also members of the haemopoietin receptor family. However, they contain a duplicated extracellular haemopoietin domain and a much longer cytoplasmic tail (Miyajima et al 1993).

Although both the receptor α-chain and β-chain cytoplasmic domains are normally required for effective signal transduction (Sakamaki et al 1992) there is some evidence that the primary role of the α-chain may be simply to dimerize the β-chain cytoplasmic domains. Chimaeric receptors that contain only β-chain cytoplasmic domains are able to signal cell proliferation (Takaki et al 1993, Eder et al 1994) and certain mutations in β_c that probably lead to ligand-independent dimerization of the β-chain also result in constitutive cell proliferation (Jenkins et al 1995).

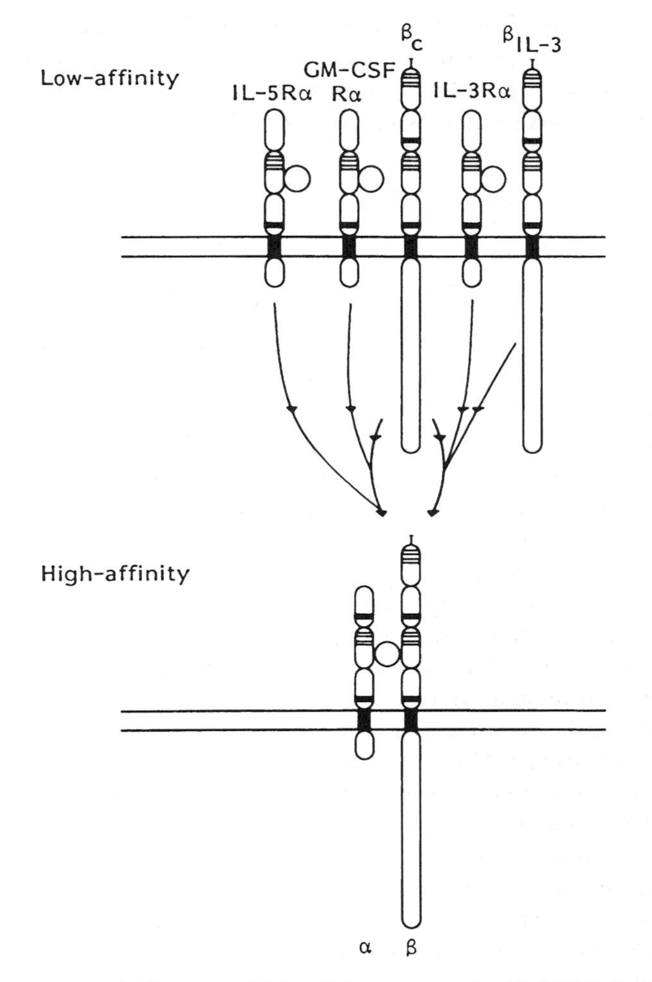

FIG. 1. Formation of functional high-affinity receptors for GM-CSF, IL-3 and IL-5. Three separate α-chains (GM-CSFRα, IL-3Rα and IL-5Rα) bind their cognate ligand with high specificity but low affinity. Each complex then binds to a β-chain to form a high-affinity complex capable of signal transduction: In human cells all three α-chains compete for a single β common chain (β_c) resulting in reciprocal cross-reactivity of binding (inhibition of high-affinity receptor formation). In mouse cells all three α-chains again compete for β_c but the IL-3Rα chain can also form a functional high-affinity receptor with the alternate β-chain, β_{IL-3}. This results in the phenomenon of hierarchical receptor *trans*-modulation in mouse cells. (From Metcalf & Nicola 1995.)

The use of a common β-chain in the three human receptors but the alternative use of β_{IL-3} in murine IL-3 receptors raises several questions. Do β_{IL-3} and β_c transduce identical cellular signals? Is there a differential distribution of β_{IL-3} and β_c amongst haemopoietic cells? Are there as yet undiscovered alternates of β_c that can be used by

GM-CSF, IL-3 or IL-5? Is the use of common and specific β-chains required for high-affinity conversion the basis for apparent cross-reactivity of the binding of GM-CSF, IL-3 and IL-5 in human cells but hierarchical receptor down-modulation in mice (Walker et al 1985, Lopez et al 1991)?

The role of β_c and β_{IL-3} in biological function and receptor *trans*-modulation

In order to address these questions, we and others have generated strains of mice in which either the β_c or β_{IL-3} genes have been functionally inactivated by homologous recombination in embryonic stem cells (Robb et al 1995, Nicola et al 1996, Nishinakamura et al 1995). Mice in which both copies of the β_c gene were disrupted ($\beta_c^{-/-}$) displayed a phenotype very similar to GM-CSF$^{-/-}$ mice with essentially normal haemopoiesis but lung pathology typical of the human disease alveolar proteinosis. In addition these mice displayed very low levels of circulating and tissue eosinophils as might be expected for mice unable to respond to IL-5. Bone marrow cells from $\beta_c^{-/-}$ mice failed to form colonies of any type in semi-solid cultures *in vitro* in response to GM-CSF or IL-5 but formed the normal numbers and types of colonies in response to IL-3. These data suggested first that GM-CSF and IL-5 could not use β_{IL-3} or any other β-chain than β_c in haemopoietic cells, and second, that β_{IL-3} present in these cells could mediate all of the colony-stimulating activities associated with the action of IL-3.

Receptor binding experiments on $\beta_c^{-/-}$ bone marrow cells with [^{125}I]GM-CSF and [^{125}I]IL-3 showed that low-affinity but not high-affinity receptors for GM-CSF could be detected, again confirming that no alternate affinity-converting β-chain for the GM-CSF receptor exists in murine bone marrow cells. On the other hand, high-affinity receptors for IL-3 were present showing that β_{IL-3} was at least as effective as β_c in forming such receptors (Robb et al 1995).

$\beta_{IL-3}^{-/-}$ mice were phenotypically normal and their bone marrow cells responded normally to maximal doses of GM-CSF, IL-3 and IL-5 *in vitro*. However, in dose–response experiments, these bone marrow cells showed a fourfold hypo-responsiveness to IL-3 compared with controls. In parallel with these results, receptor binding experiments revealed that, while GM-CSF receptors were of the same affinity as control +/+ mice, IL-3 receptors from $\beta_{IL-3}^{-/-}$ mice were of nearly 10 times lower affinity. These data suggested that, whereas β_c could mediate all of the biological functions of IL-3, it formed a lower-affinity IL-3 receptor than could be formed by β_{IL-3}. Moreover, since +/+ bone marrow cells showed a higher affinity receptor and increased responsiveness to IL-3 than did $\beta_{IL-3}^{-/-}$ bone marrow cells (which display only β_c), normal +/+ bone marrow cells must preferentially use β_{IL-3} over β_c (Nicola et al 1996).

Since $\beta_c^{-/-}$ cells express only β_{IL-3} and $\beta_{IL-3}^{-/-}$ cells express only β_c, we were able to use high affinity binding of [^{125}I]IL-3 to bone marrow cells of each strain followed by quantitative cell autoradiography to differentially map the expression pattern of each type of β-chain. The data showed that the two chains have a nearly identical qualitative and quantitative distribution amongst haemopoietic cells with expression on blast cells, cells of the granulocyte and macrophage lineages and a small proportion (10–15%) of cells with lymphoid morphology.

In normal bone marrow cells, IL-3 could *trans*-down-modulate high affinity receptors for GM-CSF, G-CSF and M-CSF at 37 °C while GM-CSF could *trans*-down-modulate only G-CSF and M-CSF receptors (hierarchical *trans*-modulation) (Walker et al 1985). In $\beta_c^{-/-}$ bone marrow cells, IL-3 could no longer *trans*-modulate GM-CSF receptors and GM-CSF had a significantly reduced capacity to *trans*-modulate G-CSF and M-CSF receptors. The behaviour of IL-3 was expected if the basis of *trans*-modulation was competition of IL-3 and GM-CSF receptors for limiting numbers of β_c chains. However, the effect on GM-CSF-induced *trans*-modulation of G-CSF and M-CSF receptors suggests a previously unknown effect of β_c activation on G-CSF and M-CSF receptors. Since IL-3 could still effectively *trans*-modulate G-CSF and M-CSF receptors on $\beta_c^{-/-}$ bone marrow cells, $\beta_{IL\text{-}3}$ activation again appears to be able to fully substitute for β_c activation.

In $\beta_{IL\text{-}3}^{-/-}$ bone marrow cells, GM-CSF was now able to very effectively *trans*-down-modulate IL-3 receptors, which it cannot do in +/+ cells, and IL-3 became an even more efficient *trans*-modulator of GM-CSF receptors than in +/+ cells. These data suggest that the hierarchical nature of receptor *trans*-modulation results from the ability of IL-3 receptors to compete for the two alternative β-chains (β_c and $\beta_{IL\text{-}3}$) while GM-CSF receptors can use only β_c. When the IL-3-specific β-chain is deleted the two receptors compete equally for β_c and hierarchical *trans*-modulation is transformed to reciprocal *trans*-modulation (Nicola et al 1996).

Cellular signalling mediated by the human β_c chain

Because these data suggested that murine β_c and $\beta_{IL\text{-}3}$ are functionally equivalent and since in humans there appears to be no equivalent to $\beta_{IL\text{-}3}$, we have concentrated our efforts to understand the cellular signalling pathways to the human β_c chain in the context of the human GM-CSFR.

Previous studies had suggested that the membrane-proximal region of the cytoplasmic domain of β_c was critical for cell proliferation (Sakamaki et al 1992, Sato et al 1993). This region contained a proline-rich sequence that was reasonably well conserved amongst most of the cytokine receptors (termed Box-1) and was required for binding the cytoplasmic tyrosine kinases of the JAK family. Receptor dimerization is thought to then activate JAK kinase activity leading to phosphorylation of cytoplasmic latent transcription factors STATs (signal transducers and activators of transcription) which then dimerize, translocate to the nucleus and activate expression of cytokine response genes (Ihle 1995, Mui et al 1995). However, other cytoplasmic tyrosine kinases are also activated by GM-CSF binding (Miyajima et al 1993) so that the relationship of the JAK/STAT pathway to cellular proliferation is still unclear.

The cytoplasmic region C-terminally adjacent to the critical region required for cell proliferation contains another sequence element conserved amongst many cytokine receptors (Box-2) that enhances the proliferative response to GM-CSF. However, neither of these two regions can activate the majority of tyrosine phosphorylation seen in cells that display full-length β_c. This seems to emanate from the cytoplasmic region between amino acids 624 to 778 and results in the activation of the Sos–Raf–Ras–MAPK pathway that has been well described for receptors with intrinsic tyrosine kinase activity (Sakamaki et al 1992, Sato et al 1993).

We transfected murine CTLL cells (T lymphocyte-derived cells which display no endogenous GM-CSF receptor chains) with expression constructs for the human GM-CSFRα and various truncated β_c chains and measured their capacity to proliferate long-term in human GM-CSF. The shortest β_c construct that allowed sustained proliferation contained 165 cytoplasmic residues (terminating at amino acid 626) and this contrasted with previous reports that truncations terminating at 533 could allow GM-CSF-induced proliferation (Sakamaki et al 1992, Sato et al 1993). However, subsequent studies by the same authors revealed that these shorter truncations only allowed the initiation of proliferation and that longer truncations to 778 were required for the maintenance of cell survival (Kinoshita et al 1995). Our studies therefore suggest that the cell survival signal requires cytoplasmic sequences of β_c only from amino acids 559 to 626 and does not require activation of the Ras pathway. In fact, CTLL cells expressing full-length β_c showed a fourfold reduced responsiveness to human GM-CSF for proliferation compared with the mutants truncated at position 626 or 783. This suggested that a negative regulatory signal is delivered from the cytoplasmic region between 783 and 897 as has been described for other cytokine receptors (Yi et al 1993, Klingmüller et al 1995).

In order to study cellular signals from β_c that lead to cellular differentiation we also transfected two different murine myeloid leukaemic cell lines (WEHI-3B D+ and M1) with cDNAs for hGM-CSFRα and several truncation mutants of human β_c (hβ_c). M1 cells can be induced to differentiate to macrophages by the action of LIF, IL-6, oncostatin M and, to a lesser extent, G-CSF but not by GM-CSF. WEHI-3B D+ cells can be induced to differentiate to granulocytes and macrophages by G-CSF, IL-6 and to a lesser extent by GM-CSF. In both cases cells transfected with expression vectors for hGM-CSFR and hβ_c acquired the capacity to be induced by hGM-CSF to differentiate into macrophages and were clonally suppressed in their proliferative capacity. The differentiation to macrophages was confirmed by morphological criteria, the acquisition of macrophage surface markers (Mac-1, F480, FcRγ, c-Fms and Gr-1), the formation of dispersed colonies in semi-solid cultures *in vitro* and the clonal suppression of proliferative capacity. The inserted receptors did not interfere with the response of M1 cells to LIF or WEHI-3B D+ cells to G-CSF. Interestingly, despite the fact that M1 cells do not display endogenous GM-CSF receptors, the transfected cells responded to hGM-CSF as well as they did to LIF by all criteria. Similarly, transfected WEHI-3B D+ cells responded more dramatically to hGM-CSF than they did to G-CSF or mouse GM-CSF (especially for clonal suppression) suggesting that both cell types contained all the signalling molecules required by β_c and that the strength of the response was probably determined by receptor number.

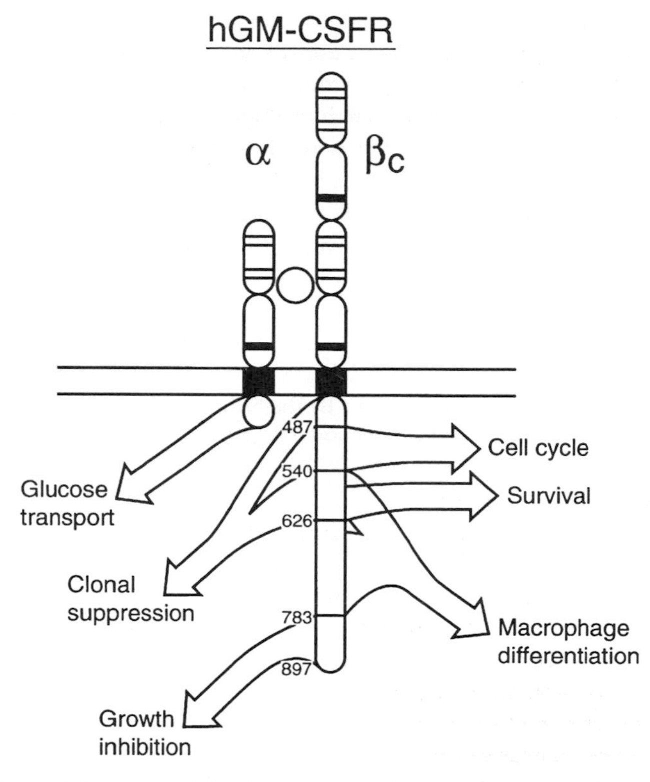

FIG. 2. The modular nature of signal transduction from the β_c cytoplasmic domain. Essential regions required for cell proliferation, cell survival, cell differentiation, clonal suppression and negative regulation of proliferation signals are shown. The amino acid numbering system is from the first amino acid of the leader sequence.

Human β_c truncated to position 783 was as effective as wild-type β_c in transducing differentiative signals in both cell types while β_cs truncated to position 541 or less were ineffective. β_c truncated to position 626 showed intermediate behaviour in that, in M1 cells, hGM-CSF induced full differentiation but at least 100-fold higher doses of hGM-CSF were needed compared with the longer β_c forms and, in WEHI-3B D+ cells, only weak differentiation was induced by hGM-CSF. These data suggest that the cytoplasmic region of β_c between 541 and 626 is essential for differentiation to macrophages in these cell lines and that the region between 626 and 783 (which includes the region that activates the Ras pathway) significantly enhances the response (Fig. 2). In M1 cells, β_c truncations that allowed macrophage differentiation in response to hGM-CSF also allowed clonal suppression while the shorter truncations (to 541 or less) did not. However, in WEHI-3B D+ cells all β_c truncations to 487 or longer showed dramatic hGM-CSF-induced clonal suppression. This suggests that, at least in WEHI-3B D+ cells, clonal suppression can be dissociated from overt macrophage differentiation and that signals for clonal suppression can emanate from either of two regions of the β_c cytoplasmic domain in the different cell lines (Fig. 2).

These results strongly suggest that the cytoplasmic domain of the human βc chain is organized in a modular fashion with different regions initiating cellular signalling pathways that result in different biological responses. Some of these modules and their responses are shown in Fig. 2. Such an organization of signalling pathways holds promise that it may be possible to devise highly specific treatments of disease by targeting only one or a few of the cellular signalling pathways normally activated in cells by cytokines and their receptors.

Acknowledgements

The original work in this paper was supported by the National Health and Medical Research Council, Australia, the Carden Fellowship Fund of the Anti-Cancer Council of Victoria, the Cooperative Research Centres' Scheme of the Australian Government, the National Institutes of Health, Grant CA2256, Bethesda, MD, USA and AMRAD Corporation, Australia. Catherine Drinkwater, Frank Kontgen, Ruili Li, Douglas Hilton and Andreas Strasser are thanked for help in the initial stages of this work and for helpful advice.

References

Dranoff G, Crawford AD, Sadelain M et al 1994 Involvement of granulocyte–macrophage colony-stimulating factor in pulmonary homeostasis. Science 264:713–716

Eder M, Ernst TJ, Ganser A et al 1994 A low affinity chimeric human α/β-granulocyte–macrophage colony-stimulating factor receptor induces ligand-dependent proliferation in a murine cell line. J Biol Chem 269:30173–30180

Golde DW, Ding DX, Rivas CI, Vera JC 1994 The colony-stimulating factors and molecular transport. Stem Cells (suppl) 12:61–66

Grant SM, Heel RC 1992 Recombinant granulocyte–macrophage colony-stimulating factor (rGM-CSF): a review of its pharmacological properties and prospective role in the management of myelosuppression. Drugs 43:516–560

Hilton DJ 1994 An introduction to cytokine receptors. In: Nicola NA (ed) Guidebook to cytokines and their receptors. Oxford University Press, Oxford, p 8–16

Ihle JN 1995 The Janus protein tyrosine kinase family and its role in cytokine signaling. Adv Immunol 60:1–35

Jenkins BJ, D'Andrea R, Gonda TJ 1995 Activating point mutations in the common β subunit of the human GM-CSF, IL-3 and IL-5 receptor suggest the involvement of β subunit dimerization and cell-type-specific molecules in signalling. EMBO J 14:4276–4287

Kinoshita T, Yokota T, Arai K-I, Miyajima A 1995 Suppression of apoptotic death in hematopoietic cells by signalling through the IL-3/GM-CSF receptors. EMBO J 14:266–275

Klingmüller U, Lorenz U, Cantley LC, Neel BG, Lodish HF 1995 Specific recruitment of SH-PTP1 to the erythropoietin receptor causes inactivation of JAK2 and termination of proliferative signals. Cell 80:729–738

Lopez AF, Vadas MA, Woodcock JM et al 1991 Interleukin-5, interleukin-3, and granulocyte–macrophage colony-stimulating factor cross-compete for binding to all surface receptors on human eosinophils. J Biol Chem 266:24741–24747

Metcalf D, Nicola NA 1995 The hemopoietic colony-stimulating factors: from biology to clinical applications. Cambridge University Press, Cambridge

Miyajima A, Mui AL-F, Ogorochi T, Sakamaki K 1993 Receptors for granulocyte–macrophage colony-stimulating factor, interleukin-3 and interleukin-5. Blood 82:1960–1974
Mui AL-F, Wakao H, O'Farrell A-M, Harada N, Miyajima A 1995 Interleukin-3, granulocyte–macrophage colony-stimulating factor and interleukin-5 transduce signals through two STAT5 homologues. EMBO J 14:1166–1175
Nicola NA (ed) 1994 Guidebook to cytokines and their receptors. Oxford University Press, Oxford
Nicola NA, Robb L, Metcalf D, Cary D, Drinkwater CC, Begley CG 1996 Functional inactivation in mice of the gene for the interleukin-3 (IL-3)-specific receptor β-chain: implications for IL-3 function and the mechanism of receptor transmodulation in hemopoietic cells. Blood 87:2665–2674
Nishinakamura R, Nakayama N, Hirabayashi Y et al 1995 Mice deficient for the IL-3/GM-CSF/IL-5 β_c receptor exhibit lung pathology and impaired immune response, while $\beta_{IL\text{-}3}$ receptor-deficient mice are normal. Immunity 2:211–222
Robb L, Drinkwater CC, Metcalf D et al 1995 Hematopoietic and lung abnormalities in mice with a null mutation of the common β subunit of the receptors for granulocyte–macrophage colony-stimulating factor and interleukins 3 and 5. Proc Natl Acad Sci USA 92:9565–9569
Sakamaki K, Miyajima I, Kitamura T, Miyajima A 1992 Critical cytoplasmic domains of the common beta subunit of the human GM-CSF, IL-3 and IL-5 receptors for growth signal transduction and tyrosine phosphorylation. EMBO J 11:3541–3549
Sato N, Sakamaki K, Terada N, Arai K, Miyajima A 1993 Signal transduction by the high-affinity GM-CSF receptor: two distinct cytoplasmic regions of the common beta subunit responsible for different signalling. EMBO J 12:4181–4189
Stanley E, Lieschke GJ, Grail D et al 1994 Granulocyte–macrophage colony-stimulating factor-deficient mice show no major perturbation of hematopoiesis but develop a characteristic pulmonary pathology. Proc Natl Acad Sci USA 91:5592–5596
Takaki S, Murata Y, Kitamura T, Miyajima A, Tominaga A, Takatsu K 1993 Reconstitution of the functional receptors for murine and human interleukin-5. J Exp Med 177:1523–1529
Walker F, Nicola NA, Metcalf D, Burgess AW 1985 Hierarchical downmodulation of hemopoietic growth factor receptors. Cell 43:269–276
Yi T, Mui AL-F, Krystal G, Ihle JN 1993 Hematopoietic cell phosphatase associates with the interleukin-3 (IL-3) receptor β-chain and down-regulates IL-3-induced tyrosine phosphorylation and mitogenesis. Mol Cell Biol 13:7577–7586

DISCUSSION

Nossal: Going back to the wild-type, to what degree are growth promotion (i.e. colony formation) and differentiation truly in opposition to each other? You have done a lot of work counting colonies, but you haven't said much about colony size. There are circumstances in which you get clonal suppression and there are those in which you see marked differentiation — are there also circumstances in which you get limited growth, i.e. rather small colonies, but differentiation to a mature cell phenotype? In the wild-type, what controls are operating in terms of degrees of division and degrees of differentiation?

Nicola: That's a difficult question, because we're using different cell lines to look at different biological responses. One of the main reasons we generated the β_c knockout mice was so we could put back our mutated β-chains into primary haemopoietic cells, in a background-free system, to look at those questions. These sorts of experiments will give the answers you are seeking.

The other point I should make is that what I showed in Fig. 2 was really the minimum region necessary for activity, between a truncation which gives a response and one that doesn't. This experiment shows that this region is required but it doesn't say anything about further upstream regions that are closer to the membrane. Consequently, the areas that give rise to cell proliferation, differentiation and even clonal suppression may all be overlapping. In my own mind, I'm not too clear about the relationship between these three processes. I suspect a lot of it may have to do with biochemical 'dose', because if you look at WEHI-3B D+ cells which give differentiated colonies in response to G-CSF, you often see larger sized colonies — at least initially. In other words, G-CSF is acting to stimulate cell proliferation and is also acting weakly to induce differentiation. Li et al (1993) have shown that if extra G-CSF receptors are put into those cells, then instead of getting that initial proliferative stimulation you actually get clonal suppression just as we're seeing with GM-CSF and M1 cells. My conclusion is that the difference between proliferation and clonal suppression may have to do with the strength of the signal rather than its actual nature.

Nossal: This is an issue that unites all three portions of this overlapping symposium. In the process of antibody formation, a B lymphocyte creates a clone of antibody-forming cells, but also B memory cells, themselves small lymphocytes. What are the rules defining the two pathways of differentiation? How many antibody-forming cells? How many memory cells? The T cell people have exactly the same dilemma: how much proliferation to effector cells and how much towards T cell memory?

Lieschke: Nick, you elegantly showed how useful the β-subunit knockout mice can be as a source of cellular reagents for looking at receptor biology *in vitro*. I wondered whether you could stand back and look at the whole animal: in particular, do you think that you've gained any insight from the two subunit knockout mice to explain why the mouse has the second β-subunit? Is the difference in affinity that you have demonstrated a satisfactory explanation? What is the evolutionary pressure that maintains this gene duplication?

Nicola: I'm not sure. One of the reasons for having two chains with the same function might be that they are under separate genetic control. Initially, we thought that the most likely scenario would be that some cells would only have the $\beta_{IL\text{-}3}$ chain and some only β_c, so that expression could be differentially regulated. We could see no difference in function, so Don Metcalf and I looked carefully and found that in the β_c knockout, high affinity IL-3 binding is only due to receptors containing $\beta_{IL\text{-}3}$, and in the $\beta_{IL\text{-}3}$ knockout, high-affinity IL-3 binding is only due to receptors containing β_c. Consequently, we were able to look at the two receptor chains separately in all bone marrow lineages by cell autoradiography. No cells had one but not the other, so I don't really know what $\beta_{IL\text{-}3}$ is doing in mice at all. Looking at affinities, for some reason the mouse prefers to use $\beta_{IL\text{-}3}$ rather than β_c, but that's only a quantitative preference.

Metcalf: There was a partial selectivity in that mature granulocytes tended to express preferentially the private β-chain for IL-3. The difference wasn't absolute; some cells exhibited common β-chains, but a subset did not. This is interesting because in the human there is no β_{IL-3}, and to everybody's surprise human granulocytes do not have receptors for IL-3. I don't know whether those two facts are linked. It does raise a question of how one is able to grow colonies of mature granulocytes from human marrow using IL-3. It suggests that some terminal maturation is possible without continuing signalling. This is also a problem with erythroid colony formation because there are no erythropoeitin receptors on the most mature nucleated erythroid cells, and yet you get mature non-nucleated erythroid cells forming *in vitro* with erythropoietin. Possibly, there is a certain 'momentum' built up in these cells that allows them to carry through some terminal maturation without continuing stimulation.

Melchers: Murine precursor B cells react not only to IL-7 but also to IL-3. Have you looked at which receptors are expressed in these cells?

Nicola: No. There are some lymphoid-like cells in bone marrow which do bind IL-3. Some of those actually have quite high levels of high-affinity receptors, but we don't know what phenotype these cells have or what they will become. We haven't looked specifically at B cell subsets.

Melchers: It's easy to grow them out *in vitro* and cross-stimulate them with either IL-7 or IL-3. That would allow you to obtain a cell population in which receptor expression could be tested.

Dexter: Getting back to Don Metcalf's point about terminal maturation in the apparent absence of receptors for the growth factors, one of the problems in those studies is the use of large numbers of cells and the presence of serum. The question always arises as to whether or not 'contaminating' mature cells, such as macrophages, are present. If they are, these cells may well be producing factors that are in turn responsible for the differentiation observed. Similarly, serum contains a variety of factors that can influence cell development. So until this experiment is done at the single cell level, in more defined conditions, we can't say whether or not there is a signal present for the terminal development. Now, this may well be relevant to Nick's paper: with the M1 or the D+ cells, is it possible that there is constitutive production of other growth factors from spontaneously differentiating cells and that these are responsible for the effects that you are seeing in terms of differentiation? I noticed that the differentiation appears to be almost totally restricted to the macrophage lineage, whereas GM-CSF will normally promote both neutrophil and macrophage development. Were your experiments done at the single cell level? If not, how can you rule out the possibility of constitutive production of other growth factors by 'spontaneously' maturing cells?

Nicola: Some of these experiments were done in bulk cultures. Most of the colony formation and clonal suppression studies were done in agar cultures with 200 cells per plate. Regardless of those issues, we have clearly shown that there is an 'all or none' effect once you go from one truncation to another. Irrespective of what mechanism you propose, even if it is some sort of secondary factor production, you can say that this absolutely requires signalling from a certain part of the β-chain. Unless you look at the differentiation of a single cell, which is very difficult, it's almost impossible to circumvent the possibility that some signalling phenomenon is giving rise to the secretion of a factor which then binds back to the same cell in an autocrine way, and that this is what actually induces it to differentiate. This might even be a problem if we were to look at single cells.

Dexter: But how do you explain the preference for macrophages, where you would expect neutrophils also to be produced?

Nicola: I suspect that a large component of differentiation may be the switching on of available pathways, and also that some cells are 'pre-committed'. My guess would be that M1 cells and WEHI-3B D+ cells are pre-committed to the macrophage pathway, because even with G-CSF, which is a strong granulocytic stimulus, mostly macrophage-like cells are produced. Most manipulations with M1 cells seem to induce macrophage differentiation.

Dexter: The reason I'm asking this is that I was intrigued to hear that the α-chain in some way facilitated glucose transport. Whose data were these?

Nicola: They were from David Golde's group, and the observations were initially made with oocytes (Ding et al 1994).

Dexter: Are these results in some way related to the observation we made with respect to the glucose concentration effect on lineage choice?

Nicola: I doubt it, unless it's a synergistic type of response. With the α-chain alone in M1 cells, at doses which would give you good levels of α-chain receptor occupancy (up to 500 ng/ml human GM-CSF), there was no detectable differentiation or induction of clonal suppression, or any of those other events I described.

Dexter: But you didn't measure glucose uptake in the cells.

Nicola: We've looked at that using a Cytosensor™ microphysiometer. We were not looking at glucose transport *per se*, but we were looking for any response that changes the acidification rate of the medium, and metabolism of glucose should have this effect. We've looked at this in M1 cells. Dale Cary has looked with α-chain alone and seen no response at all to human GM-CSF. I don't know how cell-type specific the α-chain-mediated induction of glucose transport is; all I can say is that this process alone doesn't affect any of the parameters that I mentioned today.

Metcalf: The question was raised earlier as to how much influence a regulator or a ligand has on the specific response of a cell to signalling and, more specifically, whether or not it can dictate which pathway is entered. I think the major restriction is the genetic programme operational in the cell at the time of this stimulation. A mature neutrophil will respond to CSF, but it responds by functional activation — there is no possibility of the cell proliferating. One also sees the same sort of phenomenon when inappropriate receptors are put into various haemopoietic cells. You can put erythropoietin receptors into granulocyte–macrophage progenitors, and then use erythropoietin to stimulate the formation of typical macrophage colonies. It is not possible with this stimulus for a macrophage precursor exhibiting erythropoietin receptors to generate erythroid cells. The responding cell dictates qualitatively and quantitatively what responses will occur and it is frustrating that we have no idea of the mechanisms responsible for such differences.

Burgess: All these truncation mutants have JAK/STAT binding capabilities, and yet quite different responses are induced. Is there a role for the JAKs and STATs? In particular, in the negative response element, do you envisage a phosphatase binding site?

Nicola: Jim Ihle's group has shown that a haemopoietic cell phosphatase binds to the β_c chain (Yi et al 1993). In the erythropoietin receptor, this seems to be at the C-terminal end, although I don't think it has been shown for the β_c chain. The most likely explanation of what I have shown is that this region binds and activates the phosphatase which then inhibits kinase activities, whatever they are. I suspect that this is a self-limiting system which activates kinases and then also activates the phosphatases to eventually turn itself off. According to our data, JAKs and STATs could fit in any of the processes I've described, because the regions indicated in Fig. 2 delimit truncations which function from those that do not, but function may still require upstream regions of the receptor involved in JAK/STAT activation. The surprising thing about this is that although we're seeing a lot of promiscuity in the types of JAKs and STATs that are being activated by different cytokine receptors, if you ask the question with knockout mice, the actual cytokine systems that require a particular JAK or STAT are very specific. For example, if you knock out STAT-1, it is only the response of those animals to interferon that is affected, and not to the other cytokines that have been shown to activate the same STAT (Durbin et al 1996, Meraz et al 1996).

Paul: In this system, is it known where the STAT binding sites are on the receptor? Or is the STAT binding directly to a JAK?

Nicola: In this whole cytokine system there are clearly STAT binding sites (the phosphorylated tyrosines) further down the cytoplasmic domain. What has recently been shown is that in many cases one can eliminate the STAT binding sites and still see phosphorylation and activation in the STATs, presumably mediated by JAKs. This shows that there are other ways of activating the STATs other than having to have them bind the cytoplasmic domain of the receptor, and there is also activation through the MAP kinase pathway. There is a bewildering complexity and redundancy in all of these signalling systems. I'm not sure how we're going to work out what's essential and what isn't.

Rajewsky: I protest against the use of the word 'redundancy' in this context. For almost all lymphokines studied in any depth, germline knockout experiments have established a unique, irreplaceable function.

Nicola: All I'm saying is that people are able to measure many different cytokines which activate the same JAKs. I don't know whether those are physiologically significant. The knockout studies are suggesting that they are probably not.

References

Ding DX, Rivas CI, Heaney ML, Raines MA, Vera JC, Golde DW 1994 The alpha subunit of the human granulocyte–macrophage colony-stimulating factor receptor signals for glucose transport via a phosphorylation-independent pathway. Proc Natl Acad Sci USA 91:2537–2541

Durbin JE, Hackenmiller R, Simon MC, Levy DE 1996 Targeted disruption of the mouse STAT1 results in compromised innate immunity to viral disease. Cell 84:443–450

Li J, Koay DC, Xiao H, Sartorelli AC 1993 Regulation of the differentiation of WEHI-3BD$^+$ leukemic cells by granulocyte colony-stimulating factor receptor. J Cell Biol 120:1481–1489

Meraz MA, White JM, Sheehan KC et al 1996 Targeted disruption of the *Stat1* gene in mice reveals unexpected physiologic specificity in the JAK-STAT signaling pathway. Cell 84:431–442

Yi T, Mui AL-F, Krystal G, Ihle JN 1993 Hematopoietic cell phosphatase associates with the interleukin-3 (IL-3) receptor β chain and then down-regulates IL-3-induced tyrosine phosphorylation and mitogenesis. Mol Cell Biol 13:7577–7586

General discussion I

Selection versus instruction?

Nossal: I'd like to tempt the immunologists to critique these two wonderful papers. To kick things off, I have three general questions. First, what are we to make of the redundancy of receptors and factors, shared chains of receptors, multiple receptors on a given cell, the multiple actions of a single growth factor, and the marked synergy of factors? Second, I think every one of us has to be engaged with this question of instruction versus selection. For instance, what are the roles of instruction and selection in the GM-double progenitor going to G or M, the stem cell potential to go to myeloid or erythroid development, or the T cell going to Th1 or Th2? And the third question relates to the intriguing possibility of homeobox genes as master genes, initiating a whole cascade of transcription. Is that a useful paradigm for looking at the initiation of a particular line of haemopoietic differentiation?

Paul: You alluded to the whole question of whether some or all of these differentiation effects were due to an instructive event, by which I assume you mean that the ligand causes the transcription of a set of genes that determine development of a cell in a certain direction versus a spontaneous activation of such genes followed by selection for growth of cells in which the genes have been activated. Let me ask this differently: is there a selection that you can envisage that is based on something other than growth? If we are distinguishing selection and instruction, instruction seems to me to be the induction of a programme of differentiation through a direct activation or set of transcription elements. What do we mean by selection?

Nossal: In all three of the systems we're talking about—haemopoiesis, and T cell and B cell development—there is an overlapping process of growth and differentiation. Almost everything these cells want to do as they're making their decision involves some replication, some clonogenesis, but at the same time, the cells of the expanding clone must differentiate, i.e. develop more specialized function. For example, we know that IL-6 can drive a B cell towards large-scale antibody synthesis. The interesting question, in the context of a lot of transcription going on and a lot of growth, is the very one you have posed—does the cytokine itself unleash a programme of differentiation? Does it do so by critically influencing the transcription, for instance of some *myoD*-like 'master' gene? If so, what genes might be involved? Or is it indeed, as Mike Dexter proposed, that this cell is going through some kind of internal, pre-programmed cycle? In this case, the cytokine will only be doing the job of selection.

Paul: Let's say, just for a moment, that it is selection that is important. Presumably, selection results in the survival or growth of a cell that has spontaneously differentiated, for example, to express a growth factor receptor. I assume that is the argument. We know that, for example, in the switching of a B cell to IgE production, the evidence is quite strong that IL-4, acting through STAT-6, causes transcription of germline ε. It is new transcription that we know is induced by IL-4. You might say that is a really quite trivial event at the very end of the differentiation process. We may ask whether such simple end-stage events illuminate what is happening at the beginning of haemopoietic cell differentiation, or whether they are entirely different phenomena.

Nossal: That is by no means trivial, because the same cell could have chosen to make IgG 2a and thus have a completely different effect on the immune system. I would describe that as a pretty clear instruction-style model.

Goodnow: Pre-B cell differentiation provides a good example of how difficult it can be to separate instruction and selection. Certain growth factors carry pro-B cells to the stage where they functionally rearrange their heavy chain genes, which then extinguishes expression of growth factor receptors and sets the cell in motion to depend on a different set of survival-promoting niches and switches on different homing molecules. This instructs the cell to change its selective requirements. In those kinds of differentiation steps, attempts to distinguish selection and instruction are a 'chicken and egg' exercise.

Melchers: In lymphocyte differentiation, the natural programme of the cells is death unless they are selected. Precursor B cells induced to differentiate via VDJ and VJ chain rearrangements into immature B cells have a tremendous tendency to die by apoptosis. One could argue that the primary role of cytokines, cell–cell contacts and antigen–receptor engagement is to prevent the cells from dying. In a radical experiment that we did, we cultured precursor B cells from a Rag-2 knockout mouse-cells which would normally die, and probably do *in vivo*. If these cell are grown *in vitro*, even in the presence of hydroxyurea which rules out cell division, they do not die so fast that, within six days, they would not switch in the presence of CD40-specific antibodies and IL-4 to ε. Thus it seems that one can drive these cells all the way to an εH chain-producing cell without any cell division if you give them the right ligands. However, we are cautious not to over-interpret this experiment: we concluded that first of all cytokines drive the whole process and, second, all you have to do is allow some kind of differentiation and avoid apoptosis. I think it is a question of semantics to call that either selection or instruction.

Metcalf: I think it is both instruction and selection. If you go back to the pre-haemopoietic embryo, there is an initial commitment for certain cells to enter haemopoiesis, and this shuts down permanently other options the cell might previously have had when it was totipotential. We know very little about the molecular biology of that commitment process. I find the absolute irreversibility of this step puzzling: how can this be sustained through repeated cell divisions? With the populations we're working on we begin with minute numbers of precursor cells. Consequently, there is an unavoidable element of proliferation in all our experiments. It is only in the unusual type of experiment that Fritz Melchers is able to do where it can be formally documented that cell division is not needed for certain changes. It has always been considered an important biological question whether or not a cell needs to go through mitosis to change its phenotype radically. From present information, the answer seems to be, no. It is the irreversibility I find puzzling, but the changes seem likely to be induced responses to outside signalling.

Nossal: But it is not quite as straightforward as that. For example, that self-same haemopoietic stem cell transplanted to a different region of the embryo could make brain. There is flexibility.

Metcalf: Such a cell must be at the pre-commitment stage, but the changes must still be dependent on external signalling.

Nossal: I wouldn't be as dogmatic as you about saying that a haemopoietic stem cell has been committed irrevocably to be a blood cell. There may still be some flexibility we are unaware of until we find the experimental conditions to test it.

Hodgkin: A couple of times we have touched on the question of how to relate division and proliferation to differentiation. I wanted to mention some experiments we've been doing tracking and comparing division and differentiation in B cells and the switch to IgG1 and IgE that Bill Paul mentioned. When we stimulate cells *in vitro* with IL-4 and CD40 ligand, we never see IgG1 expression until the cells have divided three or four times. We don't see IgE expression until the cells have divided six or seven times. Although the commitment to IgG1 or IgE expression might be made very early on by exposure to IL-4 in the undivided cell, the actual expression of that commitment is manifested very predictably at a certain division cycle down the track. It seems to me (at least from these experiments) that there is some relationship between the differentiation and the division cycle number. This can also be seen another way: after a few days in culture B cells become extraordinarily heterogeneous in terms of their cell surface phenotype. Many different cell surface phenotypes can be identified. At face value it appears the activation has caused the B cells to differentiate into many cell types. However, when you incorporate cell division number as a variable you see the population is following a single differentiation path. It now looks much cleaner, much clearer.

Nossal: This appearance of specialized secretory capacity after a given number of cell divisions begins to look like some kind of internal program, doesn't it.

Hodgkin: Yes, there's a number of transition states in between each cell division that show different patterns of expression of cell surface markers.

Paul: I assume you are measuring membrane expression of ε or $\gamma 1$, or alternatively secretion.

Hodgkin: Yes. We measured membrane expression.

Paul: We know that event requires not only the targeting of Iε or Iε but also an exceedingly poorly understood turn-on of a switch recombination mechanism. The germline transcription of Iε and $\gamma 1$ presumably can occur much more rapidly— these might not require any division at all.

This may only be a trivial point, but systems involving a straight choice between death or survival will be inherently difficult to analyse; when the cell has alternative choices both of which allow survival, it's a lot easier to ask the question. Perhaps that's why the seemingly trivial end-stage issue of ε versus γ2a expression is such a nice system. Since the cell does make alternative choices, it is fairly clear that that is a transcriptionally driven event and not a selective event.

Williams: Some years ago, Don Metcalf did experiments in which he took yolk sac stem cells, which are known to be quite distinct in their differentiation capacity from adult stem cells, and transplanted them (Moore & Metcalf 1970). Recently, we have taken another look at this. These cells in the yolk sac formed primitive erythroblasts and perhaps some macrophages. We've been able to reconstitute adult animals with these cells, and have looked carefully *in vivo* at comparisons between their differentiation capacity. Remarkably, once you take them out of the yolk sac microenvironment their differentiation capacity resembles that of adult stem cells, even though their proliferative capacity is very distinct (M. Yoder, unpublished observation). Our conclusion is that this is very much an instructional process of the microenvironment.

Mike Dexter, several years ago you published a paper in which you splenectomized mice and gave them massive doses of G-CSF (Molineux et al 1990). These animals became anaemic, which I would interpret, again, as being an instructional event. Would you?

Dexter: No; it is almost certainly because of spatial constraints in the bone marrow because of the over-production of mature granulocytes. If you look at the production of erythroid progenitor cells in those animals (and you can do this almost irrespective of the growth factor that you give) there's really no change. In other words, we don't have any evidence that giving supra-levels of growth factors is influencing the generation of committed cells from the multipotent stem cells, although they can obviously influence the generation of mature cells from the progenitor cells. I would go so far as to say that *in vivo*, there is no evidence to indicate that giving growth factors in excess or taking them away has an influence on lineage commitment of the multipotent stem cells, in terms of seeing major or even significant changes in the production of committed progenitor cells. I would argue that you are taking the stem cells from a restrictive environment — of the type proposed some years ago by Dov Zipori of the Weizmann Institute, Israel.

Williams: That is simply semantics.

Dexter: No it is not semantics.

Williams: You say it's restricted, but you could also say that the microenvironment is instructive in the sense that it is only providing certain cues for the cells to behave in the way that they are behaving — that is semantics.

Dexter: Or that it might be restricting access to various growth factors, or extracellular molecules or perhaps even blocking the ability of the cells to respond to such molecules. It may even be restricting access to glucose and thus modulating, indirectly, energy levels.

Strasser: Phil Hodgkin mentioned experiments where, if you have a system where you have proliferation and differentiation, you only see the differentiated phenotype later on. That could easily be explained by the common finding that differentiation and proliferation are mutually exclusive. There are several examples where if you provide an abnormal proliferative stimulus, such as the ectopic expression of Myc, you can override a rather dominant differentiative stimulus. Many of these questions will probably be answered once all the receptor pathways are linked up to the control of the cell cycle. You have to hold the cell cycle if you want to differentiate — you can't do it in S phase.

Zinkernagel: Because we culture cells in glass vials and plastic dishes we tend to forget that these processes take place in the context of a three-dimensional anatomical structure. Perhaps this is significant, particularly in the bone marrow. We often treat the bone marrow as a neutral type of environment that can be replaced by glucose and a cocktail of growth factors. This is probably not the case. The evidence from the anatomy of the immune system has taught us that these interactions occur in very precise locations where gradients of factors may play a vital role. Whether you call this regulation, selection or instruction is not so important, as long as we can define these distinct steps and understand the interrelation of anatomical structure, the messenger molecules and the physiology of the cell involved.

Dexter: I began my talk by arguing that growth factors are just one component of the regulatory environment. We know, for example, that adhesion molecules can influence the response of haemopoietic cells to their environmental milieu. If you take all the adhesion molecules and the extracellular matrix molecules into account, which are also signalling to these cells, then you may well be right in arguing that the reductionist approach we have adopted may give misleading conclusions.

Nicola: I have thought about the selection/instruction issue quite a bit. If there is a cell that will give rise to two different phenotypes, 'A' and 'B', how do you distinguish between selection and instruction? The only experiment that would convince me would be to have 100% purity of the cells, have no cell division or cell death take place in the system, and to be able to add something that would lead to every cell becoming 'A' in one case or 'B' in another. This would eliminate selection and confirm instruction.

Nossal: Take away your condition of no cell division, and the B cell CD40 ligand/IL-4 situation comes dashed close. Here, every B cell can be stimulated and must make its isotype choices. This, however, involves division, which raises the possibility of autocrine loops and other feedback mechanisms within the developing clone.

Morstyn: On a more practical level, the transplanter would obviously like to have a large number of very early haemopoietic cells. As Mike Dexter said, this doesn't seem to be possible. How realistic is a vision of mimicking the protein–protein interactions with small molecules that act intracellularly to produce a lot of early cells by stimulating proliferation without differentiation? The other possibility is that we are still missing a ligand for a known receptor. What is the more likely direction to meet this goal of getting more of these early cells? Do we have to understand all the intracellular mechanisms and interfere with them, or is there a missing ligand?

Nicola: I'd like to believe that there is a ligand missing. Although interfering with signalling pathways in a specific way is very exciting, realistically it will be a long time before we have the tools that we need to expand stem cell numbers in this manner. I am encouraged by the fact that stem cells do seem to proliferate without differentiating in the body, and so there should be some way for us to find ligands or combinations of cytokines and other factors that will reproduce the protective niche that people talk about.

Melchers: I agree with Don Metcalf that *in vivo* cell proliferation normally occurs in the haemopoietic system, so if you want to look at the *in vivo* situation you won't get around considering proliferation. However, I disagree with Phil Hodgkin and Andreas Strasser about the possibility of a B cell being able to differentiate all the way to a plasma cell. It is clear from experiments we did many years ago that all of this differentiation to an Ig-secreting plasma cell can be induced without division, with the same time scale (Karasuyama & Melchers 1988). It takes several days before the switch to other isotypes occurs and for a typical plasma cell to develop, but it occurs with the same time schedule as would occur *in vivo*. This is a very compelling parallelism in the temporal programme of late stage differentiation of B cells. None the less, it is dangerous to compare that differentiation with any other cellular differentiation. I have very little to say on instructionist versus selectionist hypotheses, except that I would emphasize that cells have a non-proliferative programme that can be manipulated by external stimuli into different directions. This is reminiscent of the cell cycle that Mitchison used to talk about — the metabolic cycle that isn't really linked to mitosis. We have very imperfect techniques for measuring and synchronizing this metabolic cycle. It would be very interesting to be able to dissect both the mitotic and the non-mitotic cycles.

Nossal: A very clear indication of exactly what you are saying is provided by the single cell studies that I did with Bussard on Ly1 B cells (Nossal et al 1970). When placed into culture, these cells are in G0 and are not secreting antibody. Under the influence of various growth factors, they secrete IgM but *without* an intervening division. As to whether my central question was semantics or not, I think we got the answer from Bill Paul when he suggested that we drop the word 'instruction' and instead look at whether a particular cytokine, interacting with that particular receptor, unleashes a new programme of gene transcription. I would like to leave on the table the question I have posed several times, namely, might there be a hierarchy of genes? Does anyone still believe in master genes? I think that most people feel that the 'master–slave' hypothesis from the MyoD story was considerably over-simplified, but does the group still find some merit in it?

References

Karasuyama H, Melchers F 1988 Establishment of mouse cell lines which constitutively secrete large quantities of interleukin 2, 3, 4 or 5, using modified cDNA expression vectors. Eur J Immunol 18:97–104

Molineux G, Pojda Z, Dexter TM 1990 Haemopoiesis in normal and splenectomized mice treated with granulocyte colony-stimulating factor. Blood 75:563–569

Moore MAS, Metcalf D 1970 Ontogeny of haemopoietic systems: yolk sac origin of *in vivo* and *in vitro* colony forming cells in the mouse embryo. Br J Haematol 18:279–296

Nossal GJV, Bussard AE, Lewis H, Mazie JC 1970 *In vitro* stimulation of antibody formation by peritoneal cells. I. Plaque technique of high sensitivity enabling access to the cells. J Exp Med 131:894

The molecular control of granulocytes and macrophages

Donald Metcalf

The Walter and Eliza Hall Institute of Medical Research, Post Office, Royal Melbourne Hospital, Victoria 3050, Australia

Abstract. The proliferation *in vitro* of granulocytes and macrophages can be regulated by the four colony stimulating factors (CSFs), stem cell factor and interleukin 6, with Flk ligand having a weaker action. Combinations of these glycoprotein regulators produce superadditive proliferative responses. The CSFs also influence commitment, maturation and mature cell functional activity and these various responses are initiated by distinct regions of the individual receptors. The injection of single CSFs into experimental animals or patients reproducibly enhances granulocyte or monocyte formation or function despite the existence of complex interacting networks of regulatory molecules. Verification of the importance of the CSFs for the regulation of basal haemopoiesis has been obtained by analysis of mice in which the genes encoding the CSFs or their receptors have been inactivated. In a casein-induced model of acute inflammatory responses, the migration of neutrophils from the marrow and localization of these cells to the inflammatory site appear not to be CSF-dependent processes even though major increases occur in CSF levels at the inflammatory site.

1997 The molecular basis of cellular defence mechanisms. Wiley, Chichester (Ciba Foundation Symposium 204) p 40–56

Neutrophilic granulocytes and macrophages are relatively short-lived cells that are produced continuously in the bone marrow from a transit pool of committed progenitor cells. Although they differ radically in morphology and surface markers, many granulocytes and macrophages are derived from common bipotential progenitor cells and are therefore closely related. This ancestral relationship suggests that their function may either show some overlap or that the cells may be intended to function in a coordinated manner. Macrophages are prominent in mediating tissue remodelling during development and in killing and/or removing effete or damaged cells throughout adult life. They are also important cells in mediating host responses to many types of invading microorganisms. Granulocytes may exhibit similar functions, at least in response to tissue injury, but appear to be primarily involved in responses to microorganisms. As a consequence, if no significant infections are in progress, there is evidence that most granulocytes are unused and die locally in the bone marrow without leaving their site of production (Metcalf et al 1995).

Seven specific regulatory factors have been identified as stimulating the formation of granulocytes and/or macrophages, but these do not act in isolation. Eight major lineages of blood cells are produced simultaneously in the bone marrow from a common pool of multipotential stem cells under the action of at least 25 characterized regulatory factors. Granulocytes and macrophages can also exhibit receptors for some regulators whose primary actions may be on cells of other lineages. This not only results in the possibility of competition between the lineages for available stem cells but also of competition for adequate stromal cell niches in the bone marrow and of complex enhancing or inhibitory interactions between many of the regulators. Similar complexity may also exist in ensuring the orderly release of cells from the marrow, their direction to correct tissue locations and in their functional activation at such locations. This situation demands that information obtained from simplified tissue cultures of purified target cells be verified by studies using whole animals before concluding that the regulatory mechanisms identified *in vitro* are necessarily of relevance *in vivo*.

The colony stimulating factors

No progress in analysing the molecular control of granulocytic and macrophage populations was possible until the mid-1960s, when methods were developed for the clonal culture *in vitro* of granulocyte/macrophage progenitor cells, which formed colonies of maturing granulocytes and macrophages (Metcalf & Nicola 1995). The formation of these cells required stimulation by the addition of tissue fragments, extracts or tissue-conditioned medium and the clonal cultures provided a quantitative bioassay for monitoring the active agents in such material. The active agents were termed 'colony stimulating factors' (CSFs) and a 15-year period followed in which indirect evidence was accumulated that the CSFs were likely to be regulators of granulocyte and macrophage formation *in vivo*. In this period, four CSFs of murine origin were purified to homogeneity using newly-evolving methods for separative protein chemistry. The four CSFs — GM-CSF (granulocyte/macrophage CSF), G-CSF (granulocyte CSF), M-CSF (macrophage CSF) and multi-CSF (multipotential CSF; also known as interleukin 3) — proved to be glycoproteins of reasonably similar polypeptide core size (Table 1) but with a varying content of carbohydrate. Each is active in stimulating cell proliferation *in vitro* at concentrations of picograms to nanograms per millilitre. A characteristic of these CSF-induced proliferative responses is that superadditive synergistic responses are observable if combinations of two or more CSFs are used (Metcalf & Nicola 1992). An example of such synergy is shown in Fig. 1, where combination of GM-CSF with M-CSF not only increases total cell production but permits the formation of unusually large colonies possibly initiated by cells that, like stem cells, have a mandatory requirement for simultaneous stimulation by two or more haemopoietic regulators.

In common with all other haemopoietic regulators, the proliferative actions of the CSFs are not confined to cells of a single lineage (Table 1). Because most granulocytic

TABLE 1 The murine colony-stimulating factors

Type	*Glycosylated mass (kDa)*	*Protein core (kDa)*	*Cells responding to proliferative stimulation in vitro*	
			Strong actions	*Weaker actions*
GM-CSF	18–30	14.4	G, M, Eo	Meg, E, Stem
G-CSF	25	19.1	G, Stem	M
M-CSF	70, >200	21, 18 (monomer)	M	G
IL-3 (multi-CSF)	22, 29, 34	16.2	G, M, Eo, Meg, E, Mast, Stem	B

G, neutrophilic granulocytes; M, monocytes/macrophages; Eo, eosinophils; Meg, megakaryocytes; E, erythroid cells; Mast, mast cells; Stem, multipotential stem cells; B, B lymphocytes.

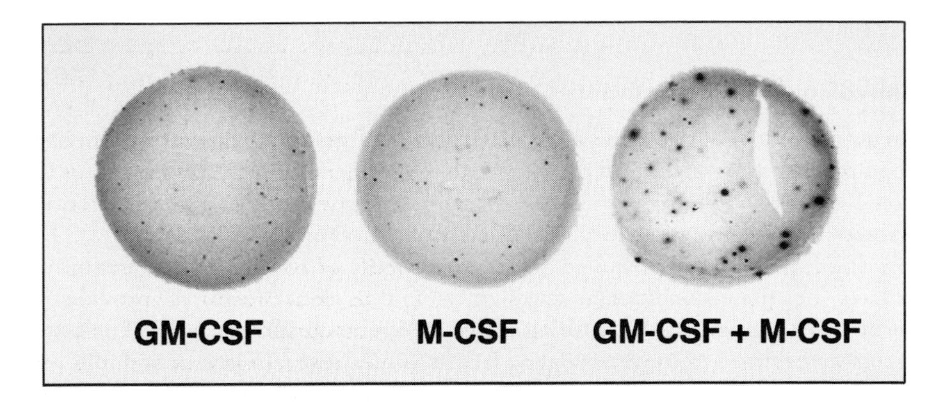

FIG. 1. Wholemount stained preparations of cultures of 25 000 bone marrow cells from C57BL/6 mice showing the major size enhancement of colonies when GM-CSF (10 ng/ml) is combined with M-CSF (10 ng/ml). Seven day cultures in 35 mm dishes.

and macrophage cells co-express specific receptors for all four CSFs, it was anticipated that the CSFs would be structurally related. This proved not to be the case when amino acid sequences were established for the four polypeptides, although relatedness has subsequently become more firmly established from their similar four-α-helical 3D configuration, common features in their gene structure and, for three of them, related and shared components in their receptors (Metcalf & Nicola 1995).

The polyfunctional actions of the CSFs

The CSFs are not simply proliferative stimuli: they also have the capacity to influence differentiation commitment decisions at the stem cell to progenitor cell stage, to influence the initiation of maturation, to maintain cell viability by ensuring adequate membrane transport and to influence the level of functional activity exhibited by mature cells. Subsequent *in vivo* studies have added further functions to the CSFs, such as influencing cell release from the marrow (Metcalf & Nicola 1995).

Acceptance of each of these functions as genuine and direct actions of the CSFs did not occur without dispute. Central to much of this controversy was the undoubted ability of the CSFs to influence cell survival (Metcalf & Merchav 1982, Begley et al 1986) by maintaining membrane transport integrity and preventing death by apoptosis (Williams et al 1990). This permitted the criticism that all of the proposed special functions of the CSFs might merely be the passive consequence of maintenance of cellular viability.

The CSF concentration does influence the number of cells developing within a colony during the culture period and studies using paired daughter cells verified this as a direct action based on shortening of cell cycle times (Metcalf 1980). Subsequent studies have shown that M-CSF initiates production of cyclin D1 that is necessary for passage of cells from the G1 to S phase of the cell cycle and confirms that the major actions of CSFs are likely to be shortening of the G1 period and achieving the critical transition from G1 to S (Roussel & Sherr 1993). Analysis of GM-CSF and G-CSF receptors has identified two regions, Box-1 and Box-2, that are necessary for eliciting mitotic signalling by interaction with JAK kinases. Conversely, viability of progenitor cells can be sustained *in vitro* by overexpression of *bcl-2* or by supplementation of the medium with ATP, but neither manoeuvre results in cell division (Vaux et al 1988, Whetton & Dexter 1983).

Analysis of the formation by individual stem cells of committed progenitor cell progeny has documented the stochastic nature of this process. However, the probability of occurrence of particular events in a stochastic process is capable of being altered by extrinsic signalling. In studies in which stem cells were stimulated to divide by stem cell factor (SCF; also known as Steel factor) plus CSFs, some evidence was obtained that the proportion of various progenitor cells generated could be influenced by the particular CSF used (Metcalf 1993). More convincing evidence of an influence of CSF on differentiation commitment was obtained from an analysis of CSF action on bipotential progenitor cells where M-CSF was able to commit such cells to macrophage formation and GM-CSF to granulocyte formation (Metcalf & Burgess 1982). Even stronger evidence of CSF-induced differentiation commitment has emerged from an analysis of CSF action on suitable myeloid leukaemic cell lines. G-CSF and GM-CSF can clearly suppress self-generation by such cells and induce commitment to the formation of maturing granulocytes or macrophages (Metcalf 1982, Maekawa et al 1990). Analysis of receptor structure has indicated the existence of a specific cytoplasmic domain of these receptors that is necessary for such differentiation induction (see Nicola et al 1996, this volume).

A CSF, acting on a single progenitor cell in fully defined medium, can stimulate the formation of granulocytic or macrophage colonies in which full cellular maturation occurs. Scepticism regarding a mandatory role for CSFs in initiating maturation was aroused by the observation that some maturation can occur in CSF-deprived cultures, particularly in cultures of CSF-dependent multipotential cell lines overexpressing *bcl-2* when such cells are cultured in the absence of CSF (Fairbairn et al 1993). However, it has been documented that a C-terminal region of the G-CSF receptor is essential for the occurrence of maturation in some G-CSF-responsive granulocytic cell lines (Fukunaga et al 1993). Similarly, in congenital neutropenia, a disease characterized by maturation arrest at the promyelocyte stage, in some cases the defect has been associated with absence or mutation of the same C-terminal region of the G-CSF receptor (Dong et al 1994).

These conflicting observations suggest that immature granulocytic and macrophage cells may be capable of a certain level of maturation in the absence of continuing CSF stimulation but that CSF signalling is capable of initiating maturation as an active event and may be the more usual, or more effective, method for initiating this process.

The actions of CSFs in maintaining cell viability *in vitro* are not disputed nor are their actions in enhancing functional activity of mature cells. Although some evidence has implicated the α-chain of the GM-CSF receptor as regulating the membrane transport of some metabolites (Ding et al 1994), other evidence has identified a region in the β-chain as being necessary for cell survival (Kinoshita et al 1995).

It remains unclear why a post-mitotic cell responds to CSF signalling merely by exhibiting increased functional activity whereas a progenitor cell with similar receptor numbers responds by proliferation. It appears that the actual response of these target cells is determined, or restricted, by gene programmes available or activated in the responding cell and, to this degree, regulators such as the CSFs are capable only of eliciting a set of signals whose actual consequences are then dictated by the cell itself.

Other regulators active *in vitro* on granulocytic populations

Subsequent to the discovery of the CSFs, three other regulators have been found that are able to directly stimulate (predominantly) granulocytic colony formation *in vitro* (Metcalf & Nicola 1995). Of these, SCF is the most potent and is able to stimulate the formation of more and larger granulocytic colonies than G-CSF. In cultures of murine bone marrow cells interleukin 6 (IL-6) stimulates the development of granulocytic colonies that resemble those stimulated by G-CSF, although with a curiously flat dose–response curve. The Flk ligand has a much weaker capacity to stimulate granulocyte colony formation and only does so at relatively high concentrations.

In vivo assessment of the actions of regulators active on granulocytic or macrophage populations

Injection of CSFs elicits responses in granulocyte/macrophage populations that agree qualitatively with the range of actions observed for these CSFs *in vitro* (see Metcalf & Nicola 1995). However, there are major discrepancies in the magnitude of the responses observed, particularly if assessed merely by changes in peripheral blood cell populations. IL-3 and GM-CSF are the most active agents *in vitro* in stimulating granulocyte colony formation, as assessed by colony numbers or size, and even SCF has a stronger action than G-CSF. However, animals injected with G-CSF develop higher neutrophil levels than those injected with GM-CSF and little change is observed following the injection of IL-3 or SCF. Responses in humans have followed a comparable pattern. Similarly, M-CSF is the strongest agent *in vitro* for stimulating macrophage proliferation yet, *in vivo*, monocyte numbers vary little following the injection of M-CSF, and at most, a relative monocytosis develops. Such apparent discrepancies are in part misleading because blood white cell levels provide only a limited view of the changes that can occur in response to the injection of a regulator. For example, if CSFs are injected intraperitoneally, the largest rise in peritoneal cell numbers is the GM-CSF-induced rise in macrophages, although again responses to M-CSF are minor. G-CSF also induces little change in local peritoneal neutrophil numbers following intraperitoneal injection. The population responding most strongly to the injection of IL-3 is splenic mast cells, which can multiply 100-fold.

The injection into mice of SCF, IL-6 or Flk ligand produces only minor responses in granulocytic or macrophage populations.

A response that could not be studied *in vitro* is the CSF-induced rise in stem and progenitor cell numbers in the peripheral blood due to release of selected cells from these populations in the bone marrow. Again G-CSF proved to be outstanding in eliciting 100-fold rises (Dührsen et al 1988) while GM-CSF is somewhat less active and IL-3 or SCF less active again. All CSFs increase the level of haemopoiesis in the spleen and, again, G-CSF is outstanding in causing a massive relocation of erythropoiesis from the now granulocytic marrow to the spleen (Molineux et al 1990).

Although the injection of single CSFs can elicit reproducible responses *in vivo*, such responses cannot necessarily be ascribed solely to the direct action of the injected CSF. There are many other haemopoietic regulators present in the injected animals and one or more of these, such as SCF, might interact with the injected CSF to produce the response observed.

These various *in vivo* observations raised a number of issues. While the major responses to G-CSF supported the contention that G-CSF was an important *in vivo* regulator of granulocyte formation, the less than impressive responses to M-CSF cast doubt on its role. In contrast to its strong local effect, and the converse situation with G-CSF, the relatively weak blood cell responses to GM-CSF indicated that some regulators might be designed for local tissue regulation whereas others function as humoral regulators with strong systemic effects. Overall, the view was expressed that, because CSFs are only readily detectable in the circulation in aplasia and severe infections, the CSFs might have little role in the regulation of basal haemopoiesis and instead only be of relevance in emergency situations.

To resolve these questions, it has been necessary to study mice in which the genes for particular regulators or their receptors have been inactivated by natural mutations or by homologous recombination. Such studies are now reasonably comprehensive and,

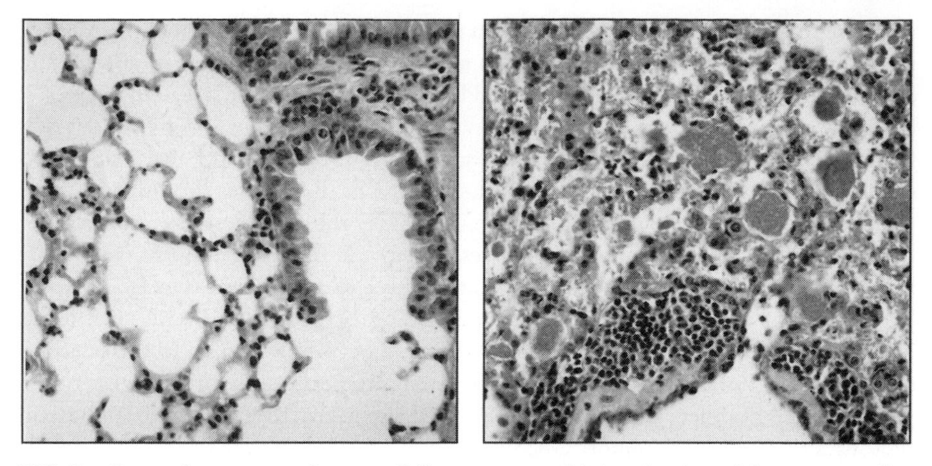

FIG. 2. Lungs from a normal mouse (*left*) or a mouse with inactivation of the gene encoding the β_c chain of the GM-CSF receptor (*right*). Note in the knockout lung the accumulation of surfactant and proteinaceous material in the alveoli and the peribronchial lymphoid accumulations. Both are typical consequences of loss of GM-CSF action.

with the exception of IL-3, have documented that CSFs are of importance for basal haemopoiesis. Inactivation of the G-CSF gene results in severe neutropenia, a major reduction in marrow neutrophils and an inability to adequately mobilize neutrophils on demand (Lieschke et al 1994a). An inability to produce M-CSF (as occurs in the spontaneous mutant *op/op* mouse with an inactivating frameshift in the M-CSF gene [Yoshida et al 1990]) is associated with a major depletion of macrophages in various tissues and a dramatic deficiency in macrophage-derived osteoclasts leading to osteopetrosis and failure of teeth eruption (Wiktor-Jedrzejczak et al 1992). Inactivation of the GM-CSF gene or the gene encoding the β-chain of the GM-CSF receptor does not lead to numerical changes in granulocyte/macrophage populations but results in functional inactivation of lung macrophages with surfactant and protein accumulation (Dranoff et al 1994, Stanley et al 1994, Robb et al 1995) (Fig. 2). This state mimics the disease alveolar proteinosis. Combined deficiency of GM-CSF and M-CSF accentuates this state and results in premature death from pneumonia, although such mice are not wholly depleted of macrophages (Lieschke et al 1994b). Deficiency in SCF formation (*Sl* mice) or SCF receptor (W^v mice) also leads to a neutropenia, evident either in the blood or marrow (Ruscetti et al 1976, Chervenick & Boggs 1969). However, deficiency in IL-3 or IL-6 has yet to be reported to result in any impairment of granulocyte/macrophage populations.

None of these individual inactivations has resulted in complete absence either of granulocytes or macrophages. Potentially a combined deficiency in G-CSF plus SCF might result in total absence of granulocyte formation but it seems evident that there must be at least one major regulator of macrophage formation remaining undiscovered.

Role of CSFs in migration of cells to inflammatory sites

In vitro studies have shown both G-CSF and GM-CSF to be positively chemotactic for neutrophils (Wang et al 1987, 1988). Although injected CSF of both types results in elevated neutrophil levels in the blood and a delayed release of stem and progenitor cells from the marrow, these observations do not necessarily prove that CSF, produced in some local inflammatory site, is responsible for the localization of neutrophils to that site.

To investigate this question, we have developed a mouse model in which an acute inflammatory response occurs intraperitoneally. Within three hours of the intraperitoneal injection of 2 ml of 0.2% casein solution, neutrophil numbers in the peritoneal cavity rise from zero up to 20×10^6 cells. Histological examination showed that this was not due to local migration from the surface of abdominal organs but to transit of neutrophils through small venules and capillaries into the loose fatty tissue of abdominal organs. Although no consistent change occurred during this period in numbers of neutrophils in the peripheral blood, there was a consistent fall in marrow neutrophils totalling two- to 10-fold more than the number of cells entering the peritoneal cavity (Fig. 3). Analysis using animals labelled with tritiated thymidine showed that the release and migration of the neutrophils was selective and involved only the oldest (most mature) of the marrow neutrophils.

Levels of GM-CSF rose in the peritoneal fluid to a maximum value of 2 ng/ml 2 h after the injection of casein and G-CSF levels rose progressively, reaching 14 ng/ml at 3 h. Somewhat lower rises in CSF concentrations occurred in the serum and these lagged slightly behind those in peritoneal fluid. For both CSFs, a concentration of 2 ng/ml in cultures of marrow cells results in the formation of maximum colony numbers, indicating that the concentrations observed in casein-injected mice are biologically relevant. This raised the possibility that one or other CSF might be actively involved in mediating the cellular migration to the peritoneal cavity. However, analysis of the response to injected casein of mice with homozygous inactivation of the genes encoding GM-CSF, G-CSF or the β-chain of the GM-CSF receptor showed in each case that such mice were able to develop a rise in neutrophil numbers in the peritoneal cavity (Fig. 3), associated with the usual fall in marrow neutrophil numbers.

These observations appear to exclude either GM-CSF or G-CSF from a mandatory role in achieving this type of neutrophil inflammatory response. It seems reasonable to assume, however, that the major rises in local CSF concentrations in this inflammatory response are not purposeless and further studies are needed to determine whether such rises are designed to achieve functional activation of the neutrophils once they have arrived at the inflammatory site.

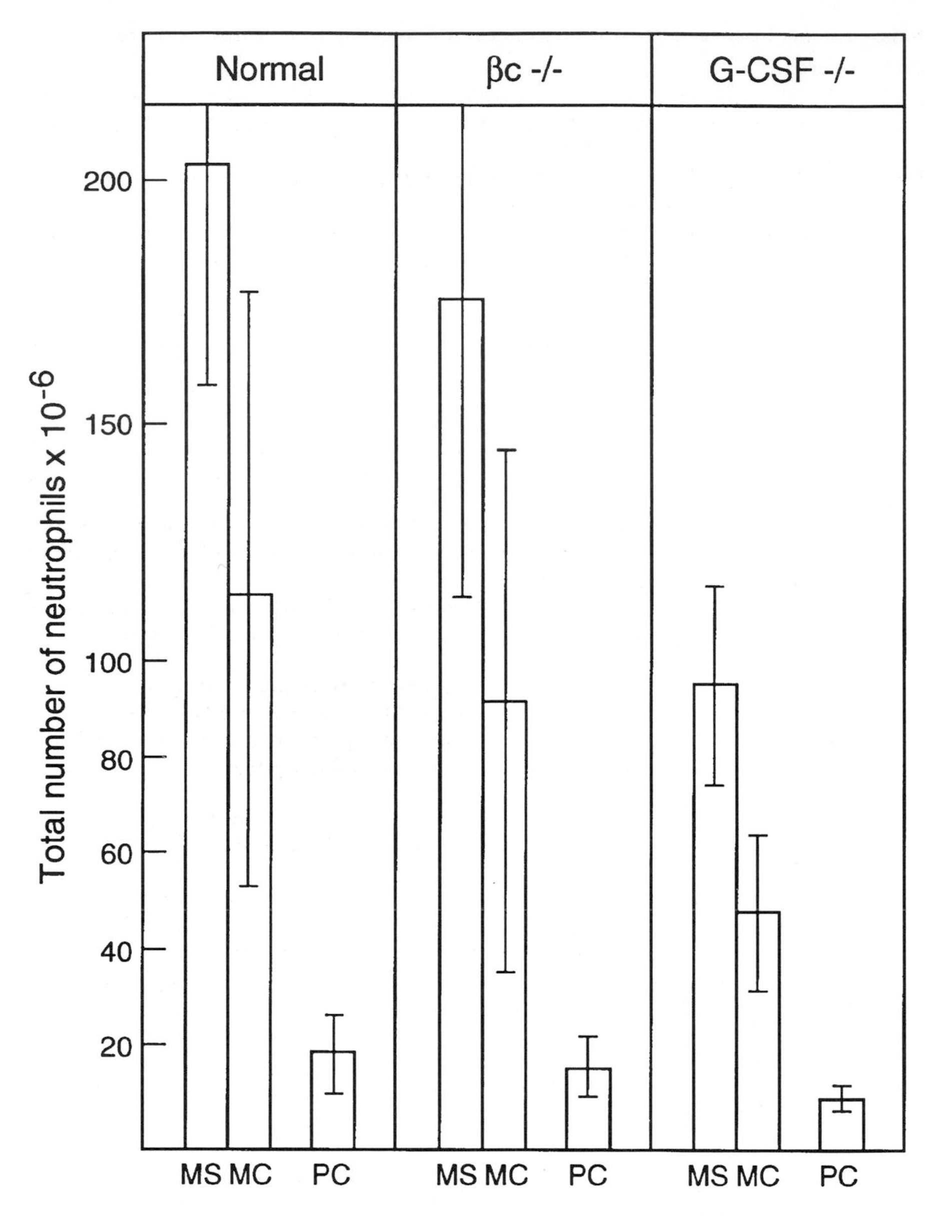

FIG. 3. Total numbers of neutrophils in the bone marrow three hours after the intraperitoneal injection of saline (MS) or casein (MC) compared with neutrophil numbers in the peritoneal cavity after casein injection (PC). Mean values ±SD from four mice per group. Note the similar responses in normal mice, and mice with inactivation of the gene encoding the β_c chain of the GM-CSF receptor ($\beta_c{}^{-/-}$) or the G-CSF gene (G-CSF $^{-/-}$). Overall, cell numbers are reduced in G-CSF$^{-/-}$ mice but the pattern of the response is as in normal mice.

References

Begley CG, Lopez AF, Nicola NA et al 1986 Purified colony stimulating factors enhance the survival of human neutrophils and eosinophils *in vitro*: a rapid and sensitive microassay for colony stimulating factors. Blood 68:162–166

Chervenick PA, Boggs DR 1969 Decreased neutrophils and megakaryocytes in anemic mice of genotype W/W^V. J Cell Physiol 73:25–30

Ding DX-H, Rivas CI, Heaney ML, Raines MA, Vera JC, Golde DW 1994 The α-subunit of the human GM-CSF receptor signals for glucose transport via a phosphorylation-independent pathway. Proc Natl Acad Sci USA 91:2537–2541

Dong F, Hoefsloot LH, Schelen AM et al 1994 Identification of nonsense mutation in the granulocyte-colony-stimulating factor receptor in severe congenital neutropenia. Proc Natl Acad Sci USA 91:4480–4484

Dranoff G, Crawford AD, Sadelain M et al 1994 Involvement of granulocyte–macrophage colony-stimulating factor in pulmonary homeostasis. Science 264:713–716

Dührsen U, Villeval J-L, Boyd J, Kannourakis G, Morstyn G, Metcalf D 1988 Effects of recombinant human granulocyte-colony stimulating factor on hemopoietic progenitor cells in cancer patients. Blood 72:2074–2081

Fairbairn LJ, Cowling GJ, Reipert BM, Dexter TM 1993 Suppression of apoptosis allows differentiation and development of a multipotent haemopoietic stem cell in the absence of added growth factors. Cell 74:823–832

Fukunaga R, Ishizaka-Ikeda E, Nagata S 1993 Growth and differentiation signals mediated by different regions on the cytoplasmic domain of granulocyte-stimulating factor receptor. Cell 74:1079–1087

Kinoshita T, Yokota T, Arai K-I, Miyajima A 1995 Suppression of apoptotic death in hematopoietic cells by signalling through the IL-3/GM-CSF receptors. EMBO J 14:266–275

Lieschke GJ, Grail D, Hodgson G et al 1994a Mice lacking granulocyte colony-stimulating factor have chronic neutropenia, granulocyte and macrophage progenitor cell deficiency, and impaired neutrophil mobilization. Blood 84:1737–1746

Lieschke GJ, Stanley E, Grail D et al 1994b Mice lacking both macrophage- and granulocyte–macrophage colony-stimulating factor have macrophages and co-existent osteopetrosis and severe lung disease. Blood 84:27–35

Maekawa T, Metcalf D, Gearing D 1990 Enhanced suppression of human myeloid leukemic cell lines by combinations of IL-6, LIF, GM-CSF and G-CSF. Int J Cancer 45:353–358

Metcalf D 1980 Clonal analysis of proliferation and differentiation of paired daughter cells: action of granulocyte–macrophage colony-stimulating factor on granulocyte–macrophage precursors. Proc Natl Acad Sci USA 77:5327–5330

Metcalf D 1982 Regulator-induced suppression of myelomonocytic leukemic cells: clonal analysis of early cellular events. Int J Cancer 30:203–210

Metcalf D 1993 The cellular basis for enhancement interactions between stem cell factor and the colony stimulating factors. Stem Cells (suppl 2) 11:1–11

Metcalf D, Burgess AW 1982 Clonal analysis of progenitor cell commitment to granulocyte or macrophage production. J Cell Physiol 111:275–283

Metcalf D, Merchav S 1982 Effects of GM-CSF deprivation on precursors of granulocytes and macrophages. J Cell Physiol 112:411–418

Metcalf D, Nicola NA 1992 The clonal proliferation of normal mouse hemopoietic cells: enhancement and suppression by CSF combinations. Blood 79:2861–2866

Metcalf D, Nicola NA 1995 The hemopoietic colony-stimulating factors: from biology to clinical applications. Cambridge University Press, Cambridge

Metcalf D, Lindeman GJ, Nicola NA 1995 Analysis of hematopoiesis in *max* 41 transgenic mice that exhibit sustained elevations in blood granulocytes and monocytes. Blood 85:2364–2370

Molineux G, Pojda Z, Dexter TM 1990 A comparison of hematopoiesis in normal and splenectomized mice treated with granulocyte colony-stimulating factor. Blood 75:563–569
Nicola NA, Smith A, Robb L, Metcalf D, Begley CG 1996 The structural basis of the biological actions of the GM-CSF receptor. In: The molecular basis of cellular defence mechanisms. Wiley, Chichester (Ciba Found Symp 204) p 19–32
Robb L, Drinkwater CC, Metcalf D et al 1995 Hematopoietic and lung abnormalities in mice with a null mutation of the common β subunit of the receptors for granulocyte–macrophage colony-stimulating factor and interleukins 3 and 5. Proc Natl Acad Sci USA 92:9565–9569
Roussel MF, Sherr CJ 1993 Signal transduction by the macrophage colony-stimulating factor receptor. Curr Opin Hematol 1:11–18
Ruscetti FW, Boggs DR, Torok BJ, Boggs SS 1976 Reduced blood and marrow neutrophils and granulocytic colony-forming cells in Sl/Sl^d mice. Proc Soc Exp Biol Med 152:398–402
Stanley E, Lieschke GJ, Grail D et al 1994 Granulocyte–macrophage colony-stimulating factor-deficient mice show no major perturbation of hematopoiesis but develop a characteristic pulmonary pathology. Proc Natl Acad Sci USA 91:5592–5596
Vaux DL, Cory S, Adams JM 1988 *bcl*-2 gene promotes haemopoietic cell survival and cooperates with c-*myc* to immortalise pre-B cells. Nature 335:440–442
Wang JM, Colella S, Allavena P, Mantovani A 1987 Chemotactic activity of human recombinant granulocyte–macrophage colony-stimulating factor. Immunol 60:439–444
Wang JM, Chen ZG, Colella S et al 1988 Chemotactic activity of recombinant human granulocyte colony-stimulating factor. Blood 72:1456–1460
Whetton AD, Dexter TM 1983 Effect of haematopoietic cell growth factor on intracellular ATP levels. Nature 303:629–631
Wiktor-Jedrzejczak W, Ratajczak MZ, Ptasznik A, Sell KW, Ahmed-Ansara A, Ostertag W 1992 CSF-1 deficiency in the *op/op* mouse has differential effects on macrophage populations and differentiation stages. Exp Hematol 20:1004–1010
Williams GT, Smith CA, Spooncer E, Dexter TM, Taylor DR 1990 Haemopoietic colony stimulating factors promote cell survival by suppressing apoptosis. Nature 343:76–79
Yoshida H, Hayashi S-I, Kunisada T et al 1990 The murine mutation osteopetrosis is in the coding region of the macrophage colony stimulating factor gene. Nature 345:442–444

DISCUSSION

Nossal: Let me cross-examine you on the multi-organ derivation of CSFs in health. You told us that most organs make CSFs, in varying amounts, but you didn't dwell on the question of which cells actually make them. As an immunologist, I am tempted to ask you whether they might be made by some of the cells that are common to all organs. From this point of view, two cell types stand out as possibilities: the dendritic cells and the vascular endothelium. For instance, where you showed muscle making plentiful CSFs, were these made by myocytes, or were they produced by some cell within muscle derived from the bone marrow? Or were they made by vascular endothelium?

Metcalf: There is no shortage of literature on the ability of particular cell types to make CSF. Almost every cell that's been examined *in vitro* has the ability to produce one or other of the CSFs, particularly after suitable priming. The question is: do these same cell types also produce CSF *in vivo* or is CSF production an induced consequence of culture *in vitro*? I think the truth lies somewhere in between these two options. Immunologists have to disabuse themselves of the possibility that lymphocytes are an important source of agents such as GM-CSF or G-CSF, although there is good evidence that they may be an important source of IL-3, whenever it is produced *in vivo*, and of IL-5. Endothelial cells, stromal cells, fibroblasts, macrophages and lymphocytes are all well-known producers of all of the CSFs, and I suspect that most cell types in the body can do the job if induced.

Lieschke: Aside from the knockout literature, the most direct way that the question of which cells produce CSFs has been assessed is by *in situ* hybridization and immunohistochemistry. For GM-CSF, placental cells are definitely a site of GM-CSF mRNA production (Kanzaki et al 1991). Secondly, given the phenotype of GM-CSF-deficient mice, it is interesting that bronchiolar epithelial cells have been shown to be a site of GM-CSF mRNA production *in vivo* (Tazi et al 1993), consistent with the capacity of respiratory epithelial cells to synthesize GM-CSF *in vitro* (Smith et al 1990, Churchill et al 1992). There are fewer studies of G-CSF production sites *in vivo*. However, we have some unpublished preliminary observations (Lieschke 1996). In knocking out the G-CSF gene in mice, we placed the *lacZ* reporter gene under control of the endogenous G-CSF promoter, hoping to identify sites of endogenous G-CSF production by histochemical tissue staining. Results from *lacZ* staining of tissues from endotoxin-treated G-CSF-deficient mice are somewhat confusing. There is an endogenous β-galactosidase-like activity in macrophages, which results in background staining in one of the cell types we would most have had expected to produce G-CSF (Dannenberg & Suga 1981). None the less, there are many more *lacZ*-positive cells in the spleens of LPS-treated G-CSF-deficient mice than in those of control wild-type LPS-treated animals. These cells appear to be parenchymal and from their nuclear morphology are most probably endothelial cells and macrophages.

Metcalf: An initially surprising study was published by Cluitmans et al (1995) who reported that, despite the clear evidence that G-CSF is important in controlling granulocyte production, no G-CSF was made by normal human bone marrow cells. We didn't believe this until we looked at our own data and realized we had made the same observation — even using repeated cycles of PCR, we cannot detect any G-CSF mRNA. It may be necessary therefore for G-CSF to be transported to the bone marrow. The finding is still a little bizarre because you can grow cells out of bone marrow and show that they can produce G-CSF, just as endothelial cells do. This may be another example of misleading data resulting from the use of cultured (induced) cells.

On this question, some years ago Hultner et al (1982) showed that the injection of C3H marrow cells into endotoxin-resistant C3H/HeJ mice, allowed them to exhibit the usual elevation of CSF levels following the injection of endotoxin. This implies that bone marrow-derived cells are major sources of CSF in this response. Whether in fact this implicates lymphoid cells, macrophages or even endothelial cells, I don't know, but it was an interesting observation.

Nossal: As far as transport of CSF is concerned, Mike Dexter has shown that extracellular matrix proteins in the glycocalyx of stromal cells can 'catch' growth factors and concentrate them locally. Is that still your position, Mike?

Dexter: Yes, the heparan sulfate proteoglycans are assuming increasing importance, not only for the haemopoietic cell growth factors but also, for example, for the fibroblast growth factors. I think that story holds up. With respect to the bone marrow not producing any detectable G-CSF, I'm assuming that those studies were performed using normal steady-state bone marrow. But what happens with a damaged marrow? There one would expect perhaps more growth factors to be produced. Have people looked at this?

Metcalf: We have not examined the production of G-CSF by damaged marrow populations. Like you, I would assume that under conditions of perturbation G-CSF production might increase.

Dexter: I was also intrigued by your suggestion that perhaps IL-8 was one of the players involved in monitoring the distribution of mature granulocytes. What about MIP-1α? It has effects quite similar to IL-8 in terms of releasing cells from the marrow within a short time after injection.

Metcalf: We haven't looked at any of these chemotactic or releasing agents. There is now a large family of them, and it is unclear within that family which are the important molecules in various situations. A number of them can share the IL-8 receptor. I was intrigued by the IL-8R knockout study of Cacalano et al (1994) because it seemed relevant to the type of model we had been working with.

Williams: Concerning your studies on peritoneal exudates, last year Dinauer and I knocked out the *gp91* phox gene, which is required for generation of superoxide in granulocytes and macrophages, and is the cause of X-linked chronic granulomatous disease (CGD) in children (Pollock et al 1995). One of the subtle changes in those animals that you would be interested in, is that when we inject their peritoneal cavity with thioglycollate, we get a typical granulocyte response. Usually these granulocytes are replaced by macrophages over the next 3–4 d, many of which are full up with debris from granulocytes. The animals that don't make superoxide in their granulocytes (i.e. harvested from CGD mice) have an exaggerated granulocyte response, and the elicitation of macrophages is greatly blunted. This infers that part of that physiological process has to do with self-destruction of granulocytes by the generation of oxidants.

Early in your talk you made a comment that struck me. If you are relying *in vivo* upon shortening of G1 to generate an amplification of cells, you may require direct input from the stem cell pool for differentiated cells. I was recently at a conference at which the statement was made that progenitor cells as we think of them make no contribution at all to functional or differentiated cells in the bloodstream. The statement was based on the observation that if you use highly purified stem cell populations you see no difference in short-term reconstitution in transplant experiments in humans. Would you comment on the concept of how important progenitor populations are for *in vivo* constitution?

Metcalf: It's logical to assume that they are. They are on a production line and you could say the same thing about a myeloblast or a promyelocyte: how do we know that that cell actually generates mature neutrophils? It is hard to have an animal that just has neutrophils and nothing else. It is not an easy argument to refute. You have to involve progenitor cells because they are the immediate progeny of stem cells. That statement stems from the puzzling ability in some very clean laboratories to actually resuscitate a lethally irradiated mouse with small numbers of pure stem cells, whereas in a dirty laboratory you will need to supplement that population with quite large numbers of more mature progenitors. If it were to be documented that the earliest appearance of mature neutrophils was entirely normal there might be a case, but the stem cells in the meantime have generated progenitor cells. It therefore requires an assumption that all neutrophils appearing day 2 after irradiation must have come from the injected stem cells. This is a non-sensical argument.

Shortman: I don't understand how you fit the observation that if you stimulate granulocyte production in bone marrow, you do see an increase in granulocytes in the peritoneum, with your original model. The model says there's a huge excess production of granulocytes anyway, and the limiting factor is getting them through that door out of the bone marrow and into the bloodstream. According to the model you would not have predicted that there would have been any increase—the limitations lie elsewhere.

Metcalf: You were obviously looking carefully at my last slide! In mice that have had repeated G-CSF injections and have developed a bone marrow full of neutrophils, there was some increase in peritoneal neutrophils. Many of these marrow neutrophils will not leave the marrow unless a special demand is made. However, a day or two previously, these mice had quite elevated blood neutrophil levels, and it is astonishing that more of these did not lodge in the peritoneal cavity. This doesn't seem to happen, so it seems possible to have high circulating neutrophil levels and almost no cells actually entering a tissue until you provoke this entry with chemotactic signals. There is something in crude casein preparations that can initiate such signals but I'm not sure what it is. The same results can be achieved using endotoxin-unresponsive C3H/HeJ mice, and endotoxin turns out to be quite a poor elicitor of the inflammatory response.

Shortman: When these cells were called into the peritoneum by other means, there was no increase in blood level. Was this because there was a faster transit through the blood?

Metcalf: We haven't measured neutrophil transit times during an inflammatory response. White cell levels in mice tend to be very variable, but overall there was at most a twofold elevation of blood neutrophils following the injection of casein.

Nossal: You mentioned that in the healthy animal only the fully formed totally mature granulocytes leave the marrow. If you fill up the marrow by means of G-CSF injections, to what degree is that orderly process perturbed? Is the fact that you now drive numerous progenitor cells out of the bone marrow and into peripheral blood simply because you have flushed out all the mature granulocytes as well, or have you in some fairly serious way perturbed that gating by injecting the G-CSF?

Metcalf: We haven't checked the kinetics of neutrophil release from a G-CSF-primed animal. The release of progenitor cells from the marrow is a highly selective process in response to G-CSF. All of the cells released are out of cycle, and there is a selective release of less mature progenitor cells. With progenitor cells of different lineages there is a curious bias in frequency that fluctuates from day to day. This is a highly complex process and is not fully developed until 5–7 d after CSF injections begin, whereas the release of mature neutrophils during an inflammatory response is a process that is finished within a couple of hours.

Strasser: You compared the death of unused granulocytes in the bone marrow with the death of autoreactive lymphocytes in the thymus. I think that is a false comparison, because in the thymus and among the developing B cells in the marrow two fundamentally different forms of death are occurring. One is death by activation; the other is death by neglect. Granulocyte death is by neglect, not by activation — it is the lack of a survival signal that causes them to die.

Metcalf: It is an assumption that they die by neglect. It would be intriguing if future work showed that there were such things as auto-reactive granulocytes.

Dexter: I'm also unsure about this term 'neglect', because you can pander those mature neutrophils as much as you want and they still die. Their death is actually programmed in them.

Tarlinton: A situation perhaps more similar to that described by Don Metcalf occurs in the periphery. After the B cells leave the marrow, far fewer are recruited into the long term circulating pool than are exported. This might involve a selective process requiring an antigen-dependent activation after export from the bone marrow.

Metcalf: Are you saying that there is an over-production of B cells, but the death of that excess population — self reactive or otherwise — occurs before the cells become antigen-responsive?

Tarlinton: No, what I am trying to say is that only a very small fraction of the amount exported from the bone marrow becomes incorporated into the recirculating pool. That incorporation appears to involve an interaction with antigen, either positively, or, as Chris Goodnow would suggest, negatively.

Nossal: We are talking about two different phenomena. On the one hand, we have the question of localized death within the marrow, which stathmokinetic studies can show, and which may or may not be antigen dependent. On the other hand, we have the brief survival of the cells that leave the marrow, with only a small proportion entering the recirculating lymphocyte pool as long-lived cells. Again, this phenomenon, i.e. conversion to a long-lived phenotype, may or may not be influenced by antigen exposure.

Goodnow: In the thymus the numbers are known more accurately. When people have looked in the transgenic mouse where most of the T cells are eliminated in the thymus or fail to be positively selected, although essentially all the T cells are eliminated late in the double-positive it doesn't change the absolute number of thymocytes. Thus the real number-controlling mechanism has nothing to do with positive or negative selection. Instead, it probably has much more to do with the factors that control the pool of granulocytes in marrow and is probably antigen-independent and involved in determining numbers of stromal niches.

Melchers: But there is a transgenic mouse expressing one and the same monoclonal antibody as antigen-specific receptor in all B-lineage cells, the one that Fred Alt has looked at. On a Rag-2 knockout genetic background all B lineage cells express the same receptor, and only this receptor. Still, 98% of the B cells die as immature cells in the marrow. Isn't that proof that the process isn't antigen specific?

Goodnow: Yes, I'm sure it is the same with the B cells — it is just that the numbers are a bit more slippery.

Metcalf: Are we therefore saying that in neither the thymus nor the marrow can we ascribe what Macfarlane Burnet called the 'slaughter of the innocents' to an antigen-driven process? Are they dying for some other reason? This would bring the three populations quite close together: the body, for whatever reason, is generating more cells than are needed.

Nossal: So the real question is not whether negative selection (*à la* von Boehmer) exists or doesn't — because it most assuredly does — but rather, it concerns the proportion of total thymic cell death ascribable to negative selection. As matter of fact, Don Metcalf studied this nearly 40 years ago. If antigenic selection had been the only regulator, animals bearing up to 30 functioning thymic grafts should have had an excess of peripheral lymphocytes, but they did not! In a sense we're all agreeing that there are phenomena that we don't understand very well which control these cell numbers. In addition, there are quite specific antigen-dependent phenomena, such as negative selection in the thymus, which kill T cells and limit thymic output.

Rajewsky: I wanted to comment on what Fritz Melchers said. You can't deduce from these kinds of data that in a normal mouse, cells would not die from negative selection.

Melchers: I wasn't even addressing negative selection. It has been shown beautifully that the expression of the antigen in the primary organ deletes the immature B cells expressing the monoclonal antibody specific for this antigen almost entirely. I was addressing the question of whether there is some positive selection. I got the impression from Don Metcalf that he was worried about how that small population of granulocytes got into the peripheral tissues. What makes them go into the periphery, while the majority die? The same problem exists for B cells: what makes the few come out? At least in the monoclonal antibody-expressing mouse on a Rag-2 knockout genetic background there doesn't appear to be a positively selecting antigen on the horizon that would do that.

Nossal: Dennis Osmond, while on sabbatical in my lab, showed very clearly that the exit of B cells from the marrow was random, and it wasn't, as Don described for granulocytes, that the most mature left. In fact, after the last reduction division of the small immature B cell, departure from the marrow was stochastic, whether or not the IgD had started to appear.

Rajewsky: With respect to a potential positive selection in these monoclonal models: the receptors that are being used are actually taken from a cell which was originally also selected into the periphery. The fact that cells come out in such models may simply reflect that fact.

Melchers: But you could turn that argument around and say that if that was the case they should all come out.

References

Calacano G, Lee J, Kikly K et al 1994 Neutrophil and B cell expansion in mice that lack the murine IL-8 receptor homolog. Science 265:682–684

Churchill L, Friedman B, Schleimer RP, Proud D 1992 Production of granulocyte–macrophage colony stimulating factor by cultured human tracheal epithelial cells. Immunology 75:189–195

Cluitmans FHM, Esendam BHJ, Landegent JE, Willemze R, Falkenberg JHF 1995 Constitutive *in vivo* cytokine and hematopoietic growth factor gene expression in the bone marrow and peripheral blood of healthy individuals. Blood 85:2038–2044

Dannenberg AM, Suga M 1981 Histochemical stains for macrophages in cell smears and tissue sections: β-galactosidase, acid phosphatase, nonspecific esterase, succinic dehydrogenase, and cytochrome oxidase. In: Adams DO, Edelson PJ, Koren MS (eds) Methods for studying mononuclear phagocytes. Academic Press, New York, p 375–396

Hultner L, Staber FG, Mergenthaler H-G, Dormer P 1982 Production of murine granulocyte–macrophage colony-stimulating factors (GM-CSF) by bone marrow-derived and non-hemopoietic cells *in vivo*. Exp Hematol 10:798–808

Kanzaki H, Crainie M, Lin H et al 1991 The *in situ* expression of granulocyte–macrophage colony-stimulating factor (GM-CSF) mRNA at the maternal–fetal interface. Growth Factors 5:69–74

Lieschke GJ 1996 Physiological role of haemopoietic growth factors in mice revealed by targeted gene disruption. PhD Thesis, University of Melbourne, Melbourne, Victoria, Australia

Pollock JD, Williams DA, Gifford MAC et al 1995 Mouse model of X-linked chronic granulomatous disease, an inherited defect in phagocyte superoxide production. Nat Genet 9:202–209

Smith SM, Lee DKP, Lacy J, Coleman DL 1990 Rat tracheal epithelial cells produce granulocyte/macrophage colony stimulating factor. Am J Respir Cell Mol Biol 2:59–68

Tazi A, Bouchonnet F, Grandsaigne M, Boumsell L, Hance AJ, Soler P 1993 Evidence that granulocyte–macrophage colony-stimulating factor regulates the distribution and differentiated state of dendritic cells/Langerhans cells in human lung and lung cancers. J Clin Invest 91:566–576

General discussion II

Differences between membrane-bound and secreted isoforms of stem cell factor

Williams: We have heard from Don Metcalf that there may be some differences in the way a cell responds to the two isoforms of the same factor, the membrane-associated and secreted forms of stem cell factor (SCF; also known as Steel factor). A major focus for our laboratory is to understand these two isoforms (Majumdar et al 1994). Although the *Steel* mutant mouse, which is deficient in SCF, is widely thought not to have immune deficiency, this is an oversight—there is overwhelming evidence both *in vitro* and *in vivo* that these animals have abnormal thymocyte development (R. Kapur & D. A. Williams, unpublished results).

We have approached this question by means of traditional transgenic and cell mutation approaches, and also by the use of *in vitro* systems. In the latter, we generated homozygous SCF-deficient stromal cell lines, and then generated transfectants expressing isoforms of the SCF cDNAs representing membrane-associated and secreted forms. We then asked a very simple question: when we take either c-Kit$^+$ or CD34$^+$ cells and put them on these stable transfectants, do the cells behave differently and can we get at the biochemical nature of that difference (Toksoz et al 1992)? We examined the response of Mo7E cells to secreted or membrane-associated forms of SCF. There is a definite difference in how long c-Kit (the receptor for SCF) remains phosphorylated (activated). Activation of c-Kit is prolonged in response to membrane-associated SCF and shortened in response to secreted SCF (Miyazawa et al 1995). Addition of recombinant soluble factor to cultures containing membrane-associated SCF shortens the activation of c-Kit. We were interested to see whether these differences in c-Kit activation had implications downstream, and what their physiological relevance was. To look at this, we generated the same type of lines expressing murine forms of SCF and used purified primary c-Kit$^+$ murine bone marrow cells. One of the downstream signalling events we looked at was MAP kinase activation. Again, we see a very distinct difference in the kinetics of MAP kinase activation depending on whether the c-Kit$^+$ bone marrow cells are exposed to soluble versus membrane-associated SCF. Somewhat counterintuitively, the longer c-Kit is activated the more transient the MAP kinase activation is. Thus when the c-Kit$^+$ cells are exposed to membrane-associated SCF they display activation of c-Kit for prolonged periods, and transient activation of MAP kinase is observed. Parallel cultures of these cells were used to examine the biological effects of SCF isoforms. After 3 d exposure to stromal cells expressing either secreted or membrane-associated SCF *in vitro*, we assayed cells for high proliferative potential colony-forming cells (HPP-CFCs), a primitive haemopoietic progenitor cell. There is a distinct difference in the number of HPP-CFC colonies surviving, with more evident after 72 h in cultures expressing membrane-associated SCF versus soluble SCF. What may be even more interesting is the stem cell phenotypes of these cells. If you take these primary HPP-CFC colonies and re-plate them, there is a significant difference in the re-plating potential. If you believe that the HPP-CFC colony re-plating assay measures the proliferative nature of the cell, c-Kit$^+$ primary cells exposed to membrane-associated SCF appear to maintain a higher proliferative capacity than cells exposed to soluble factor. Interestingly, these differences are due to signalling through the same receptor with different isoforms of the same protein. The MAP kinase activation we see parallels very nicely work done in PC12 cells. PC12 cells exposed to nerve growth factor (NGF) display terminal differentiation and cessation of proliferation. This differentiation is associated with a prolonged MAP kinase activation. In contrast, PC-12 cells exposed to epidermal growth factor (EGF) have a short MAP kinase activation and these cells continue to proliferate without differentiation (Traverse et al 1992).

Cory: Are any of these results explicable in terms of the short half-life of the soluble form of SCF?

Williams: In Mo7E cells, where we have looked at this carefully with pulse–chase experiments, cells exposed to soluble SCF demonstrate rapid c-Kit receptor internalization and degradation. But if the receptor is exposed to the membrane-associated isoform of SCF, it remains on the cell for a prolonged time. The results are definitely related to internalization and degradation of the receptor–ligand complex.

Nicola: Were those results obtained at a single dose of soluble SCF or is that the maximum effect you get when you titrate the soluble SCF?

Williams: The results are from around 20 ng/ml of SCF. If recombinant SCF is added, the same response is seen over a wide range of concentrations. It is relatively hard to quantitate membrane-associated protein in the same way; instead, we looked at a number of clones producing membrane-associated proteins, to make sure that this is not a dose effect. We see the same effect—at least over different ranges of mRNA production from the transfectants.

Burgess: Concerning the kinetics of the MAP kinase activation by the membrane-bound form of the ligand, my understanding is that c-Kit remains activated for a long time after the MAP kinase is down-regulated. Presumably you have to infer that there is another control point between Kit and MAP kinase, which may be a phosphatase. I don't know that there is another system which describes an intermediate control point. Can you tell us about the other proteins in this chain? Is Ras still loaded, is Raf activated?

Williams: We haven't looked at that yet. Your point about phosphatases is probably right on the mark. At least from a standpoint of physiologically relevant phosphatases, I know that Alan Bernstein has crossed moth-eaten and W^v and sees abrogation of the phenotype of moth-eaten when he crosses in a hypo-functioning receptor (Paulson et al

1996). At least in that system, one would argue that the phosphatase is probably an important component of the pathway.

Melchers: If you take a soluble isoform of SCF and insolubilize it artificially, does it have the same effect as the membrane-bound form?

Williams: This system is reductionistic but not to the point of only one factor, so the stromal cells are making all the normal factors and adhesive ligands that they normally make. Obviously, the story may be more complicated than just SCF itself.

References

Majumdar MK, Feng L, Medlock E, Toksoz D, Williams DA 1994 Identification and mutation of primary and secondary proteolytic cleavage sites in murine stem cell factor DNA yields biologically active, cell-associated protein. J Biol Chem 269:1237–1242

Miyazawa K, Williams DA, Gotoh A, Nishimaki J, Broxmeyer HE, Toyama K 1995 Membrane-bound Steel factor induces more persistent tyrosine kinase activation and longer life-span of c-kit gene encoded protein than its soluble form. Blood 85:641–649

Paulson RF, Versely S, Siminovitch KA, Bernstein A 1996 Signalling by the W/Kit receptor tyrosine kinase is negatively regulated *in vivo* by the protein tyrosine phosphatase Shp1. Nature Genet 13:309–315

Toksoz D, Zsebo KM, Smith KA et al 1992 Hematopoiesis in long-term bone marrow cultures by murine stromal cells selectively expressing the membrane-bound and secreted forms of the human homolog of the Steel gene product, stem cell factor. Proc Natl Acad Sci USA 89:7350–7354

Traverse S, Gomez N, Paterson H, Marchall C, Cohen P 1992 Sustained activation of the mitogen-activated protein (MAP) kinase cascasde may be required for differentiation of PC12 cells: comparison of the effects of nerve growth factor and epidermal growth factor. Biochem J 288:351–355

CSF-deficient mice — what have they taught us?

Graham J. Lieschke

Whitehead Institute for Biomedical Research, 9 Cambridge Center, Cambridge, MA 02142–1479, USA

Abstract. Haemopoietic growth factor-deficient mice have been particularly instructive for defining the usual physiological role of these factors. Mice now exist lacking the granulopoietic factors G-CSF, GM-CSF, M-CSF (CSF-1), SCF, several other factors influencing haemopoiesis (including erythropoietin, interleukins 5 and 6), combinations of these factors (GM- & M-CSF; G- & GM-CSF; G- & GM- & M-CSF) and several CSF receptor components. Most of these mice were generated by targeted gene disruption, others are spontaneously arising mutants. The phenotypes of these mice indicate that the granulopoietic factors have both unique and redundant roles *in vivo*. Some factors are uniquely important in baseline myelopoiesis. Experimental infection of CSF-deficient mice indicates unique roles for some factors in emergency 'overdrive' haemopoiesis. Recovery from myeloablation evaluates the role of CSFs in emergency 'restoring normality' haemopoiesis. Redundancy also exists in the capacity of CSFs to support complete granulocyte development *in vivo*. Some factors are not involved in all the *in vivo* roles suggested by the range of their actions demonstrable *in vitro*. Some CSFs have indispensable roles in non-haemopoietic tissues. Some factors have *in vivo* roles not anticipated from previous studies. Mice deficient in several factors have identified compensating roles for factors by revealing exacerbated and additional phenotypic features, and may unmask additional *in vivo* roles.

1997 The molecular basis of cellular defence mechanisms. Wiley, Chichester (Ciba Foundation Symposium 204) p 60–77

An important achievement of the study of haemopoiesis *in vitro* has been the identification of a set of haemopoietic growth factors that influence the survival, proliferation, differentiation and function of granulocytes, macrophage cells and their precursors. They include those originally named colony stimulating factors (CSFs): granulocyte-, granulocyte/macrophage- and macrophage-CSF (G-CSF, GM-CSF and M-CSF [also designated CSF-1]). Other factors involved in these processes include interleukin (IL)-3 (also known as multi-CSF) , IL-5, IL-6, IL-11, stem cell factor (SCF; also known as Steel factor) and Flt-3/Flk-2 ligand. However, an important issue is the extent to which the demonstrable *in vitro* activities of these factors correlate with their actual physiological function *in vivo*. Previous insights into the *in vivo* role of these factors derive from several approaches: creating transgenic mice over-expressing a factor; reconstituting lethally irradiated mice with marrow cells infected with a recombinant retrovirus overexpressing a factor; and administrating pharmacological doses of factor (reviewed in Lieschke & Dunn 1992, Lieschke & Burgess 1992a,b). While these approaches demonstrate the capacity for supraphysiological levels of some factors to stimulate granulopoiesis and monocyte/macrophage development *in vivo*, they do not define their usual physiological role. Other insights come from measurement of serum or tissue levels of endogenously produced factor (e.g. Kawakami et al 1990), or demonstration of sites of mRNA production. These approaches correlate factor production with *in vivo* scenarios, but still incompletely define the dynamic physiological role of factors.

The study of factor-deficient mice is one particularly direct way of defining the unique and overlapping physiological roles of haemopoietic growth factors. As a result of gene targeting studies or spontaneously occurring mutations, mice now exist lacking the granulopoietic factors G-CSF (Lieschke et al 1994a), GM-CSF (Dranoff et al 1994, Stanley et al 1994), M-CSF (Wiktor-Jedrzejczak et al 1990, Yoshida et al 1990), SCF (Zsebo et al 1990), several other factors influencing haemopoiesis (including erythropoietin [Wu et al 1995], IL-5 [Kopf et al 1996], IL-6 [Kopf et al 1994, Poli et al 1994, Dalrymple et al 1995]), combinations of these factors (Lieschke et al 1994b, Nilsson et al 1995, Seymour et al 1995, Grail et al 1996) and several CSF receptor components (Nishinakamura et al 1995, Robb et al 1995). The phenotype of IL-3-deficient mice has not been reported, although $\alpha_{IL\text{-}3}$ mutant mice with impaired IL-3 signalling have been characterized (Ichihara et al 1995). Studies of these factor-deficient mice have permitted the physiological role of these factors to be re-evaluated, and such reappraisals are now being published (Wiktor-Jedrzejczak 1993, Metcalf 1995).

This paper reviews the insights about the physiological roles of the CSFs and other factors revealed from studying CSF-deficient mice. The focus is on granulopoietic factors since this has been an area of particular interest in our work. However, where analogous or contrasting examples can be drawn for other haemopoietic lineages, they are mentioned.

Granulopoietic CSFs and baseline haemopoiesis

A useful distinction can be drawn *in vivo* between baseline and emergency granulopoiesis. Baseline granulopoiesis refers to steady-state conditions in which neutrophil production and clearance rates are equal and the numbers of neutrophils and their precursors are constant in all compartments. One manifestation of this is the maintenance of blood neutrophil levels within a relatively narrow normal range (Fig. 1). Although the regulatory mechanisms controlling this are not defined, particularly those comprising the negative feedback loop of this apparently homeostatic process, the actions of the granulopoietic CSFs *in vitro* to stimulate the production of large numbers of mature cells from a much smaller number of progenitor cells (called colony forming cells [CFCs] or alternatively, colony forming units [CFUs]) made them likely candidates as the proliferative stimulus.

The haemopoietic profiles of CSF-deficient mice have identified those CSFs which are primarily responsible *in vivo* for balanced single-lineage haemopoietic cell production (Table 1). Under basal conditions, G-CSF-deficient mice have chronic neutropenia, with neutrophil levels 20–35% of age-matched wild-type controls (Lieschke et al 1994a). This indicates that G-CSF is critically important for the maintenance of the normal quantitative balance of granulopoiesis. Partially SCF-deficient Sl/Sl^d mice have a 40% reduction in blood neutrophil levels and 70–90% reduction in blood monocytes (totally SCF-deficient Sl/Sl mice die before birth from erythropoietic failure), identifying SCF as another physiological regulator of balanced neutrophil and monocyte/macrophage production (Ruscetti et al 1976, Shibata & Volkman 1985), although it has more critical effects on erythropoiesis (Russell 1979).

Similarly, young M-CSF-deficient mice have monocytopenia, although the degree has varied between several reports (Wiktor-Jedrzejczak et al 1982, Lieschke et al 1994b, Nilsson et al 1995) and in mice of different ages (Begg et al 1993). Baseline tissue macrophage populations are also reduced at some sites (Cecchini et al 1994). IL-5-deficient mice have a 10% reduction in blood eosinophil levels (Kopf et al 1996). These observations identify these factors as critically important for normally balanced *in vivo* monocyte and eosinophil production, respectively. Adequate erythropoiesis is totally dependent on erythropoietin since erythropoietin-deficient mice die *in utero* from erythropoietic failure (Wu et al 1995).

In mice lacking G-CSF, SCF, M-CSF or IL-5, the peripheral blood picture is accompanied by reductions in at least some subtypes of the relevant morphologically recognizable and/or clonable haemopoietic progenitor cells. Interestingly, IL-6-deficient mice, which do not have significant basal neutropenia (Dalrymple et al 1995), have a 10% reduction in total numbers of marrow CFU-GM with a compensating increase in total numbers of splenic CFU-GM, and a larger relative reduction in marrow and splenic pools of earlier precursors such as CFU-S (Bernad

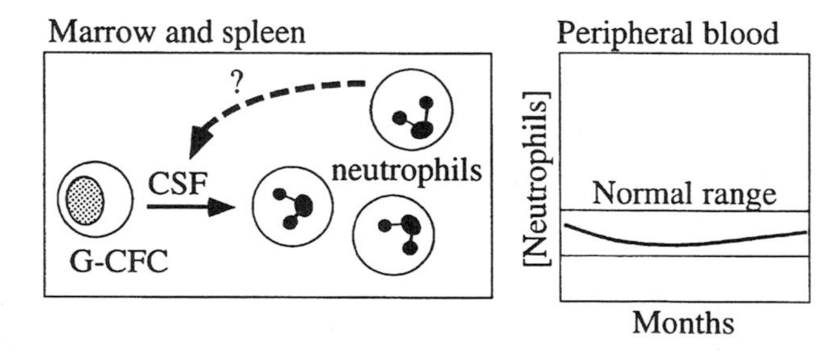

FIG. 1. Baseline granulopoiesis. Under the influence of colony-stimulating factor(s) (CSF), granulocyte progenitor cells (G-CFC, granulocyte colony-forming cells) form mature neutrophils. In the peripheral blood, neutrophil levels remain within a narrow normal range over time. (Adapted from Lieschke 1996.)

TABLE 1 Effects of haemopoietic growth factor deficiency on baseline myelopoiesis

	Neutrophil or monocyte/macrophage population (relative number of cells)				
Factor deficiency	*Peripheral blood*	*Marrow*	*Spleen*	*Tissues*	*Reference(s)*
Stem cell factor (*Sl/Sl^d*)[a]	↓Neutrophils ↓Monocytes	↓Cellularity ↓Granulopoiesis ↓GM-CFCs ↓M-CFCs	Not tested	Peritoneal macrophages—normal	Ruscetti et al (1976) Shibata & Volkman (1985)
G-CSF	↓Neutrophils ↓Monocytes[c]	Normal cellularity ↓Granulopoiesis ↓G- & M-CFCs	Normal cellularity Normal granulopoisis	Not tested	Lieschke et al (1994a)
GM-CSF	Normal	Normal	Possible ↑GM-CFCs Macrophages—normal	Liver phagocytes—normal	Stanley et al (1994) Lieschke (1996)
M-CSF (CSF-1) (<12 week)[b]	↓Monocytes	↓Cellularity ↓M-CFCs	↑Cellularity ↑CFCs	↓Macrophages in tissues	Wiktor-Jedrzejczak et al (1982) Begg et al (1993) Cecchini et al (1994)
IL-5	↓Eosinophils	↓Eosinophils	Not tested	Peritoneal eosinophils—normal	Kopf et al (1996)
IL-6	↓Leukocytes but normal neutrophils	↓Cellularity ↓CFU-GM	↓Cellularity ↑CFU-GM	Not tested	Bernad et al (1994) Dalrymple et al (1995)

[a]Partial deficiency only (Flanagan et al 1991).
[b]Results tabulated for young (<12 week) mice; haemopoiesis in M-CSF-deficient mice corrects by 22 weeks of age (Begg et al 1993).
[c]Older mice only (e.g. 27–29 week).
CFCs, colony forming cells; CFU, colony forming unit.

et al 1994), indicating that the size of the blood neutrophil compartment can be maintained in the absence of IL-6, despite alterations in the distribution and size of the progenitor and earlier precursor pools. Although erythropoietin is mandatory for terminal erythroid development, CFU-E and BFU-E precursors are demonstrable in significant numbers in the fetal livers of erythropoietin-deficient mice (Wu et al 1995), but in absolute terms the total numbers are reduced despite an increased frequency amongst nucleated cells. This indicates that erythroid lineage commitment *in vivo* is not exclusively directed by erythropoietin.

Given the potent capacity of GM-CSF to support granulocyte/macrophage colony development *in vitro*, an unexpected finding was that GM-CSF-deficient mice had unperturbed haemopoiesis with normal levels of circulating neutrophils, monocytes and eosinophils, and normal numbers of marrow progenitor cells (Stanley et al 1994). A modest elevation of splenic progenitor cells may have been related to subclinical infections in the diseased lungs of some GM-CSF-deficient mice. Hence, GM-CSF is either not implicated in baseline granulopoiesis *in vivo*, or its role can be replaced by other factors completely.

Granulopoietic CSFs and emergency haemopoiesis

Emergency granulopoiesis refers to situations where there is an acute requirement for augmented neutrophil production. Two emergency situations which may be driven by CSF production can be distinguished (Fig. 2): (1) suprabasal neutrophil production, in which the production of extra neutrophils is superimposed on otherwise normal basal granulopoiesis (such as occurs during acute bacterial infections); and (2) the restoration of normal neutrophil levels from a depleted state, such as occurs following myeloablation by a cytotoxic drug. Although CSFs are the obvious candidates to drive granulopoiesis under these two conditions, different CSFs may be important in each circumstance. In fact, a possible physiological distinction between different granulopoietic factors *in vivo* that is difficult to discriminate from *in vitro* studies would be the employment of different factors to drive granulopoiesis under basal and different emergency conditions. However, if granulopoiesis were indeed regulated by a typical negative-feedback homeostatic loop, the CSFs involved in restoring normality would by definition include those involved in baseline granulopoiesis.

Studies in CSF-deficient mice identified critical roles for some CSFs in suprabasal overdrive emergency granulopoiesis (Table 2). *Listeria*-infected G-CSF-deficient mice failed to develop a significant neutrophilia and had an impaired capacity to control the infection (Lieschke et al 1994a), indicating at least that a G-CSF-driven marrow is mandatory for a normal haematological response to this infection, but also suggesting that G-CSF is involved in driving the neutrophilic response. These data are consistent with observations linking elevated serum G-CSF levels with neutrophilia in some bacterial infections (Kawakami et al 1990). (Interestingly, *Listeria*-infected G-CSF-deficient mice also had a delayed and obtunded monocyte response to *Listeria* infection, indicating that the marrow macrophage progenitor cell deficiency of these

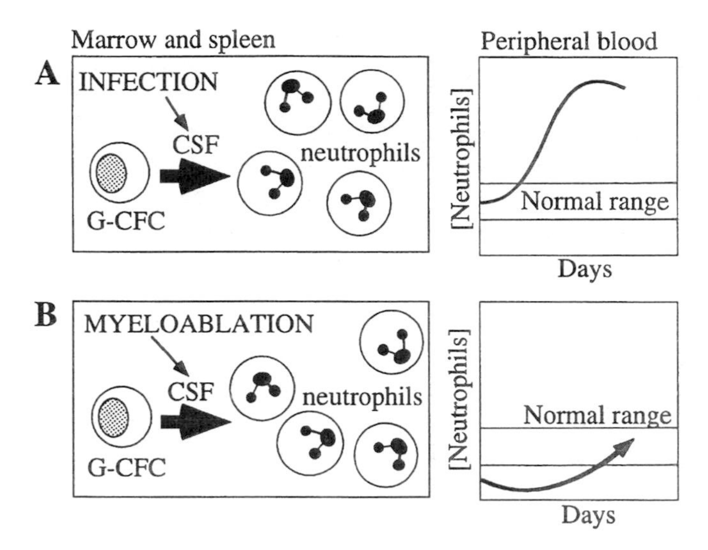

FIG. 2. Emergency granulopoiesis. Under the influence of colony-stimulating factor(s) (CSF), granulocyte progenitor cells (G-CFC, granulocyte colony-forming cells) form mature neutrophils. In overdrive emergency granulopoiesis, such as occurs during infection (A), suprabasal levels of blood neutrophils result from augmented granulopoiesis driven by CSFs generated in response to infection. In emergency granulopoiesis restoring normality after a myeloablative insult (B), CSF production drives granulopoiesis to restore blood neutrophils to normal levels. (Adapted from Lieschke 1996.)

mice is of functional relevance.) The infection-vulnerability of G-CSF-deficient mice has physiological importance, since the long-term survival of G-CSF-deficient mice held in a conventional animal house is impaired, and sporadic bacterial infections contribute to this increased mortality (G. Lieschke & D. Grail, unpublished results).

IL-6 is also essential as an emergency stimulus in the haematological response to *Listeria*, since *Listeria*-infected IL-6-deficient mice fail to develop a neutrophilia in the first 48 h after infection and are vulnerable to this pathogen (Kopf et al 1994, Dalrymple et al 1995). Under baseline conditions IL-6-deficient mice have normal numbers of peripheral blood neutrophils and approximately normal total body numbers of CFU-GM (although the relative distribution of CFU-GM between the marrow and spleen is altered), and hence this lack of a 48 h neutrophil response would appear to reflect primarily an impaired emergency response rather than an altered basal state at the onset of infection. However, for a full description of the basal state influencing the 48 h neutrophil response to *Listeria* infection, it would be useful to enumerate the number of readily mobilizable neutrophils in IL-6-deficient mice such as has been done in G-CSF-deficient mice (Lieschke et al 1994a), since in this short period, mobilization of neutrophils from marginated and marrow stores may be important in influencing the peripheral blood picture. This is because the generation of new neutrophils from progenitor cells takes some time.

TABLE 2 Effects of haemopoietic growth factor deficiency on emergency myelopoiesis

	Type of myelopoietic emergency				
	Experimentally induced infection		*Other scenarios*		
Factor deficiency	*Effects in peripheral blood*	*Effects on control of infection*	*Scenario*	*Features*	*References*
G-CSF	↓Neutrophilia delayed monocytosis	Vulnerable to *Listeria*	Incidental infections	Increased mortality	Lieschke et al (1994a) G. J. Lieschke & D. Grail (unpublished data)
GM-CSF	Not tested	Not tested	Incidental lung infection	Able to form abscess Rarely succumb	Stanley et al (1994)
			Endotoxin challenge	↑Tolerance, normal blood leukocyte response	Basu et al (1994)
			Haemopoietic reconstituion after lethal irradiation	Normal with wild-type and GM-CSF$^{-/-}$ cells	Dranoff et al (1994)
M-CSF (CSF-1)	Not tested	Not tested	Endotoxin challenge	↑Tolerance	Szperl et al (1995)
IL-5	↓Eosinophilia	Normal progress of *Mesocestoides corti* infection	Not tested	Not tested	Kopf et al (1996)
IL-6	↓Neutrophilia	↑Susceptibility to *Listeria*	Recovery after 5-FU	Faster marrow CFU-GM recovery Normal recovery of marrow cellularity & blood leukocytes	Bernad et al (1994) Dalrymple et al (1995)

5-FU, 5-fluorouracil; CFU-GM, colony-forming unit — granulocyte/macrophage.

IL-5 plays a role in suprabasal overdrive emergency eosinophil production analogous to that of G-CSF in neutrophil production, since IL-5-deficient mice have both a modest basal blood and marrow eosinophil deficit, and a markedly impaired eosinophil response to infection with the parasite *Mesocestoides corti* (Kopf et al 1996).

There have been fewer studies in CSF-deficient mice of emergency granulopoiesis restoring normality after myeloablation. Interestingly, in contrast to the marked deficit in overdrive emergency granulopoiesis of IL-6-deficient mice, after a dose of 5-fluorouracil IL-6-deficient mice show normal kinetics of blood and marrow leukocytes and enhanced marrow CFU-GM recovery (Bernad et al 1994), indicating that IL-6 is not essential for this form of emergency granulopoiesis. However, after phenylhydrazine-induced anaemia, IL-6 mice (which have increased total numbers of both splenic and marrow BFU-E) have lower blood erythrocyte levels despite a more marked reticulocytosis, and increased susceptibility to a second dose of phenylhydrazine, implicating IL-6 in emergency erythropoiesis (Bernad et al 1994).

Initial observations have not yet identified any critical role for GM-CSF in emergency haemopoiesis. GM-CSF-deficient mice display greater variation in peripheral blood granulocyte levels than wild-type mice, and this includes sporadic elevated neutrophil levels (Stanley et al 1994), indicating that suprabasal neutrophil levels can be generated in the absence of GM-CSF. Dranoff et al (1994) were able to reconstitute GM-CSF-deficient mice by GM-CSF-deficient marrow cells, but they did not describe the kinetics of the reconstitution. Acute leukopenia following endotoxin administration occurs in GM-CSF-deficient mice (Basu et al 1994), indicating that GM-CSF is not essential for this response.

Redundancy amongst granulopoietic CSFs *in vivo*

Although all CSF-deficient mice studied to date have revealed unique roles for each CSF, they have also indicated that substantial redundancy exists in the capacities of CSFs to support haemopoietic development *in vivo*. Mice deficient in the factors of predominant importance for lineage-specific haemopoiesis (G-CSF, M-CSF and IL-5) still have morphologically mature neutrophils, monocytes and macrophages, and eosinophils, respectively, indicating that in each case the repertoire of factors remaining is able to fully support mature leukocyte development, although the quantitative balance is perturbed in the absence of a critical regulator. In most cases, the identity of these other factors is unknown, and were they to be identified, confirmation would be required that these other factors are usually utilized in these capacities in a fully factor-replete animal.

Redundancy (i.e. an overlapping role) should be distinguished from compensation (i.e. an augmented role resulting from recruitment to rectify a deficit). It is more difficult to implicate a factor as *compensating* for the absence of another *in vivo* than it is to ascribe to it the capacity to act in an overlapping role, since compensation requires the demonstration not only of necessity but also of supranormal levels of production. Several experimental approaches are available to attempt to draw this distinction. One is to assess the production of remaining factors directly *in vivo* by looking in a CSF-deficient mouse for increased gene transcription, tissue factor production or measuring serum factor levels: increased gene expression and factor production over that seen in wild-type mice indicates compensation. Another less direct approach is to assess the capacity of tissues and organs to release or be induced to make CSFs in *ex vivo* preparations of conditioned media (e.g. Wiktor-Jedrzejczak et al 1990, Stanley et al 1994, Metcalf 1995). A further approach is to generate mice deficient in two or more factors and to compare the effect of the absence of additional factors with the consequences of single factor deficiency. This approach (discussed further below) is a particularly direct way of evaluating the necessity of a second factor in the face of deficiency of another.

Other haemopoietic effects of CSF deficiency

Several of the granulopoietic CSFs have essential roles in other haemopoietic lineages indicated by deficits in the factor-deficient mice (Table 3). SCF-deficient mice have macrocytic anaemia and impaired megakaryocytopoiesis (Russell 1979, Ebbe et al 1973); IL-6 deficient mice have perturbed erythropoiesis and impaired megakaryocytopoiesis (Bernad et al 1994); young M-CSF-deficient mice have lymphopenia (Wiktor-Jedrzejczak et al 1982); and IL-5-deficient mice have reduced peritoneal B-1 ($CD5^+$) cells (Kopf et al 1996).

Non-haemopoietic roles of CSFs

Granulopoietic CSFs have primary pleiotropic actions that are not always restricted to the haemopoietic system, some of which were unsuspected until the factor-deficient mice were generated (Table 3). To date, no primary non-haemopoietic phenotypic consequences have been recognized in G-CSF-deficient or IL-5-deficient mice. In contrast, SCF, M-CSF and IL-6 have essential primary roles in non-haemopoietic systems.

SCF-deficient mice (*Sl/Sl*) have been known to lack germ cells and pigment cells since their discovery (Bennett 1956), defining an essential role for this factor in gametogenesis and melanogenesis. These processes are also impaired in heterozygous animals and partially SCF-deficient Sl/Sl^d animals (Russell 1979).

M-CSF-deficient (*op/op*) mice were first recognized by their osteopetrotic bone disorder (Marks & Lane 1976), which is due to impaired osteoclast number and function. A pregnancy defect in M-CSF-deficient mice identified a requirement for maternal M-CSF production for normal fertility (Pollard et al 1991), supporting previous observations implicating M-CSF in placental development. Unexpectedly, however, M-CSF is also required for normal mammary development during pregnancy (Pollard & Hennighausen 1994).

A suspected role for IL-6 in stimulating osteoclast development and function was confirmed by analysis of IL-6-deficient mice, which, although they have normal

TABLE 3 Other effects of haemopoietic growth factor deficiency

Factor deficiency	*Effects on non-myeloid haemopoietic lineages*	*Non-haemopoietic effects*	*Reference(s)*
Stem cell factor	Erythropoietic failure *in utero* (*Sl*/*Sl*) Macrocytic anaemia (Sl/Sl^{d}) ↓Marrow megakaryocytes (Sl/Sl^{d})	↓Germ cells ↓Pigment cells	Bennett (1956) Ebbe et al (1973) Russell (1979)
G-CSF	None described	None described	Lieschke et al (1994a)
GM-CSF	None described	Pulmonary disease: *alveolar proteinosis* Impaired surfactant clearance	Dranoff et al (1994) Stanley et al (1994) Ikegami et al (1996)
M-CSF (CSF-1)	Lymphopenia (3–4 week)	Osteopetrosis ↓Female fertility ↓Mammary development in pregnancy	Marks & Lane (1976) Pollard et al (1991) Pollard & Henninghausen (1994)
IL-5	↓Peritoneal $CD5^{+}$ B-cells	None described	Kopf et al (1996)
IL-6	↑Marrow and spleen BFU-E ↑Vulnerability to phenylhydrazine-induced anaemia ↓Marrow CFU-Meg	Increased bone turnover and vulnerability to oestrogen-induced bone loss ↓Acute phase response	Bernad et al (1994) Poli et al (1994) Kopf et al (1994) Fattori et al (1994)

BFU-E, burst-forming unit—erythroid; CFU-Meg, colony-forming unit—megakaryocytic.

amounts of trabecular bone, exhibit elevated rates of bone turnover, and are protected from the effects of oestrogen-depletion induced by ovariectomy to cause loss of bone mass (Poli et al 1994).

Unexpectedly, GM-CSF-deficient mice develop pulmonary alveolar proteinosis, sometimes complicated by infection, ascribing to GM-CSF an irreplaceable important role in pulmonary physiology (Dranoff et al 1994, Stanley et al 1994). The disorder does not appear to be secondary to infection because surfactant excess is morphologically evident in young (21 d) mice reared in a clean environment, without evidence of inflammation or infection (Dunn & Lieschke 1996). The fundamental primary problem is likely to be a defect in surfactant lipoprotein clearance, because transcription of surfactant protein A, B and C mRNAs is normal, but numerous surfactant lipid-laden macrophages are found in the lungs and studies of surfactant metabolism indicate impaired clearance (Dranoff et al 1994, Stanley et al 1994, Ikegami et al 1996).

As the phenotypes of CSF-deficient mice are studied in greater detail, it is likely that additional primary and secondary non-haemopoietic features will be recognized.

Insights from mice deficient in several CSFs

Mice deficient in two or more CSFs provide further information about the interacting physiological roles of these factors. Depending on the individual phenotypes of the single-factor-deficient mice and the consequence of double-factor deficiency, different conclusions about their *in vivo* roles may be made.

GM-CSF- and M-CSF-deficient mice

Mice deficient in both GM-CSF and M-CSF were the first double-CSF-deficient mice reported (Lieschke et al 1994b). Compared with GM-CSF-deficient mice, GM-CSF- and M-CSF-deficient mice have particularly severe lung disease with dense surfactant lipoprotein accumulation, increased vulnerability to incidental pulmonary infection, and reduced survival due in part to fatal bacterial pneumonic infections. These observations unequivocally implicate M-CSF in ameliorating the lung disease of GM-CSF-deficient mice, and because mice deficient in M-CSF alone do not have pulmonary disease, identify M-CSF as an available but non-essential regulatory factor in pulmonary pathology. Mice deficient in these two factors still have circulating monocytes and phagocytically active macrophages in their diseased lungs and at other sites of infection, indicating that full monocyte/macrophage development can be supported by other factors. The haemopoietic profile of GM-CSF- and M-CSF-deficient mice resembles that of mice deficient in M-CSF alone (Lieschke et al 1994b, Nilsson et al 1995) and, similarly to them, substantially corrects over time (Nilsson et al 1995), indicating that GM-CSF is not essential for monocyte/macrophage development in the absence of M-CSF and suggesting (but not proving) that GM-CSF is not responsible for the gradual correction of haemopoiesis in M-CSF-deficient mice.

G-CSF- and GM-CSF-deficient mice

Seymour et al (1995) have reported some preliminary observations about mice deficient in both G-CSF and GM-CSF. The degree of neutropenia in these mice is comparable to that of mice deficient in G-CSF alone, indicating that granulopoiesis in G-CSF-deficient mice is not exclusively dependent on GM-CSF. Mice deficient in both these factors have 60% cumulative mortality in the first year of life (G. Lieschke, D. Grail & J. Seymour, unpublished results). Since multifocal pneumonia with pulmonary abscesses is prevalent in mice that die, G-CSF appears important in ameliorating the severity of GM-CSF lung disease. A comprehensive characterization of these mice is underway.

Mice deficient in other factor combinations

To investigate the role of IL-3 in supporting haemopoiesis in G-CSF-deficient mice and *vice versa*, we have interbred IL-3-deficient mice (generated by G. Dranoff, Dana Farber Cancer Institute, Boston, MA, USA) with G-CSF-deficient mice to generate mice deficient in both these factors. IL-3- and G-CSF-deficient mice are viable and fertile (G. Lieschke & G. Dranoff, unpublished results), but haematological data are not yet available.

Similarly, tightly linked GM-CSF- and IL-3-disruptive mutations have been generated by sequential gene targeting and resultant GM-CSF- and IL-3-deficient mice have been generated (personal communications: G. Dranoff, V. Tybulewicz).

Mice deficient in the β_c chain of the GM-CSF, IL-3 and IL-5 receptors would be expected to combine the phenotypes of GM-CSF and IL-5 deficiency (but not to be IL-3 deficient since IL-3 signalling would still be possible using the β_{IL-3} chain). Indeed, such mice have pulmonary alveolar proteinosis resembling that of GM-CSF-deficient mice and concomitant blood and marrow eosinophil deficiency (Nishinakamura et al 1995, Robb et al 1995). The degree of eosinophil depletion in the blood (5–18% of normal) and bone marrow (20% of normal) of β_c-deficient mice is greater than that observed in IL-5-deficient mice alone (40% of normal in the bone marrow), suggesting that GM-CSF may support eosinophil development in the absence of IL-5. However, there were differences between baseline control groups and in the strains of mice used, and so this conclusion must be viewed with caution.

Initial analysis of mice deficient in three factors (G-CSF, GM-CSF and M-CSF) has commenced (Grail et al 1996, Seymour et al 1995) and is expected to provide further insights into the interacting physiological roles of these three factors.

Conclusion

Initial analyses of CSF-deficient mice have revealed definitive insights into the physiological role of these factors in haemopoiesis and in other tissues. Continuing studies of integrated physiological responses in CSF-deficient mice are likely to reveal further distinctions and overlaps in the roles of these factors *in vivo*, particularly in physiological processes secondarily dependent on the actions of these factors and the haemopoietic cell populations they support.

Acknowledgements

Most of the author's work was conducted at the Ludwig Institute for Cancer Research (Melbourne Tumour Biology Branch, Victoria, Australia). I thank my collaborators for their contributions to this work: at the Ludwig Institute for Cancer Research, Ashley Dunn, Dianne Grail, George Hodgson, Edouard Stanley, Sunanda Basu, Cathy Quilici, John Seymour, Kerry Fowler, Darryl Maher and Jonathan Cebon; at the University of Melbourne, Christina Cheers, Yi Zhan; at the Walter and Eliza Hall Institute for Medical Research, Donald Metcalf; and at the Whitehead Institute for Biomedical Research, Glenn Dranoff and Prakash Rao. G. J. L. is a Howard Hughes Medical Institute Postdoctoral Physician Fellow.

References

Basu S, Marino M, Savoia H et al 1994 Increased tolerance to endotoxin by granulocyte macrophage colony stimulating factor deficient mice. Blood (suppl 1) 84:510(abstr 2026)

Begg SK, Radley JM, Pollard JW, Chisholm OT, Stanley ER, Bertoncello I 1993 Delayed hematopoietic development in osteopetrotic (*op/op*) mice. J Exp Med 177:237–242

Bennett D 1956 Developmental analysis of a mutation with pleiotropic effects in the mouse. J Morphol 98:199–234

Bernad A, Kopf M, Kulbacki R, Weich N, Koehler G, Gutierrez-Ramos JC 1994 Interleukin-6 is required *in vivo* for the regulation of stem cells and committed progenitors of the hematopoietic system. Immunity 1:725–731

Cecchini MG, Dominguez MG, Mocci S et al 1994 Role of colony stimulating factor-1 in the establishment and regulation of tissue macrophages during postnatal development of the mouse. Development 120:1357–1372

Dalrymple SA, Lucian LA, Slattery R et al 1995 Interleukin-6-deficient mice are highly susceptible to *Listeria monocytogenes* infection: correlation with inefficient neutrophilia. Infect Immun 63:2262–2268

Dranoff G, Crawford AD, Sadelain M et al 1994 Involvement of granulocyte–macrophage colony-stimulating factor in pulmonary homeostasis. Science 264:713–716

Dunn AR, Lieschke GJ 1996 Granulocyte–macrophage colony-stimulating factor (GM-CSF)-deficient mice. In: Durum SK, Muegge K (eds) Cytokine knockouts. Humana Press, Totowa, NJ

Ebbe S, Phalen E, Stohlman F Jr 1973 Abnormalities of megakaryocytopoiesis in Sl/Sl^d mice. Blood 42:865–871

Grail D, Quilici C, Lieschke GJ, Seymour J, Hodgson G, Dunn AR 1996 Insights into the regulation of haematopoiesis revealed through the creation of mice deficient in one or more cytokines. Proceedings of the 8th Lorne Cancer Conference, Lorne, Victoria, Australia, 8–11 February 1996(abstr 417)

Ichihara M, Hara T, Takagi M, Cho LC, Gorman DC, Miyajima A 1995 Defective interleukin-3 (IL-3) response of the A/J mouse is caused by a branch point deletion in the IL-3 receptor α subunit gene. EMBO J 14:939–950

Ikegami M, Ueda T, Hull W et al 1996 Surfactant metabolism in transgenic mice after granulocyte–macrophage colony-stimulating factor ablation. Am J Physiol 270:650L–658L (Lung Cell Mol Physiol 14)

Kawakami M, Tsutsumi H, Kumakawa T et al 1990 Levels of serum granulocyte colony-stimulating factor in patients with infections. Blood 76:1962–1964

Kopf M, Baumann H, Freer G et al 1994 Impaired immune and acute-phase responses in interleukin-6 deficient mice. Nature 368:339–342

Kopf M, Brombacher F, Hodgkin PD et al 1996 IL-5-deficient mice have a developmental defect in $CD5^+$ B-1 cells and lack eosinophilia but have normal antibody and cytotoxic T cell responses. Immunity 4:15–24

Lieschke GJ 1996 Physiological role of haemopoietic growth factors in mice revealed by targeted gene disruption. PhD Thesis, University of Melbourne, Melbourne, Victoria, Australia

Lieschke GJ, Burgess AW 1992a Drug therapy: granulocyte colony-stimulating factor and granulocyte–macrophage colony stimulating factor. 1. N Engl J Med 327:28–35

Lieschke GJ, Burgess AW 1992b Drug therapy: granulocyte colony-stimulating factor and granulocyte–macrophage colony stimulating factor. 2. N Engl J Med 327:99–106

Lieschke GJ, Dunn AR 1992 Physiologic role of granulocyte colony-stimulating factor: insights from *in vivo* studies. In: Abraham NG, Konwalinka G, Marks P, Sachs L, Tavassoli M (eds) Molecular biology of haematopoiesis, vol 2. Intercept, Andover, p 201–216

Lieschke GJ, Grail D, Hodgson G et al 1994a Mice lacking granulocyte colony-stimulating factor have chronic neutropenia, granulocyte and macrophage progenitor cell deficiency, and impaired neutrophil mobilization. Blood 84:1737–1746

Lieschke GJ, Stanley E, Grail D et al 1994b Mice lacking both macrophage- and granulocyte–macrophage colony-stimulating factor have macrophages and co-existent osteopetrosis and severe lung disease. Blood 84:27–35

Marks SC, Lane PW 1976 Osteopetrosis, a new recessive skeletal mutation on chromosome 12 of the mouse. J Hered 67:11–18

Metcalf D 1995 The granulocyte–macrophage regulators: reappraisal by gene inactivation. Exp Hematol 23:569–572

Nilsson SK, Lieschke GJ, Garcia-Wijnen CC et al 1995 Granulocyte–macrophage colony-stimulating factor is not responsible for the correction of hematopoietic deficiencies in the maturing *op/op* mouse. Blood 86:66–72

Nishinakamura R, Nakayama N, Hirabayashi Y et al 1995 Mice deficient for the IL-3/GM-CSF/IL-5 β_c receptor exhibit lung pathology and impaired immune response, while β_{IL3} receptor-deficient mice are normal. Immunity 2:211–222

Poli V, Balena R, Fattori E et al 1994 Interleukin-6 deficient mice are protected from bone loss caused by estrogen depletion. EMBO J 13:1189–1196

Pollard JW, Hennighausen L 1994 Colony stimulating factor 1 is required for mammary gland development during pregnancy. Proc Natl Acad Sci USA 91:9312–9316

Pollard JW, Hunt JS, Wiktor-Jedrzejczak W, Stanley ER 1991 A pregnancy defect in the osteopetrotic (*op/op*) mouse demonstrates the requirement for CSF-1 in female fertility. Dev Biol 148:237–283

Robb L, Drinkwater CC, Metcalf D et al 1995 Hematopoietic and lung abnormalities in mice with a null mutation of the common β subunit of the receptors for granulocyte–macrophage colony-stimulating factor and interleukins 3 and 5. Proc Natl Acad Sci USA 92:9565–9569

Ruscetti FW, Boggs DR, Torok BJ, Boggs SS 1976 Reduced blood and marrow neutrophils and granulocytic colony-forming cells in *Sl/Sl*^{d} mice. Proc Soc Exp Biol Med 152:398–402

Russell ES 1979 Hereditary anemias of the mouse: a review for geneticists. Adv Genet 20:357–459

Seymour JF, Lieschke GJ, Stanley E et al 1995 Amyloidosis, hematopoietic and pulmonary abnormalities in mice deficient in both G-CSF and GM-CSF. Blood (suppl 1) 86:593(abstr 2359)

Shibata Y, Volkman A 1985 The effect of hemopoietic microenvironment on splenic suppressor macrophages in congenitally anemic mice of genotype *Sl/Sl*^{d}. J Immunol 135:3905–3910

Stanley E, Lieschke GJ, Grail D et al 1994 Granulocyte–macrophage colony-stimulating factor-deficient mice show no major perturbation of hematopoiesis but develop a characteristic pulmonary pathology. Proc Natl Acad Sci USA 91:5592–5596

Szperl M, Ansari AA, Urbanowska E et al 1995 Increased resistance of CSF-1-deficient, macrophage-deficient, TNF-α-deficient, and IL-1 α-deficient *op/op* mice to endotoxin. Ann N Y Acad Sci 762:499–501

Wiktor-Jedrzejczak W 1993 *In vivo* role of macrophage growth factors as delineated using CSF-1 deficient *op/op* mouse. Leukemia (suppl 2) 7:117S–121S

Wiktor-Jedrzejczak W, Ahmed A, Szczylik C, Skelly RR 1982 Hematological characterization of congenital osteopetrosis in *op/op* mouse. J Exp Med 156:1516–1527

Wiktor-Jedrzejczak W, Bartocci A, Ferrante AW et al 1990 Total absence of colony-stimulating factor 1 in the macrophage-deficient osteopetrotic (*op/op*) mouse. Proc Natl Acad Sci USA 87:4828–4832 (erratum: Proc Natl Acad Sci USA 88:5978)

Wu H, Liu X, Jaenisch R, Lodish HF 1995 Generation of committed erythroid BFU-E and CFU-E progenitors does not require erythropoietin or the erythropoietin receptor. Cell 83:59–67

Yoshida H, Hayashi S-I, Kunisada T et al 1990 The murine mutation osteopetrosis is in the coding region of the macrophage colony-stimulating factor gene. Nature 345:442–444
Zsebo KM, Williams DA, Geissler EN et al 1990 Stem cell factor is encoded at the *Sl* locus of the mouse and is the ligand for the c-*kit* tyrosine kinase receptor. Cell 63:213–224

DISCUSSION

Nossal: Which factor do you think is promoting the appearance of mature granulocytes in the triple knockout mice?

Lieschke: The best candidates would be SCF, IL-6 or possibly IL-11. There are other possibilities, but I am less inclined to think that they are involved. For example, the Flt-3/Flk-2 ligand clearly does affect granulopoiesis but it's not very good as a granulopoietic colony-stimulating factor in its own right (Hudak et al 1995). Defective granulopoiesis in *Listeria*-infected IL-6-deficient mice (Dalrymple et al 1995) focuses interest on IL-6 as possibly being a more important granulopoietic factor *in vivo* than had previously been thought.

Metcalf: The evidence is quite good that SCF regulates neutrophil production. The papers written on this are more than 20 years old and have been forgotten. W/W^v mice lacking the SCF receptor or *Sl* mice unable to produce SCF efficiently both have a fairly profound neutropenia (Chervenick & Boggs 1969, Ruscetti et al 1976). Until somebody does a G-CSF/SCF double knockout, we don't know whether together these factors can account for the stimulation of all neutrophil production.

Burgess: Graham, why do you think it is that the GM-CSF knockout mice are resistant to LPS?

Lieschke: LPS kicks off a complex physiological response including production of many cytokines and related molecules which, if uncontrolled, leads to septic shock and may culminate in death. Since GM-CSF-deficient mice are resistant to LPS, GM-CSF is involved either in initiating the cascade of acute-phase reactants and pro-inflammatory cytokines, or in amplifying it. An interesting therapeutic possibility is suggested by this observation: GM-CSF antagonism may be a useful intervention to interrupt the deteriorating physiology of patients with severe Gram-negative septic shock. Other related approaches tested in such patients have been to antagonise IL-1 or TNF, but these have been notoriously unsuccessful. Perhaps one of the reasons why these approaches didn't work is that there are several important molecules initiating the pro-inflammatory cytokine cascade, or several important amplifying molecules. Therapies simply directed at intervening with the effects of one cytokine may not be enough. But the LPS resistance of GM-CSF-deficient mice mean that GM-CSF must be included in the list of potential initiators and amplifiers of the cytokine cascade leading to septic shock.

Burgess: Given that the TNF concentration rises in the serum of the GM-CSF$^{-/-}$ mice without any significant phenotype, does this mean that GM-CSF is downstream of TNF?

Lieschke: I don't think that the knockout mice necessarily indicate where in the stream a factor lies. Although one obvious explanation for the concurrence of absent GM-CSF production, high TNF levels and obtunded physiological responses in LPS-treated GM-CSF-deficient mice is indeed that TNF exerts its effects through GM-CSF, there are other possible explanations. For example, GM-CSF may be upstream of TNF but the serum TNF response may not be critically dependent on GM-CSF alone because of an alternative LPS-activated TNF-producing pathway. However, GM-CSF may uniquely also induce an ancillary 'priming' or 'facilitating' pathway which has to be activated concurrently for high TNF levels to mediate adverse physiological effects, and it is this that is missing in LPS-treated GM-CSF mice. So although factor-deficient mice can show that molecules interact in eliciting a physiological response, they do not necessarily precisely define the temporal connection of the interaction.

Dexter: What happens in the IL-3 knockouts? Although the results haven't been published, I know one has been made by Victor Tybulewicz (personal communication). When I talked to him he mentioned that he couldn't see any effects on haemopoiesis or resistance to infections. I think that someone else has done this knockout with similar results. Much more interesting would be double or triple knockouts involving IL-3.

Lieschke: Both Victor Tybulewicz (personal communication) and Glenn Dranoff (personal communication) have made IL-3-deficient mice. As far as we know from their brief mentions at meetings, there is no significant phenotype. In a collaboration, Glenn and I have interbred the G-CSF- and IL-3-deficient mice to generate mice deficient in both these factors. All I can tell you at the moment is that G-CSF- and IL-3-deficient mice are viable and can be interbred. We have not yet studied their haemopoietic system. Now that we know we can maintain this colony by interbreeding the double-deficient animals, we will be able to generate the cohorts of animals that are matched in time and age that are optimal for these haematological studies.

Morstyn: There were reports that G-CSF suppresses TNF production and endotoxin is cleared more rapidly in G-CSF-treated animals. My prediction is that G-CSF-deficient knockout mice would not be protected against endotoxin treatment in the way the GM-CSF mice seem to be. Have you looked at this in these mice?

Lieschke: Those experiments are in progress.

Nossal: Before people get too excited about G-CSF as a therapy for septic shock, we have to remember that there is such a thing as a complement cascade. It is going to be a tough problem to beat. There are at least three bodily systems going haywire in septic shock: cytokines, coagulation and complement.

Nicola: There are data that suggest that GM-CSF is essential for some types of dendritic cell formation. What happens to the dendritic cells in the GM-CSF knockouts?

Lieschke: Glenn Dranoff's group counted dendritic cells in their GM-CSF animals and found no deficit (Dranoff et al 1994). Together with Ken Shortman's group, we also enumerated lymph node thymic and splenic dendritic cells in a large cohort of animals, and found no marked deficit (K. Shortman, unpublished data). This was interesting, because at that time GM-CSF was viewed as an obligatory factor for their production *in vitro*. This shows that there must be other factors able to support dendritic cell formation.

Rajewsky: What does the GM-CSF knockout tell us about alveolar proteinosis in the human?

Lieschke: The disease these mice get has all the morphological features of the human disease. This raises the likelihood that some forms of congenital alveolar proteinosis may be a consequence of GM-CSF deficiency or mutations in the GM-CSF receptor. Whether the acquired form of the human disease represents a relative deficiency of GM-CSF and an acquired somatic mutation, and whether acquired alveolar proteinosis would respond to GM-CSF are all interesting possibilities that are being actively pursued in Melbourne and by others. It would be very exciting if the accumulation of this excess lipoprotein in the alveoli could be ameliorated or cleared by GM-CSF administration, even if the fundamental aetiology was something else. This has essentially been an untreatable and almost invariably fatal clinical condition.

Metcalf: At this moment, there is a patient with alveolar proteinosis in the adjacent hospital who is receiving GM-CSF treatment for his disease.

Burgess: The symptoms certainly resolve in the short term after GM-CSF treatment, but the effects haven't been followed up long-term.

Lieschke: Huffman et al (1996) created a dominant GM-CSF transgene driven by a surfactant protein promoter that directed transgene expression to the respiratory epithelium and used this as a model of therapy for alveolar proteinosis in GM-CSF-deficient mice. It completely ameliorated the alveolar proteinosis in otherwise GM-CSF-deficient mice. This model is a genetic test of whether administering GM-CSF ameliorates the proteinosis which develops in the face of GM-CSF deficiency. It is very tempting to extrapolate that it would be a useful therapy for the acquired form of the disease as well.

References

Chevrenick PA, Boggs DR 1969 Decreased neutrophils and megakaryocytes in anemic mice of genotype W/W^v. J Cell Physiol 73:25–30

Dalrymple SA, Lucian LA, Slattery R et al 1995 Interleukin-6-deficient mice are highly susceptible to *Listeria monocytogenes* infection: correlation with inefficient neutrophilia. Infect Immun 63:2262–2268

Dranoff G, Crawford AD, Sadelain M et al 1994 Involvement of granulocyte–macrophage colony-stimulating factor in pulmonary homeostasis. Science 264:713–716

Hudak S, Hunte B, Culpepper J et al 1995 Flt3/Flk2 ligand promotes the growth of murine stem cells and the expansion of colony-forming cells and spleen colony-forming units. Blood 85:2747–2755

Huffman JA, Hull WM, Dranoff G, Mulligan RC, Whitsett JA 1996 Pulmonary epithelial cell expression of GM-CSF corrects the alveolar proteinosis in GM-CSF-deficient mice. J Clin Invest 97:649–655
Ruscetti FW, Boggs DR, Torok BJ, Boggs SS 1976 Reduced blood and marrow neutrophils and granulocyte colony-forming cells in *Sl*/*Sl*^{d} mice. Proc Soc Exp Biol Med 152:398–402

Clinical benefits of improving host defences with rHuG-CSF

George Morstyn, MaryAnn Foote and *Steve Nelson

*Amgen Inc., 1840 Dehavilland Drive, Thousand Oaks, CA 91320–1789 and *LSU Medical Center, New Orleans, LA, USA*

Abstract. Recombinant human granulocyte colony-stimulating factor (rHuG-CSF) has been shown to stimulate the production and function of neutrophils *in vitro* and *in vivo*. Clinical studies in patients receiving myelosuppressive chemotherapy showed earlier neutrophil recovery, a reduction in infectious complications of neutropenia, and the use of fewer antibiotics. Its use has also been established for mitigating the infectious complications associated with severe chronic neutropenia (SCN). Data are emerging that neutropenia also contributes to the risk of infections in patients with acquired immunodeficiency syndrome (AIDS) and in neonates with presumed sepsis, and that rHuG-CSF may be a useful adjunct therapy in these patients. More recent studies have focused on enhancing neutrophil number and function in patients with infections not associated with neutropenia. These studies were approached cautiously because of the suggestion that neutrophils might non-selectively amplify the body's inflammatory response in the immunocompetent host and lead to inadvertent tissue injury. Preclinical models have provided a strong rationale for clinical studies to determine whether rHuG-CSF lessens the severity or duration of serious infections or their complications in patients with suboptimal outcome from antibiotics. These studies suggest that elevation of neutrophil levels in these settings is not only safe but has clinical benefit.

1997 The molecular basis of cellular defence mechanisms. Wiley, Chichester (Ciba Foundation Symposium 204) p 78–87

Host defences against infection include physical barriers (e.g. intact skin, lysozyme in tears and saliva), specific immune effectors (e.g. cells capable of exhibiting memory to agents), and innate immune effector cells (e.g. neutrophils). Neutrophils respond to exogenous infective agents and recognize them without previous exposure. Neutrophils are the first cells to respond to chemotactic stimuli produced by infected tissue, such as the chemokines, and this causes them to move to the site of infection. This paper reviews some of the work that has been done to study the clinical effects of recombinant human granulocyte colony-stimulating factor (rHuG-CSF) on various non-neutropenic infectious disease states.

Role of endogenous G-CSF

Mice lacking endogenous G-CSF have chronic neutropenia and impaired neutrophil mobilization, indicating that G-CSF is indispensable for maintaining the normal quantitative balance of neutrophil production during steady state granulopoiesis *in vivo* (Lieschke et al 1994). It is possible that G-CSF also has a role in the leukocytosis seen during infections. In response to an infection or to neutropenia, the amount of circulating endogenous G-CSF in the blood has been shown to increase (Metcalf 1988, Watari et al 1989, Kawakami et al 1990, Morstyn et al 1991, Cebon et al 1994). The highest endogenous G-CSF levels are found in neutropenic patients and are correlated with fever (Cebon et al 1994). Because serum or plasma levels of G-CSF have been found to increase significantly during bacterial sepsis and other febrile illnesses, it is believed that G-CSF may play a critical role in the endogenous regulation of the host immune response.

Preclinical studies of rHuG-CSF in non-neutropenic infection models

The observation that endogenous G-CSF levels are increased during infection, coupled with the documented ability of rHuG-CSF to increase the number of neutrophils in the peripheral blood system and the importance of the neutrophil in the body's response to infection, led to *in vivo* investigations of the effect of rHuG-CSF in infectious diseases. A number of animal models of non-neutropenic infection have demonstrated the efficacy of rHuG-CSF.

Pneumonia model

Administration of rHuG-CSF was associated with a reduction in viable bacterial counts and a significant improvement in survival in several animal models of pneumonia, including studies with splenectomized mice and ethanol-treated rats.

Splenectomized mice exposed to an aerosol of *Streptococcus pneumoniae* were administered rHuG-CSF 24 h before exposure and for 3 d after exposure. The rHuG-CSF-treated animals had a 70% survival rate compared with 20% for the control animals ($P < 0.001$) (Herbert et al 1990). Use of rHuG-CSF decreased the recovery of viable *S. pneumoniae* from the lungs and tracheobronchial lymph nodes of the splenectomized and sham-operated mice compared with control animals.

Pre-treatment of rats with rHuG-CSF before intraperitoneal administration of ethanol and intratracheal challenge with *Klebsiella pneumoniae* increased the recruitment of neutrophils into the lungs of control rats and attenuated the adverse effects of ethanol on neutrophil entry into the lung at 4 h (Nelson et al 1991). Only 2% of viable bacteria remained in the rHuG-CSF-treated animals compared with 79% in control animals. In ethanol-treated animals, rHuG-CSF decreased the proliferation of *K. pneumoniae* from 255% to 118% ($P < 0.01$). 72 h after inoculation, only 8% of the animals in the rHuG-CSF-treated group had died compared with 100% in the control group (Fig. 1) (Nelson et al 1994).

The only study showing potential negative effects of rHuG-CSF in the lung was reported by Freeman et al (1996) in which rats exposed to intrabronchial *Escherichia coli* and treated with rHuG-CSF had increased circulating and pulmonary

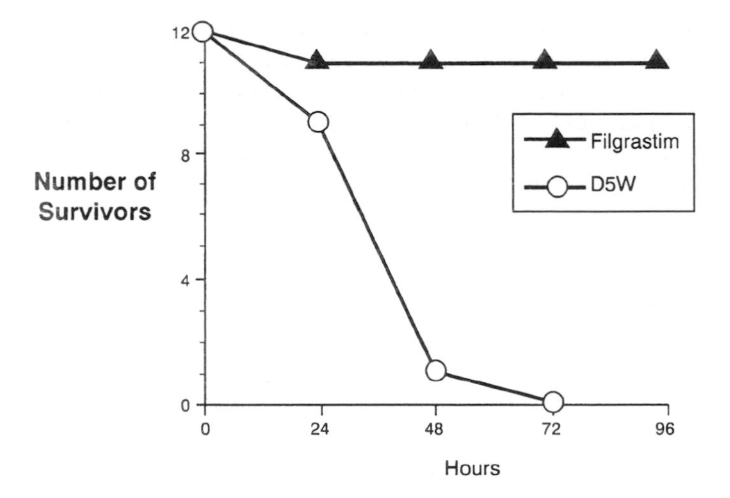

FIG. 1. Pre-treatment with rHuG-CSF (Filgrastim) enhanced the survival of ethanol-treated rats challenged with *Klebsiella pneumoniae* (Nelson et al 1994). D5W, 5% dextrose in water.

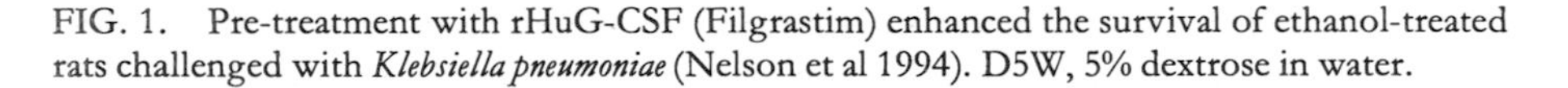

neutrophils. The rats had worsened pulmonary injury during both infectious and non-infectious challenges, suggesting a pro-inflammatory effect of rHuG-CSF. Of note, the massive amount of inoculum used in this model may be more reflective of a model of acute lung injury than pneumonia. A previous study by the same group using a canine *E. coli* peritonitis model showed that rG-CSF improved survival without worsening lung injury (Eichacher et al 1994). Clearly, human studies are required to determine how predictive any of these models are for the clinical situation.

Septic shock

In dogs with an implanted *E. coli*-infected peritoneal clot, pre-treatment with recombinant canine (rc)G-CSF at 0.1 or 5 μg/kg/d produced a dose-dependent neutrophilia and reduction in mortality (22% and 10%, respectively) compared with control animals (46%, $P = 0.06$) (Eichacher et al 1994). Pre-treatment with rHuG-CSF (50 or 250 μg/kg/d) protected mice against bacterial endotoxin-induced mortality (Görgen et al 1992). Mortality in mice treated with rHuG-CSF was 83% compared with 33% and 0% for those pre-treated with 50 μg/kg and 250 μg/kg, respectively. The reduction in mortality was accompanied by suppression of bacterial endotoxin (LPS)-induced serum tumour necrosis factor (TNF). Pre-treatment with rHuG-CSF reduced the accumulation of water and albumin leakage in the lungs of guinea-pigs challenged with LPS, suggesting that rHuG-CSF may attenuate the lung responses to intravenous LPS (Kanazawa et al 1992).

Pre-treatment with rHuG-CSF also prevented sepsis-induced leukopenia, increased the influx of neutrophils into the infected site, and decreased the incidence of bacteraemia in ethanol-treated rats challenged with subcutaneous *E. coli* (Lang et al 1993). A porcine model of *Pseudomonas* sepsis showed that administration of rHuG-CSF before bacterial challenge reduced temperature elevations, mean pulmonary arterial pressure, plasma levels of TNF and endotoxin compared with control animals that did not receive rHuG-CSF (Haberstroh et al 1995). There were no deaths in the rHuG-CSF-treated group, but two due to pulmonary failure in the control group.

A study was done to determine whether rHuG-CSF given with antibiotics would influence the morbidity and mortality in animals with pneumonia and sepsis (Smith et al 1995). Unlike other studies, this one was designed to simulate clinical sepsis, with rHuG-CSF administered after sepsis was evident, not at the time the bacteria were introduced into the animal's body. Rabbits were infected with penicillin-sensitive, Gram-negative bacteria, and then randomized to receive penicillin with placebo or 6.5 to 8 μg/kg/d rHuG-CSF for 5 d. The rabbits treated with rHuG-CSF had significantly increased neutrophil levels by day 4 of the study compared with rabbits treated with placebo. There was a trend toward improved survival in the rHuG-CSF-treated animals, with most of the survival benefit seen in the first 24 h of treatment.

Clinical studies with rHuG-CSF in non-neutropenic infection

The role of rHuG-CSF is well established in the chemotherapy setting (Bronchud et al 1988, Gabrilove et al 1988a,b, Morstyn et al 1989, Crawford et al 1991) as is the chronic administration for severe chronic neutropenia (SCN) (Dale et al 1993, Welte et al 1990). Recently, early studies of the use of rHuG-CSF have been reported in non-neutropenic settings. The difficulties in designing studies in these settings are: (1) that antibiotics are effective in many settings and there is a need to show that rHuG-CSF added to antibiotic therapy produces better clinical outcomes; (2) that rHuG-CSF works best in prevention models in which neutrophil levels are increased before exposure to the infectious agent, and because it is often not possible to predict when a patient will become infected, the preventive use of rHuG-CSF has limited clinical applicability; and that (3) when the studies were initiated there was concern that rHuG-CSF-induced neutrophilia, in theory, might make the inflammatory consequences of infection worse. To date, these clinical data suggest potential benefits of rHuG-CSF.

Normal volunteer studies

To develop a rationale for the potential benefits of rHuG-CSF in addition to elevating neutrophils, a study was done in normal volunteers. Healthy human volunteers were given 4 ng/kg bacterial endotoxin (LPS) which induced neutropenia and fever (van Deventer 1995). After administration of rHuG-CSF 5 μg/kg/day, the volunteers had increased neutrophil counts and increased levels of BPI, a neutrophil-derived endotoxin inhibitor. Compared with volunteers receiving placebo, rHuG-CSF completely blocked neutrophil recruitment to the lungs. In another study in a similar group, cytokine release in the blood from individuals pre-treated with 480 μg rHuG-CSF was stimulated by LPS; treatment with rHuG-CSF decreased the levels of LPS-induced TNF and increased the release of soluble TNF-receptor and interleukin (IL)-1ra (Kuhlberg & van der Meer 1996). Similar results were reported in another study when cytokine release was stimulated by agents other than LPS (Hartung et al 1995).

Pneumonia

Studies have been done in the setting of community-acquired pneumonia (CAP). Lower respiratory tract infections are associated with significant morbidity and mortality. Since the clearance of microorganisms from this area is dependent upon a functioning neutrophil system, patients with neutropenia or compromised neutrophil function (due to alcoholism, trauma, or chronic illness) are particularly susceptible to pneumonia. Pneumonia, in itself, can impair neutrophil function and lead to chronic disease.

In a phase I, open-label trial of CAP, 31 patients were treated with rHuG-CSF (Filgrastim) at doses as great as 600 μg/d for up to 10 d in conjunction with antibiotics (De Boisblanc et al 1996). Doses >150 μg/d increased neutrophil counts up to three times the normal value and were not associated with any clinically evident toxicity. Overall, pulmonary symptoms and signs improved or remained normal for the majority of the patients regardless of the dose. The study was too small to determine whether there was any clinical benefit, and clinical outcomes (i.e. duration of fever, antibiotic use, hospitalization) had no apparent relation to dose. However, this study appeared to establish a safe dose that was effective and a phase III study was initiated.

In the phase III trial, treatment with rHuG-CSF plus antibiotics was compared with antibiotic treatment alone in 756 patients with moderate CAP (Nelson et al 1996). Preliminary analysis of the study showed that the primary endpoint, time to resolution of morbidity (i.e. absence of clinical signs and symptoms), was not met. The analysis, however, did demonstrate a benefit in patients receiving rHuG-CSF in resolving radiographic signs of pneumonia, as well as clinical benefit, i.e. reduction of incidence of end-organ failure and incidence of acute respiratory distress syndrome (ARDS).

Clinical response to rHuG-CSF in patients with pneumonia and severe sepsis was reported by Wunderink et al (1996). Mortality at day 15 and day 29, and resolution of end-organ failure on day 29 were assessed in a double-blind, placebo-controlled, randomized study in which placebo or rHuG-CSF was administered at 300 μg/d intravenously for 5 d. Mortality at day 15 and day 29 was less in the rHuG-CSF-treated group (2/12 and 3/12) compared with the placebo-treated group (4/6 and 4/6) ($P=0.042$ and $P=0.0108$, respectively). These two studies suggest there may be clinical benefit of rHuG-CSF in severely ill patients with pneumonia, and additional phase III trials to study the prevention of severe complications and the reduction of severe complications and mortality are expected to continue in 1996.

Neonatal sepsis

The effects of rHuG-CSF in neonatal sepsis were studied in a randomized, placebo-controlled, phase I/II trial (Gillan et al 1994). Forty-two neonates, gestational age 26 to 42 weeks, with presumed bacterial sepsis in the first 3 d of life, were administered rHuG-CSF at 1, 5, or 10 μg/kg/d. There was a significant increase in their peripheral blood neutrophil count 24 h after administration of the 5 and 10 μg/kg/d doses, a dose-dependent increase in neutrophil storage pool, and increase in functional activity of neutrophils. In another study, rHu-CSF 5 μg/kg/d was given to 12 neutropenic neonates receiving intensive care, seven of whom had confirmed infections (Bedford-Russell et al 1995). Use of rHuG-CSF significantly increased absolute neutrophil and monocyte counts ($P < 0.01$), and in six patients, these increases were thought to have contributed to recovery. Future studies are planned with rHuG-CSF in this setting.

Conclusions

Recombinant HuG-CSF has not been observed to have any dose-limiting adverse effects. Recombinant HuG-CSF (Filgrastim) has been used to treat more than 1.2 million patients in the oncology and AIDS settings. The most common dose-independent adverse event is mild to moderate bone pain. In view of the findings that some cytokines may have a role in the development of organ damage, such as lung injury, it is important to note that rHuG-CSF has not been observed to precipitate lung inflammation or ARDS even in patients with moderate to severe pneumonia, and this safety profile forms a basis for further study of rHuG-CSF in patients with severe infections. There has been confirmation of this by preliminary results from a clinical trial investigating the use of rHuG-CSF. Animal models suggest a role in prevention and therapy as an adjunct to antibiotics. Early clinical trials support the safety and potential reduction in severe complications. Further trials are needed to define the exact role of rHuG-CSF in non-neutropenic settings.

References

De Boisblanc BP, Mason CM, Andresen J et al 1996 Phase I safety trial of filgrastim (r-metHuG-CSF) in non-neutropenic patients with severe community-acquired pneumonia. Resp Med, in press

Bedford-Russell AR, Davies EG, Ball SE et al 1995 Granulocyte colony-stimulating factor treatment for neonatal neutropenia. Arch Dis Child 72:53F–54F

Bronchud MH, Potter MR, Morgenstern G et al 1988 *In vitro* and *in vivo* analysis of the effects of recombinant human granulocyte colony-stimulating factor in patients. Br J Cancer 58:64–69

Cebon J, Layton JE, Maher D, Morstyn G 1994 Endogenous haemopoietic growth factors in neutropenia and infection. Br J Haematol 86:265–274

Crawford J, Ozer H, Stoller R et al 1991 Reduction by granulocyte colony-stimulating factor of fever and neutropenia induced by chemotherapy in patients with small-cell lung cancer. N Engl J Med 325:164–170

Dale DC, Bonilla MA, Davis MW et al 1993 A randomized controlled phase III trial of recombinant human granulocyte colony-stimulating factor (Filgrastim) for treatment of severe chronic neutropenia. Blood 81:2496–2502

Eichacher PQ, Waisman Y, Natanson C et al 1994 Cardiopulmonary effects of granulocyte-colony-stimulating factor in a canine model of bacterial sepsis. J Appl Physiol 77:2366–2373

Freeman BD, Correa R, Karzai W et al 1996 Controlled trials of rG-CSF and CD11b-directed Mab during hyperoxia and *Escherichia coli* pneumonia in rats. J Appl Physiol 80: 2066–2076

Gabrilove JL, Jakubowski A, Scher H et al 1988a Effect of granulocyte colony-stimulating factor on neutropenia and associated morbidity due to chemotherapy for transitional-cell carcinoma of the urothelium. N Engl J Med 318:1414–1422

Gabrilove JL, Jakubowski A, Fain K et al 1988b Phase I study of granulocyte colony-stimulating factor in patients with transitional cell carcinoma of the urothelium. J Clin Invest 82:1454–1461

Gillan ER, Christensen RD, Suen Y, Ellis R, van de Ven C, Cairo MS 1994 A randomized, placebo-controlled trial of recombinant human granulocyte colony-stimulating factor administration in newborn infants with presumed sepsis: significant induction of peripheral and bone marrow neutrophilia. Blood 84:1427–1433

Görgen I, Hartung T, Leist M et al 1992 Granulocyte colony-stimulating factor treatment protects rodents against lipopolysaccharide-induced toxicity via suppression of systemic tumour necrosis factor alpha. J Immunol 149:918–924

Haberstroh J, Breuer H, Lucke I 1995 Effect of recombinant human granulocyte colony-stimulating factor on hemodynamic and cytokine response in a porcine model of *Pseudomonas* sepsis. Shock 4:216–224

Hartung T, Docke W, Gantner F et al 1995 Effect of G-CSF treatment on *ex vivo* blood cytokine response in human volunteers. Blood 85:2482–2489

Herbert JC, O'Reilly M, Gamelli RL 1990 Protective effect of recombinant human granulocyte colony-stimulating factor against pneumococcal infections in splenectomized mice. Arch Surg 125:1075–1078

Kanazawa M, Ishizaka A, Hasegawa N, Suzuki Y, Yokoyama T 1992 Granulocyte colony-stimulating factor does not enhance endotoxin-induced acute lung injury in guinea pigs. Am Rev Resp Dis 145:1030–1035

Kawakami M, Tsutsumi H, Kumakawa T et al 1990 Levels of serum granulocyte colony-stimulating factor in patients with infections. Blood 76:1962–1964

Kuhlberg BJ, van der Meer JWM 1996 Infectious diseases: the role of haematopoietic growth factors. Gardiner-Caldwell Communications, Macclesfield

Lang CH, Molina PE, Abumrad NN 1993 Granulocyte colony-stimulating factor prevents ethanol-induced impairment in host defense in septic rats. Alcohol Clin Exp Res 17:1268–1274

Lieschke GJ, Grail D, Hodgson G et al 1994 Mice lacking granulocyte colony-stimulating factor have chronic neutropenia, granulocyte and macrophage progenitor cell deficiency, and impaired neutrophil mobilization. Blood 84:1737–1746

Metcalf D 1988 The molecular control of blood cells. Harvard University Press, Cambridge, MA

Mooney DP, Gamelli RL, O'Reilly M, Herbert JC 1988 Recombinant human granulocyte colony-stimulating factor and *Pseudomonas* burn wound sepsis. Arch Surg 123:1353–1357

Morstyn G, Campbell L, Lieschke G et al 1989 Treatment of chemotherapy-induced neutropenia by subcutaneously administered granulocyte colony-stimulating factor with optimization of dose and duration of therapy. J Clin Oncol 7:1554–1562

Morstyn G, Cebon J, Layton J et al 1991 Cytokines in infections and as anticancer agents. Ann Hematol 29A(abstr A96)
Nelson S, Summer W, Bagby G et al 1991 Granulocyte colony-stimulating factor enhances pulmonary host defenses in normal and ethanol-treated rats. J Infect Dis 164:901–906
Nelson S, Daifuku R, Andresen J 1994 Use of Filgrastim (r-metHuG-CSF) in infectious diseases. In: Morstyn G, Dexter TM (eds) Filgrastim (r-metHuG-CSF) in clinical practice. Marcel Dekker, New York, p 253–266
Nelson S, Farkas S, Fotheringham N, Ho H, Marrie T, Movahhed H 1996 Filgrastim in the treatment of hospitalized patients with community acquired pneumonia (CAP). Am J Resp Crit Care Med 153A(abstr 535)
Smith WS, Sunmicht GE, Sharpe RW, Samuelson D, Millard FE 1995 Granulocyte colony-stimulating factor versus placebo in addition to penicillin G in a randomized blinded study of Gram-negative pneumonia sepsis: analysis of survival and multisystem organ failure. Blood 86:1301–1308
van Deventer SJH 1995 Progress in cytokine research and implications for immunotherapy of sepsis. Clinician 13:9–13
Watari K, Asano S, Shirafuji N et al 1989 Serum granulocyte colony-stimulating factor levels in healthy volunteers and patients with various disorders as estimated by enzyme immunoassay. Blood 73:117–122
Welte K, Zeidler C, Reiter A et al 1990 Differential effects of granulocyte–macrophage colony-stimulating factor and granulocyte colony-stimulating factor in children with severe congenital neutropenia. Blood 75:1056–1063
Wunderink RG, Leeper KV, Schein RMH et al 1996 Clinical response to Filgrastim (r-metHuG-CSF) in pneumonia with severe sepsis. Am J Resp Crit Care Med 153A(abstr 123)

DISCUSSION

Nossal: I'd like to start by asking you a question prompted by the surprisingly low starting neutrophilia in the group of pneumonia patients. This suggests, for whatever reason, that this cohort did not have a very marked shift to the left. Has anyone decomposed those results in terms of a subset that showed a mobilization of neutrophils and an adequately high count? And in such a group, might the G-CSF not have had the incremental results that it had on the various clinical parameters measured?

Morstyn: The results have been decomposed. Not surprisingly, the group that does the worst is the group labelled as having 'relative leukopenia'. This is a small subset of patients who have a low normal neutrophil count even though they have an overwhelming infection. In this group, the difference between controls and the patients given G-CSF is most dramatic. There are clearly some patients that don't receive G-CSF with neutrophil counts of 40–50 000, but it is interesting that overall as a group, the absolute neutrophil level only goes up to 10–12 000/cmm. We didn't study the data on left-shift: the point was not to dissect out a group that G-CSF works less well in. Instead we wanted to see in whom the G-CSF works the best, and by any of the known prognostic factors that we looked at, including relative low white cell count, signs of sepsis, and multilobar pneumonia, the most severely affected patients always had the greatest benefit from G-CSF.

Nossal: It seems clear that as this drug reaches widespread clinical practice, clinicians will be using it in a greater variety of circumstances. For example, they will certainly be using the drug if the white cell count, in the presence of serious bacterial infection, only manages to reach 5000. In the real world, that is what is going to happen, which is why it is tremendously important to dissect your data every which way. The results you've just given are very encouraging and merit follow-up in wider trials.

Morstyn: The other side of this is that the group that we studied is at the severe end of the spectrum of pneumonia. We're not suggesting G-CSF is going to benefit the vast majority of patients with less severe disease; instead we're looking at the most-sick 150 000 patients per year in the US who have pneumonia sepsis and severe pneumonia.

Goodnow: Is it possible, in the largest group which had *Pneumococcus*, to look at IgG titres to *Pneumococcus*, to break the data down to those that had the lowest or highest antibody titres?

Morstyn: We haven't done that; I don't think it was part of the data collection for the study.

Goodnow: Might you expect in theory some sort of synergy, where you could give a cocktail of human monoclonal antibodies to some of the common organisms at the same time as G-CSF?

Morstyn: That is an interesting idea, because one of the effects of G-CSF is to up-regulate the Fc receptors; this could lead to a greater effect against organisms coated with monoclonal antibody as a result of the treatment you propose.

Burgess: You began by telling us that in patients with ARDS, the neutrophils are probably dysfunctional, but you didn't tell us about the neutrophil function of the patients treated with G-CSF.

Morstyn: A number of investigators, including David Dale, have shown that G-CSF enhances neutrophil function in normals, and we have also found improved function in patients with AIDS. We've studied neutrophil function in small pilot studies, but in a study such as the one I presented with 40 centres and 750 patients it is hard to establish neutrophil function tests on all the patients at all the centres, and so this was not investigated.

Burgess: Many of your patients will have been at risk of arthritis: did the study look at this?

Morstyn: I'm not sure how many had rheumatoid arthritis; exacerbation of rheumatoid arthritis did not appear to be a problem. There are some patients with Fetty's syndrome whose joint pain becomes worse when their neutrophil count is restored.

Burgess: Although you increased the neutrophil count in patients, is there any indication that these neutrophils are functional? In this group of severe pneumonia patients, the neutrophils are supposed to be dysfunctional.

Morstyn: There are a number of criteria used to assess neutrophil function and G-CSF has been reported to enhance many of these. We have looked at this in burn and head trauma patients, in a small study in pneumonia, and in AIDS. The neutrophils are certainly functioning, but it is hard to convince yourself that the neutrophil function is better than in control patients.

Nossal: Superficially, it is tempting to use G-CSF in localized conditions such as empyema, peritonitis and sub-phrenic abscess. Perhaps these result when normal bodily processes cannot mobilize enough defence in such a localized environment. Is there anything at all in the data that suggests that the empyema patients got any benefit?

Morstyn: I think we are preventing the spreading of the infection locally. They didn't have empyema at the start of the infection, so the hypothesis is that G-CSF prevented the infection from spreading to the lung wall, and that's why only one person on G-CSF actually developed empyema versus six in the placebo group. I don't know if anyone has tried using G-CSF to treat a patient with empyema. Some of the data suggest G-CSF could be used not only to prevent but also to treat ARDS. When we elevate neutrophil counts, I think these neutrophils are available to go to where they're needed, due to the local host responses. I haven't thought of G-CSF as an agent which you would inject locally.

Lieschke: In pneumonia, elevated serum G-CSF levels reflecting endogenous G-CSF production and active bacterial infection can coexist (Kawakami et al 1990). Were these patients screened for serum G-CSF levels at the start of the study?

Morstyn: Although we have serum from patients, we haven't measured this. In a pilot study, Steve Nelson measured levels in pneumonia patients and showed that they were high and came back to normal as the patients' infections resolved. The hypothesis here would have to be that G-CSF levels are not high enough—and the levels certainly don't elevate the neutrophil count as highly as when G-CSF is given exogenously.

Zinkernagel: Did you say that your selection of patients had an inbuilt bias towards taking the less sick patients for the treatment?

Morstyn: Yes. When we started the study we thought we would end up with a group of patients on both arms that had a mortality rate of about 10%. In fact, we ended up with a group that had a mortality rate closer to 5%. As the study progressed the mortality rate in both groups increased and we got a sicker group of patients. We think physicians were initially unwilling to put the sickest patients onto the study, even if they were eligible. The eligibility criteria excluded the patients with septic shock.

Reference

Kawakami M, Tsutsumi H, Kumakawa T et al 1990 Levels of serum granulocyte colony-stimulating factor in patients with infections. Blood 76:1962–1964

General discussion III

Test tube data and *in vivo* realities

Nossal: I would like to direct our attention to the *in vivo veritas* note which Graham Lieschke struck in his paper. I have a few questions for discussion. First, with respect to the cytokines we are talking about, what reflections do any of us have on the results that would have been predicted from *in vitro* studies and the results that have actually been obtained in real life? Second, what surprises, if any, have we had from the extensive series of single and multiple knockout mice? And thirdly, if you were now the person controlling George Morstyn's budget, and can only do a limited number of studies at any one time (even with the massive resources of Amgen), what next would you do? (For the sake of argument, let's pretend he doesn't just control the cytokines that he does, but that he controls them all. Given how well Amgen are doing, he soon might.)

Zinkernagel: The difference between the results we obtained with interleukin (IL) knockout mice and those obtained with neutralizing or blocking antibodies looking at the same question, suggest to me that more effort should be made by the stem cell researchers to make neutralizing antibodies and test them in parallel with the knockout models. The insights might actually be complementary, if not additive.

Nossal: I'll direct that one to Bill Paul, who has thought as much as any of us about comparative effects of anti-IL-4 antibodies and IL-4 knockouts, where the results are not the same.

Paul: In many respects the results are the same, with two interesting exceptions. First, anti-IL-4 antibodies profoundly inhibit the production of IgE in normal animals, and IL-4 knockout mice have very dramatic diminution of serum IgE. IgG1 levels are not affected by anti-IL-4 antibodies, but IgG1 levels of IL-4 knockouts are diminished about 10-fold. Mice in which *Stat6*, which controls many IL-4-mediated differentiation events, has been knocked out have normal IgG1 levels but don't have IgE. These differences may reflect the degree of inbreeding and the background of the mice. There's one very dramatic result, though, that points to a potential insight that comes from studying knockouts. IL-4 knockout mice have recently been prepared using BALB/c embryonal stem cells. BALB/c mice infected with *Leishmania* are uniformly susceptible. They develop a progressive disease and die, whereas most other strains develop a transient disease and recover. When BALB/c mice are treated with monoclonal anti-IL-4 antibodies, they become resistant to *Leishmania*. Nancy Noben-Trauth examined genetically pure BALB/c IL-4 knockouts. Curiously, these animals remained susceptible. In the course of the experiments, she explicitly repeated the anti-IL-4 experiment and showed that the anti-IL-4 caused control BALB/c mice to become resistant, but anti-IL-4 treatment of knockouts had no effect. This established that anti-IL-4 really worked specifically, because if its action had been non-specific, it ought to have had an effect on the IL-4 knockout mice. On the other hand, it implied that IL-4 knockouts were displaying a property that was different from the wild-type animal. Somehow they had 'adapated'. This is a cautionary tale; it does imply that mice may develop alternative regulatory systems when a dominant system is ablated.

Nossal: How does this relate to the dramatic results of the CTLA-4 knockouts?

Mathis: I think they were completely predicted from the antibody blocking experiments: addition of anti-CTLA-4 monoclonal antibody to *in vitro* model systems of T cell activation usually augments T cell proliferation.

Metcalf: One problem with knockouts is that the animal can learn how to compensate for a serious initial deficiency. The most dramatic example of this is the *op/op* mouse, which has no capacity to make M-CSF. It suffers terrible osteopetrotic problems, but at the age of three months these defects became corrected (Begg et al 1993). This shows how you can be misled if you examine animals at the incorrect age. As to whether antibody blocking experiments give the same answer as knockouts, the record with haemopoietic regulators is not bad. Early studies done using anti-erythropoietin (EPO) antibodies observed a depression of erythropoiesis similar to that seen in EPO or EPO receptor knockout mice (Garcia & Schooley 1962). Studies by Hammond et al (1991) on dogs that develop antibodies against injected human G-CSF gave essentially the same answer as the G-CSF knockout mice. However, one of the limitations with the antibody approach is that injected antibodies can't effectively penetrate all tissues, nor can they do much about membrane-displayed molecules.

We have looked carefully in G-CSF and GM-CSF knockout mice to determine whether compensation occurs by overproduction of another factor with overlapping actions, but we have found no evidence that this might occur.

Williams: Steel mice provide another example where antibody blocking experiments give different results from knockouts. The heterozygotes have the characteristic ventral body spot and the homozygotes are white. The melanocytes don't migrate correctly to the hair follicle via the dermis and epidermis. You can't study the function of melanocytes in the hair follicle if the cells have never migrated into place. You can only study these functions with antibody injections.

Rajewsky: Except if you do a conditional knockout.

Williams: That's right.

Paul: Another problem with monoclonal antibodies is they may have different effectiveness. Some antibodies are good inhibitors, some are terrible, and you often don't know this when you use them. I agree with Klaus Rajewsky's point that the future almost certainly lies with conditional knockouts. With these, one has the opportunity to make acute interventions.

Nossal: We have to remember that antibodies can be agonistic as well as antagonistic. One can be badly tricked in this way.

Burgess: It is early days in the physiology of these cytokines; there are many surprises ahead of us. For instance, at one stage we all thought that TGF-α was the molecule controlling the production of cells in the gastrointestinal tract. Then it was found that the TGF-$\alpha^{-/-}$ mice have normal cell production in the gastrointestinal tract. Our immediate response was to assume that there must be redundancy. However, perhaps TGF-α doesn't have a role in this process, even though it has a pharmacological effect. In the GM-CSF knockout mice, the major observation is the lung pathology. We haven't looked at a role for GM-CSF as a T cell cytokine adjuvant in the GM-CSF$^{-/-}$ mice. In haemopoiesis the *in vitro* clues have been magnificent. They have set us off in several useful directions, but we are going to find lots of surprises.

Rajewsky: To the list of surprises, that are perhaps not so surprising in retrospect, I would add the inflammatory bowel disease seen after IL-10 and IL-2 administration.

Dexter: I agree with Tony Burgess that the *in vitro* results have been predictive for at least most of the *in vivo* experiments. However, the biggest surprise to me concerned IL-3. *In vitro* it is one of the most powerful stimuli for most myeloid lineages. 10–15 years ago, you would have put your money on IL-3 as the haemopoietic growth factor which was going to have the most profound clinical effect. Its *in vivo* role still puzzles me, and we are no closer to this now than we were 10 years ago.

Rajewsky: At the risk of repeating myself, I think one would get a clearer answer if IL-3 was knocked out inducibly in the adult mouse, compared with the standard IL-3 knockout situation where the mouse has to grow up in the absence of IL-3 and perhaps adapts to this in a way in which we don't yet understand.

Dexter: But even in patients or in animal models where you give IL-3, its effect is marginal. Unlike G-CSF and erythropoietin, which have clear effects *in vivo* — excess red blood cells and neutrophils — IL-3 has minimal effects. I'm not sure it is even worth giving to patients.

Morstyn: IL-3 application can increase platelet counts by about 30% and can induce basophils. However, for an effective therapy I think we should be looking for drugs that are fairly dominant on one lineage. When factors have effects on multiple cells it becomes difficult to untangle the biology. It would be really nice if we were able to engineer the molecule rather than the receptor to actually give us the part of the biology we want.

Metcalf: Drug resistant tuberculosis is emerging as a significant clinical problem. I would like to raise another possible use of haemopoietic factors as pioneered by the work of Murray et al (1996). These studies improved the immunogenicity of bacillus Calmette–Guérin by engineering in a capacity for the organisms to produce one or other haemopoietic factor. This is potentially an ideal approach — to take a living organism that has been well enough tried clinically, and give it 'teeth', as a totally independent way of tackling the problem. Widespread immunization using bacillus Calmette–Guérin with 'teeth' might be the way to tackle the increasing problem of drug-resistant tuberculosis. Such an approach will require some unusual collaboration between microbiologists, immunologists and haematologists to pool the required expertise.

I have a suggestion for George Morstyn: the use of haemopoietic factor combinations clinically. It is clear that the body uses such combinations to regulate blood cell formation efficiently, and combination of two or more agents, such as G-CSF and GM-CSF, widens the range of responding cells. These concepts seem to be reaching the clinic very slowly. The sort of patients George is now working on, the high-risk pneumonia cases, to me would seem to be a quite logical group for such trials. It must surely be advantageous to combine the anti-microbial actions of granulocytes and macrophages.

Nossal: If one looks back in history, the pattern is that sheer economics have dictated that drugs must be registered one by one, and then clinical empiricism takes over. For example, cytotoxic drugs acting singly are not too great in leukaemia but gradually leukaemologists learnt how to use the drugs in combination. Isn't that likely to happen with cytokines?

Paul: Nevertheless, the first FDA approval for the Roche protease inhibitor was only for use in combination with another drug.

Nossal: But the other drug had already been assessed. The huge number of interesting cytokines makes the sort of trial Don Metcalf proposed quite difficult.

Metcalf: Except that two of these cytokines (G-CSF and GM-CSF) have already been licensed, and they are logical partners which exhibit synergy *in vivo*. It is simply a matter of two different companies talking to one another.

Morstyn: The problem with combinations of cytokines is the lack of predictiveness of the animal models. Looking at GM-CSF in infections, animal models suggest anything from benefit to harm. If there are animal models that suggest in a particular infectious disease a particular combination of cytokines will work, that would give us some direction. Before enrolling into large studies it is important to have some predictive models and, to an outsider, the existing models appear to give a total morass of conflicting data.

Burgess: With regard to vaccines, animal data now indicate that certain cytokines will act as adjuvants. Do you think a company or several companies would be ready to run adjuvant trials with GM-CSF or another cytokine?

Morstyn: Sandoz were moving in that direction. One of the problems here concerns which of the many cytokines are the best at augmenting particular immune responses.

Paul: One cytokine we really want to eliminate in inducing cellular immunity is IL-4, since it will suppress induction of Th1 responses very powerfully. The difficulty with the potential prophylactic use of anti-cytokine antibodies (or cytokines) stems from their possible side effects. It is one thing to treat a desperately ill patient with a very powerful drug; it's another to vaccinate millions of healthy people with agents that have potential for side effects. We need many good models before we can begin to move into large populations.

Nossal: This prompts me to question why G-CSF is so extraordinarily safe and relatively free of side effects, whereas many other cytokines — and certainly two that have been in extensive use, IL-2 and GM-CSF — have quite vexing side effects.

Morstyn: Interestingly, G-CSF is not produced in the resting marrow. We are using it as a humoral factor in the body. I have always thought of some of the other cytokines as more locally acting factors in normal physiology; for instance, it is unusual to see patients with high circulating levels of GM-CSF. In addition G-CSF also displays lineage dominance. We now have experience of administering G-CSF to 1.2 million patients with few reports of adverse effects. The same is true for EPO: it is a humoral factor made by the kidney and it works on the marrow.

Williams: One can measure SCF in the blood. By your logic, SCF shouldn't have profound side effects.

Morstyn: IL-1 and TNF can also be measured in the circulation, but there may not be an appropriate response. The significance isn't that we can measure G-CSF in the blood. I think that the body probably uses it as a humoral agent, made in one place (perhaps where the infection is) and which then goes back and works on the bone marrow. The body has adapted to it being given systemically.

Nicola: That is tautologous: you are effectively saying that G-CSF doesn't have side effects because the body tolerates it. The question concerns why this is so. One explanation might be that this is because neutrophils don't do as much damage as macrophages do. Do other neutrophil-activating agents show side effects? Is it as simple as neutrophils versus macrophages?

Davis: I think you are incorrectly stating what George Morstyn is saying. His hypothesis is that by definition the body will learn to tolerate molecules that can be present at high systemic levels. Factors that are generally released in short bursts and which are present at minute systemic levels would have a greater potential to be toxic.

Nicola: That doesn't say anything about the mechanism of toxicity.

Rajewsky: In shigellosis, it is the excess neutrophils recruited into the gut that do the most severe damage (Perdomo et al 1994).

Morstyn: Coming from an oncology background, where people are profoundly sick when they lack neutrophils, I have a bias that neutrophils are good, and that more neutrophils should be better! I don't know whether activating them with other agents would create a problem, but clearly the neutrophils produced in response to G-CSF have not had the harmful effects that people had predicted.

Zinkernagel: Has anyone analysed the function of the neutrophils under the G-CSF treatment?

Morstyn: There are many technical problems associated with neutrophil function studies. The studies available were done to pick up diseases such as granulomatosis—they weren't designed to show enhanced function. But to the extent that people have looked, neutrophils produced in response to G-CSF are actually enhanced in function. One of the areas we studied neutrophil function in extensively was in patients with HIV, where there was a feeling that although the patients had neutropenia this was not a very important risk factor for infection. It was hypothesized that although G-CSF produces neutrophils in the HIV setting, these do not function normally. There are now several studies in this setting showing that those neutrophils produced in response to G-CSF are in fact normally functioning.

Nossal: Don Metcalf, can I call on you to summarize today's proceedings?

Metcalf: We have had a sample of the three areas where considerable progress is being made. Most recently we have heard about the widespread clinical use of haemopoietic growth factors, which have now been administered to more than a million people. The second area that has been moving rapidly is receptor substructure. I once predicted that nobody would ever be able to deduce the complete signalling pathway from the receptor to the nucleus, and I am now at serious risk of being proved wrong. However, I still suspect that we will find it difficult to understand the complexities of the control of nuclear transcription. The third area is the knockout area: taking studies back to the whole animal. One of the things becoming evident from the knockout studies is that for macrophages, eosinophils and platelets, there are missing haemopoietic factors yet to be found.

References

Begg SK, Radley JM, Pollard JW, Chisholm OT, Stanley ER, Bertoncello I 1993 Delayed hematopoietic development in osteopetrotic op/op mice. J Exp Med 177:237–242

Garcia JF, Schooley JC 1962 An immunological study of human urinary erythropoietin. In: Jacobson LO, Doyle M (eds) Erythropoiesis. Grune & Stratton, New York, p 56–57

Hammond WP, Price TH, Souza LM, Dale DC 1991 Chronic neutropenia. A new canine model induced by human granulocyte colony-stimulating factor. J Clin Invest 87:704–710

Murray PJ, Aldovini A, Young RA 1996 Manipulation and potentiation of antimycobacterial immunity using recombinant bacille Calmette–Guérin strains that secrete cytokines. Proc Natl Acad Sci USA 93:934–939

Perdomo OJJ, Cavaillon JM, Huerre M, Ohayon H, Gounon P, Sansonetti PJ 1994 Acute inflammation causes epithelial invasion and mucosal destruction in experimental shigellosis. J Exp Med 180:1307–1319

T cell receptor biochemistry, repertoire selection and general features of TCR and Ig structure

M. M. Davis*†, D. S. Lyons†, J. D. Altman†, M. McHeyzer-Williams‡, J. Hampl*, J. J. Boniface† and Y. Chien†

**Howard Hughes Medical Institute, Beckman Center and †Department of Microbiology and Immunology, Stanford University School of Medicine, Stanford, CA 94305–5428, USA*

Abstract. T cell recognition is a central event in the development of most immune responses, whether appropriate or inappropriate (i.e. autoimmune). We are interested in reducing T cell recognition to its most elemental components and relating this to biological outcome. In a model system involving a cytochrome c-specific I-E^k restricted T cell receptor (TCR) derived from the 2B4 hybridoma, we have studied the interaction of soluble TCR and soluble peptide–MHC complexes using surface plasmon resonance. We find a striking continuum in which biological activity correlates best with the dissociation rate of the TCR from the peptide–MHC complex. In particular, we have found that weak agonists have significantly faster off-rates than strong agonists and that antagonists have even faster off-rates. This suggests that the stability of TCR binding to a given ligand is critically important with respect to whether the T cell is stimulated, inhibited or remains indifferent. It also suggests that the phenomenon of peptide antagonists might be explained purely by kinetic models and that conformation, either inter- or intramolecular, may not be a factor. We have also studied TCR repertoire selection during the establishment of a cytochrome c response, initially using an anti-TCR antibody strategy, but more recently using peptide–MHC tetramers as antigen-specific staining reagents. These tetramers work well with either class I or class II MHC-specific TCRs and have many possible applications. Lastly, we have also tried to correlate the structural and genetic features of TCRs with their function. Recent data on TCR structure as well as previous findings with antibodies suggest that both molecules are highly dependent on CDR3 length and sequence variation to form specific contacts with antigens. This suggests a general 'logic' behind TCR and Ig genetics as it relates to structure and function that helps to explain certain anomalous findings and makes a number of clear predictions.

1997 The molecular basis of cellular defence mechanisms. Wiley, Chichester (Ciba Foundation Symposium 204) p 94–104

‡Present address: Department of Immunology, Duke University Medical Center, Durham, NC 27707, USA.

Much of T cell biology hinges on the recognition of particular peptide–major histocompatibility complex (MHC) complexes by specific T cell receptor (TCR) heterodimers on $\alpha\beta$ T cells. While we have a good understanding of the genes which give rise to these receptors and their close relationship to immunoglobulins, the study of the binding characteristics of TCRs is relatively recent and one which has been fraught with technical obstacles because of the relatively low affinities that are involved. None the less, the development of expression systems which allow TCRs and MHCs to be made in a soluble form at high yields combined with the advent of surface plasmon resonance as a means of measuring weak protein–protein interactions has speeded work up considerably and there are now significant data regarding the binding of a number of TCRs to their peptide–MHC ligands (Boniface & Davis 1996). We have followed up on our initial studies in this area (Matsui et al 1991, 1994) with a recent survey of weak agonist and antagonist peptide–MHC complexes and find a good correlation between affinity/dissociation rates and the biological response of the T cell (Lyons et al 1996). Specifically, we find that in a series of peptides derived from cytochrome c, which all have identical MHC binding properties, weaker agonist responses correlate with faster dissociation/lower affinity and that antagonist peptides have still faster dissociation rates roughly equivalent to their potency. Thus the phenomenon of peptide antagonists of TCRs may have a purely kinetic basis. We have proposed a 'substrate depletion' model which may account for the blockade of T cell responsiveness that results.

We have also used the information obtained about the kinetics of TCR binding to design a specific T cell staining reagent, based on forming a 'tetramer' of peptide–MHC complexes. We have used reagents made from both class I and class II MHCs bound with different peptides to show specific staining of T cells or directly *ex vivo*. Reagents of this type should be of general use in investigating T cell memory formation and monitoring the T cell response to pathogens (such as HIV).

Lastly, the structural basis for TCR-mediated recognition of peptide–MHC, particularly the strong skewing of diversity towards the complementarity determining region (CDR) 3 regions of V_α and V_β, has long been an intriguing mystery. Recent results concerning the antibody-like nature of TCR $\gamma\delta$ recognition indicate that peptide–MHC recognition *per se* does not require a CDR3 diversity bias. Then what does? The recent literature on antibody–antigen interactions suggests that CDR3 diversification plays a key role in antigen complementarity and may be a 'nucleation site' for antigen binding, especially during the low-affinity, pre-somatic mutation phase of most antibody responses.

TCR affinity of weak agonists and antagonists

The recent discovery of 'altered peptide ligands' which produce either a qualitatively different response or are antagonists for an otherwise normal response to the original peptide has generated considerable interest and some controversy as to its mechanism. In particular, speculation has centred on whether it represents chiefly 'conformational' changes in the way a TCR binds to peptide–MHC, including the way other molecules (e.g. CD4, CD8, CD3) figure in the binding event, or whether it might be the result of some affinity or kinetic phenomenon (Sloan-Lancaster & Allen 1996). In the work I will describe here, we have used a TCR referred to as '2B4' which is derived from a T helper cell (Th) hybridoma which recognizes a C-terminal peptide of moth or pigeon cytochrome c bound to the class II MHC molecule I-E^k (Hedrick et al 1982). After engineering soluble forms of both the 2B4 TCR and the I-E^k heterodimer, replacing the transmembrane regions with signal sequences for glycan phosphatidylinositol (GPI) linkage, we have been able to select for cell lines which express high levels of TCR GPI and I-E^k GPI on their surfaces which can then be periodically cleaved with a specific enzyme to release these proteins in a soluble form. By fixing the 2B4 TCR to a BIAcore™ biosensor 'chip' using standard *N*-hydroxysuccinamide chemistry, we were able to show specific binding to soluble I-E^k molecules loaded (to 80–90%) with a moth cytochrome c peptide (MCC 88–103) but not to empty I-E^k molecules or those loaded with irrelevant peptides (Matsui et al 1994). These early studies using surface plasmon resonance showed that the potency of a series of three cytochrome peptides had a very interesting inverse correlation with the dissociation rate. Thus the most active peptide MCC (88–103) bound most stably to the 2B4 TCR when complexed to I-E^k with a $T_{1/2}$ of 12 sec while the least stimulatory (1000 ×) peptide MCC (T102S) had a $T_{1/2}$ of 2 sec. Seeking to extend these measurements from agonist peptides to antagonistic ones, we immediately ran into problems with the sensitivity of the BIAcore™ instrument as it becomes very difficult to measure affinities less than $\sim 100\,\mu$M. As all of the antagonist peptide–I-E^k complexes we tested gave no binding, we were clearly dealing with affinities less than this threshold. It therefore became necessary to develop a new competition assay which uses a recently developed method to put native I-E^k–MCC complexes onto the sensor chip and then bind these complexes with TCR in solution. This then gives a positive binding signal which can be subject to competition from antagonist peptide–MHC also in the solution phase. This has now given us accurate data on three MCC antagonists and together with improved sensitivity in a direct BIAcore™ binding assay shows that antagonist peptide–I-E^k complexes have affinities ranging from 500 μM to 3 mM (compared with 50 μM for MCC–I-E^k and 200 μM for T102S–I-E^k). In addition, the two most potent antagonists have higher affinities than the weakest one and of the decreases in affinity all seem due to faster dissociation rates ($T_{1/2} \sim 1$ sec). Thus at least in the 2B4/cytochrome c peptide system it seems that affinity and kinetic parameters are good predictors of biological activity.

Why would a faster dissociation rate correlate with first weaker agonist and then antagonist peptides? One model would be that peptide–MHC ligands that bind less stably to the TCR would disrupt the formation of TCR oligomers, which seems to be necessary for T cell activation.

Another possibility is that rapid dissociation could change the nature of the signal delivered by the T cell receptor. This is supported by the kinetic proof-reading model of McKeithan (1995) which suggests that the multiple reactions in the signal transduction machinery of the T cell could amplify even slight differences in affinity between agonist and antagonist ligands for the TCR. Of particular importance in models of this type would be the dissociation rate (Rabinowitz et al 1996) as rapid dissociation might allow only unproductive initial phosphorylation reactions, as observed by Sloan-Lancaster et al (1994) and Madrenas et al (1995), while more long-lived complexes would allow the entire signal transduction pathway to be completed, resulting in T cell activation. We would generalize these ideas further to what we term a 'substrate depletion' model which postulates that some substrate required for normal T cell activation is depleted by the engagement of TCR by antagonist peptide–MHC complexes, thus weakening the ability of agonist peptides to trigger a response. This critical substrate could be CD_3 ζ associated molecules or one or more of the other intermediates involved in TCR activation. The observation that antagonist and weak agonist peptides have faster off-rates than strong agonists (although there may be exceptions) would allow them to engage more TCR molecules via serial engagement and thus more rapidly deplete a limiting substrate, making it impossible for the T cell to be stimulated or allow it to be stimulated only weakly.

Tracking specific T cell responses

A long-standing problem in T cell biology has been the question of how to follow specific T cells in a response *in vivo*. It would be enormously useful, for example to follow a cohort of T cells through their initial activation, memory cell formation and dispersal without resorting to artificial aids such as transgenic mice. It would also be very interesting to gauge the effects of the different steps in thymic selection on a diverse group of immature T cells selected solely for their ability to recognize a particular peptide–MHC.

Our initial efforts have focused on the $CD4^+$ T cell response to pigeon cytochrome c in H-2^k mice, where it has long been known that most of the reactive T cells express a $V_{\alpha 11}$ $V_{\beta 3}$ TCR heterodimer. Making use of available monoclonal antibodies to these V regions as well as activation markers such as CD44 and L-selectin, we were able to visualize a small population of $CD4^+$ T cells that expressed $V_{\alpha 11}$ and $V_{\beta 3}$ determinants, had an activated phenotype and were rapidly expanding in the draining lymph nodes following antigenic challenge. Sequence analysis of either populations or individual T cells suggested that at the peak of the response in the $V_{\alpha 11}{}^+ V_{\beta 3}{}^-$ activated phenotype cells were essentially pure with respect to cytochrome c/I-E^k reactivity as they were highly selected for CDR3 length and sequence as well as J region usage (McHeyzer-Williams & Davis 1995).

We also noted a shift in the kinds of CDR3 sequences encountered later in the response versus earlier, suggesting a 'maturation' of the T cell repertoire with continued selection. This may be akin to affinity maturation in B cells, although we saw no evidence for somatic mutation nor would one expect high affinity TCRs to be necessary (as membrane-bound receptors that have ligands on other cell surfaces are generally of lower affinity).

There are inherent limitations in use of anti-V_β or V_α antibodies to follow specific T cells, however, as in most cases the responses are not as restricted as those of pigeon cytochrome c and in addition there are relatively few anti-V_α antibodies available. Even in the highly restricted cytochrome c system described, we could also isolate pure antigen-specific cells at peak times in the response, whereas at other time points this is not possible. The ideal reagent with which to identify $\alpha\beta$ T cells of a given specificity would be the peptide–MHC that they are specific for. Unfortunately, all TCRs characterized to date have such fast dissociation rates ($T_{1/2} < 30$ sec) that a reagent based on monomeric peptide–MHC would be useless. If one could conveniently multimerize a peptide–MHC complex such that an average two or more TCR contacts could be made simultaneously, then binding might be stable enough to make a good staining reagent. We have engineered just such a reagent by fusing a signal peptide for biotinylation on the C-termini of both class I and class II MHC molecules. These are produced in *Escherichia coli*, refolded with various peptides and then biotinylated *in vitro*. When complexed with streptavidin (which has four biotin binding sites) we are able to create a peptide–MHC 'tetramer'. Using commercially available streptavidin phycoerythrin or Texas red conjugates we have now made a variety of reagents that are able to stain T cells according to their antigen specificity, either as lines or directly *ex vivo*. Recently, in collaboration with A. McMichael and colleagues at Oxford University, we have shown that HIV peptide/HLA-A2 specific populations of $CD8^+$ T cells can be detected in the peripheral blood of asymptomatic HIV carriers. The appearance of these gag or pol specific T cells (which can be as much as 1% of all $CD8^+$ cells in the blood) correlates well with cytotoxicity, although most of the antigen-specific cells have a non-activated (DR^-) 'memory' cell ($CD45^+$) phenotype (J. D. Altman, P. A. H. Moss, P. J. R. Goulder, D. H. Barouch, M. McHeyzer-Williams, J. I. Bell, A. J. McMichael & M. M. Davis, unpublished results).

CDR3 diversification: a general strategy for TCR and Ig complementarity to antigens?

One striking feature of the gene rearrangements which create both T cell receptors and immunoglobulins is how strongly skewed the sequence diversity is towards the CDR3 loop in one or both of the chains making up a heterodimer. In the case of $\alpha\beta$ TCRs this concentration of diversity in both V_α and V_β CDR3 loops has been proposed to correlate with the recognition of peptides bound to MHC molecules (Davis & Bjorkman 1988, Chothia et al 1988) and experimental data have supported this view (Jorgensen et al 1992, Katayama et al 1995). For immunoglobulins, however, diversity is also strongest in the CDR3 region of V_H, and while not as pronounced as in most of the TCR genes, it is still considerable. Recent work on human Ig genes enables us to make reasonably accurate calculations regarding diversity in the CDRs; evidence suggests that there are only about 50 functional V_H and a similar number of V_L gene segments (Cox et al 1994, Cook et al 1994) thus limiting the naïve repertoire to 50 CDR1s and CDR2s of each type. In contrast to this, the CDR3 of heavy chain is composed of the C-terminal part of V_H, parts or all of a D region and the N-terminal part of a J_H. In between V and D, and D and J, can be N- or P-element insertions of up to 10 nucleotides. Thus a conservative estimate of the number of possible amino acid sequences that could be encoded would be 10^8. Lacking N-region diversification and a D region, the diversity of the light chain CDR3 is relatively modest — about 10^2. It has never been explained why there should be such a dichotomy between the V_H CDR3 and all the other CDR loops involved in antigen-binding. Even more of a problem is TCR δ, which has the most CDR3 diversity of any antigen receptor by far ($\sim 10^{13}$; Elliott et al 1988) and yet now appears to recognize ligands in an antibody-like manner (Schild et al 1994). Clearly there must be some chemical or structural 'logic' behind this phenomenon. To fill this yawning gap in our knowledge, we have proposed a model in which the principal antigen specificity of an Ig or TCR is derived from its most diverse CDR3 loops (M. M. Davis, Y. Chien & B. Arden, unpublished results). In the case of antibodies, we imagine that most of the specific contacts (and hence free energy) with antigen are made by the V_H CDR3 and that the other CDRs provide 'opportunistic' contacts which contribute to the energy of binding and specificity, but in a relatively minor way. Once antigen has been encountered and clonal selection anoints particular B cells, somatic mutation would then 'improve' the binding of the CDR1s and 2s to convert the typically low affinity antibodies to the higher affinity models as observed by Berek & Milstein (1988). This model predicts that a single V_H might be able to accommodate most antigens, provided that full CDR3 diversity was possible, as apparently observed by Taylor et al (1994).

While direct tests of this hypothesis are just beginning, it would seem to hold considerable promise as a general mechanism for antigen receptor specificity and as an answer to an intriguing molecular genetic puzzle.

Acknowledgements

We thank the Ciba Foundation for sponsoring this excellent Symposium and Drs Nossal, Metcalf and Miller for their inspirational contributions to science. This work was supported by the Howard Hughes Medical Institute and grants from the US National Institutes of Health.

References

Berek C, Milstein C 1988 The dynamic nature of the antibody repertoire. Immunol Rev 105:5–25
Boniface JJ, Davis MM 1996 The affinity and kinetics of T-cell receptor binding to peptide/MHC complexes and the analysis of transient biomolecular interactions. In: Herzenberg LA, Herzenberg L, Weir DM, Blackwell C (eds) Weir's handbook of experimental immunology, 5th edn. Blackwell Science, Oxford
Chothia C, Boswell DR, Lesk AM 1988 An outline structure of the T cell receptor. EMBO J 7:3745–3755
Cook GP, Tomlinson IM, Walter G et al 1994 A map of the human immunoglobulin V_H locus completed by analysis of the telomeric region of chromosome 14q. Nat Genet 7:162–168
Cox JPL, Tomlinson IM, Winter G 1994 A directory of human germline V_{κ} segments reveals a strong bias in their usage. Eur J Immunol 24:827–836

Davis MM, Bjorkman PJ 1988 T cell antigen receptor genes and T cell recognition. Nature 334:395–402

Elliott JF, Rock EP, Patten PA, Davis MM, Chien Y 1988 The adult T-cell receptor δ-chain is diverse and distinct from that of fetal thymocytes. Nature 331:627–631

Hedrick SM, Matis LA, Hecht TT et al 1982 The fine specificity of antigen and Ia determinant recognition by T cell hybridoma clones specific for pigeon cytochrome c. Cell 30:141–152

Jorgensen J, Esser B, Fazekas de St Groth B, Reay PA, Davis MM 1992 Mapping T cell receptor/peptide contacts by variant peptide immunization of single-chain TCR transgenics. Nature 355:224–230

Katayama CD, Eidelman FJ, Duncan A, Hooshmand F, Hedrick SM 1995 Predicted complementarity determining regions of the T cell antigen receptor determine antigen specificity. EMBO J 14:927–938

Lyons DS, Lieberman SA, Hampl J et al 1996 A TCR binds to antagonist ligands with lower affinities and faster dissociation rates than to agonists. Immunity 5:53–61

Madrenas J, Wange RL, Wang JL, Isakov N, Samelson LE, Germain RN 1995 Zeta phosphorylation without ZAP-70 activation induced by TCR antagonists or partial agonists. Science 267:515–518

Matsui K, Boniface JJ, Reay PA, Schild H, Fazekas de St Groth B, Davis MM 1991 Low affinity interaction of peptide–MHC complexes with T cell receptors. Science 254:1788–1791

Matsui K, Boniface JJ, Steffner P, Reay PA, Davis MM 1994 Kinetics of T cell receptor binding to peptide–MHC complexes: correlation of the dissociation rate with T cell responsiveness. Proc Natl Acad Sci USA 91:12862–12866

McHeyzer-Williams M, Davis MM 1995 Antigen-specific development of primary and memory T-cells *in vivo*. Science 268:106–111

McKeithan K 1995 Kinetic proofreading in T-cell receptor signal transduction. Proc Natl Acad Sci USA 92:5042–5046

Schild HN, Mavaddat C, Litzenberger C et al 1994 The nature of MHC recognition by $\gamma\delta$ T cells. Cell 76:29–37

Sloan-Lancaster J, Allen PM 1996 Altered peptide ligand induced partial T cell activation: molecular mechanisms and role in T cell biology. Ann Rev Immunol 14:1–27

Sloan-Lancaster J, Shaw AS, Rothbard JB, Allen PM 1994 Partial T cell signaling: altered phospho-ζ and lack of zap-70 recruitment in APL-induced T cell anergy. Cell 79:913–922

Taylor LD, Carmack CE, Huszar D et al 1994 Human immunoglobulin transgenes undergo rearrangement, somatic mutation and class switching in mice that lack endogenous IgM. Int Immunol 6:579–591

DISCUSSION

Nossal: I have a theoretical question for you. You have shown that a relatively small affinity difference (approximately 10–20 fold off-rate difference) can make a crucial difference to the cells' response. Along similar lines, three groups (including Frank Carbone's) have shown that only an approximately 10-fold difference in binding affinity makes the difference between positive and negative selection when T cells are responding to MHC and peptide in thymus organ cultures. Because those affinities are so low, the cell seems to need different co-receptor molecules to receive signals. How come this hotch-potch of further recognition phenomena doesn't obscure what is, after all, a modest difference in affinity? If I were designing the system, I would have thought that this is a highly dangerous thing to do. What if four times more of some particular adhesion molecule, up-regulated for some purpose or another, markedly increases binding and disturbs this fine-tuning? Could this turn positive selection into negative selection?

Davis: There are two phenomena occurring here: cell adhesion and receptor engagement. There are a hundred times more adhesion molecules with potential ligands on the cell then there will be peptide MHCs of the right sort that will be able to engage the T cell receptor. Already, this tells us that by far the dominant entity in terms of cell–cell contact energetically, given similar affinities, is going to be the adhesion molecule. We therefore have a situation where the cells are brought together by adhesion molecules—at least briefly—and then receptor engagement is possible. These are clearly independent events. There is evidence of signalling through adhesion molecules, but clearly the main event in T cell activation is receptor engagement. The CD4 and CD8 co-receptors add a bit to that, but they're not the main event. The binding between CD4 and class II MHC is so weak that no one has ever been able to show it. CD8 is only slightly better.

The second issue concerns how a 10-fold difference in affinity could make the difference between agonist and antagonist. McKeithan (1995) showed that in a multistep process where the propagation of the signal is dependent somehow on receptor engagement, as soon as you go off, that chain breaks. For example, if you have a multistep process then you will essentially amplify small differences in affinity or off-rate along the chain.

Zinkernagel: It is wonderful that it is now possible to measure these processes, but I would argue that even for antibodies, some of the key parameters are not really known. For example, we do not know what affinity, what off-rate or on-rate of neutralizing antibodies is biologically important in terms of protective immunity *in vivo*. This has implications for some of the theoretical considerations in your paper, because I don't agree that specificity is a low affinity phenomenon to start out with. We do not know what is important either for haptens or against viruses and what thresholds of parameters we should take to even start attempting to make interpretations and extrapolations.

Davis: It is true that all the work has so far been done in haptens; we need to do a lot more in terms of protein antigens. But from first principles, membrane molecules seeing things on other membrane molecules or aggregates, which some people believe may be the way some of these antigens are presenting, are invariably low affinity events. I would be shocked if someone were to find events with nanomolar affinities occurring on the surface of a B cell in a primary response.

Zinkernagel: These conditions apply to neutralizing antiviral antibodies: they exhibit a similar range of affinities/avidities and this is probably because of rather than despite the fact that neutralizing determinants form crystal-like, highly organized antigen patterns on viral envelopes.

Miller: Lanzavecchia and colleagues have suggested that for T cell activation to occur, a ligand has to serially engage a certain number of T cell receptors (Valitutti et al 1995). How do you fit that into your scheme?

Davis: That is part of what I was saying about antagonists. The reason the antagonists might be so effective is that they can serially engage many more T cell receptors in the same time than an agonist peptide can.

Miller: You didn't mention anything about negative signals from antagonists. What gives a negative signal, and what do we mean by the term 'negative signal'?

Davis: Part of the problem is that up to now, no one has had the slightest idea what they mean by 'negative signal'! I'm trying to make it more specific by saying that a negative signal could occur by interfering with the machinery for the positive signal. That is, by partially phosphorylating it so that it can neither propagate the full signal nor can it be used again if a perfectly good signal comes along.

Nossal: That's a very attractive way of explaining anergy, cell death and all sorts of things.

Paul: Microscopically, instead of looking at the whole process, you look at an individual receptor. The letting-go property of that receptor is probably a stochastic event, and since you can pile on a lot of molecules and occupy many receptors, you would anticipate that a four- or fivefold change in off-rate could be overcome by simply having a larger number of events. You would also have to assume that the antagonist is 'antagonizing' itself because of a very substantial fraction of interactions which might fulfil the stability requirement, even with a rapid mean off-rate. A corollary of this is that as the state of the cell changes, an antagonist could easily become an agonist. In a sense it is almost a prediction of the model that the cell can tune itself so that depending on its situation, an agonist could easily become an antagonist or vice-versa.

I want to go back to your antibody issue and Mel Cohn's view about the primary and the secondary repertoire. His view was that the existing heavy chains and light chains are in fact evolutionarily selected for the ability to construct antibodies against antigens of high degrees of danger. The correlate of that would be that the binding of pneumococcal polysaccharides, for example, would not be by CDR3, as you are suggesting is the case for the secondary repertoire, but largely by CDR1 or CDR2. What is your view on this?

Davis: I have a dim view. Evolution has been around for a long time and clearly it exerts an influence at a rather profound level: some of the populations are still alive and some aren't. I could see a skewing of things that way. The dangerous attraction of this hypothesis is that it is a way you could design a plunger: you have this plunger and you have a sort of active centre, where you create an enormous amount of diversity and you display this diversity. You only need that little bit, and the rest of it is a certain stickiness; I would see most of the V region repertoire as being a backup or as being specialized for some of these very rare or chemically unusual antigens.

Paul: What about polysaccharides? You could argue that a lot of 'dangerous' antigens are associated with polysaccharides.

Davis: Yes, they may require some special accommodation in antibody structure: it would be interesting to see. Our present goal is to construct mice that have only a single V region of each type, to see what the responses are. Perhaps they will die at the slightest provocation, but I suspect they will be able to mount a lot of responses.

Kirberg: Regarding your model of antagonism, how do you integrate 'cold target inhibition' (Jameson et al 1993) into the model? If the T cell is confronted by two targets, one with the agonist and the other with the antagonist peptide, how could the T cell be fooled away to the target with the antagonist instead of killing the target with the agonist peptide?

Davis: The substrate that we're imagining here would have to be diffusible. I think you would also have to have small units of conditional aggravation. I believe in aggregates, but not fast aggregates. It would have to be something diffusible so that something that happens on the one side of the cell in the case you are mentioning could have this global effect.

Kirberg: Couldn't one build an antagonist that works by gluing the T cell receptor to its ligand? If Valitutti et al (1995) are right that the T cell receptor is engaged several times, couldn't it be that there are antagonists which actually inhibit several engagements due to increased affinity of the T cell receptor for the antagonist–MHC ligand?

Melchers: It is not the T cell receptor that is engaged several times, it is the MHC.

Kirberg: You are right; that is what they directly showed. But if we look at Mark Davis' results together with those of Weber et al (1992) and Valitutti et al (1995), they are quite suggestive of several MHC engagements by the same T cell receptor.

Davis: The point is valid. If you could lengthen the time of engagement between a given peptide and a TCR, this would restrict the number of engagements and make it harder to activate the cell. In a recent meeting Lanzavecchia showed an experiment that addresses that point. He made a hybrid antibody that bound to an allodeterminant on a presenting cell with the other Fab directed against the TCR. He showed that this was a poorer stimulant of the TCR than the initial peptide–MHC. However, this experiment is not satisfactory in that the binding characteristics of the hybrid antibody were not really defined.

Mosmann: Perhaps we could go back to the substrate depletion model you showed. It seemed a little incomplete the way you had it drawn. If a substrate is simply converted into the first intermediate on the pathway, that's not going to help, because the next time there is any engagement it can push the pathway a little further down. Are you suggesting that the first product has an alternate decay pathway that takes it off into something that can't be used? Or is there a slow pathway that has to go along with the fast pathway?

Davis: The other possibility is partial phosphorylation, for example, as has been seen on the ζ chain by Sloan-Lancaster et al (1994) and Madrenas et al (1995). You could take that and say that instead of causing a quick decay, which is one possibility, you alter it such that it is no longer a good substrate for the upstream part of the pathway.

Mosmann: Then you are suggesting a state that would not be within the normal pathway, so how would you induce this different kind of phosphorylation?

Davis: We're dealing with physiology here. Anything can be done by cells *in vitro*!

Goodnow: You need to weave phosphatases into this, because all these phosphorylation events essentially create their own demise by recruiting phosphatases, such as the haemopoietic cell phosphatase that terminates signals from the erythropoietin receptor or the antigen receptors on B cells.

Reference

Jameson SC, Carbone FR, Bevan MJ 1993 Clone specific T cell receptor antagonists of major histocompatibility complex class I-restricted cytotoxic T cells. J Exp Med 177:1541–1550

Madrenas J, Wange RL, Wang JL, Isakov N, Samelson LE, Germain RN 1995 Zeta phosphorylation without ZAP-70 activation induced by TCR antagonists or partial agonists. Science 267:515–518

McKeithan K 1995 Kinetic proofreading in T-cell receptor signal transduction. Proc Natl Acad Sci USA 92:5042–5046

Sloan-Lancaster J, Shaw AS, Rothbard JB, Allen PM 1994 Partial T cell signaling: altered phospho-ζ and lack of zap-70 recruitment in APL-induced T cell anergy. Cell 79:913–922

Valitutti S, Müller S, Cella M, Padovan E, Lanzavecchia A 1995 Serial triggering of many T cell receptors by a few peptide–MHC complexes. Nature 375:148–151

Weber S, Trunecker A, Oliveri F, Gerhard W, Karjalainen K 1992 Specific low-affinity recognition of major histocompatibility complex plus peptide by soluble T-cell receptor. Nature 356:793–796

Immunology and immunity studied with viruses

Rolf M. Zinkernagel

Institute of Experimental Immunology, University Hospital of Zurich, Schmelzbergstr. 12, CH-8091 Zurich, Switzerland

Abstract. Immunity to viruses is used to define important biological parameters of immunology. Specificity, tolerance and T and B cell memory were analysed with murine model infections. The key parameters of antigen kinetics, localization and patterns of T and B cell response induction in maintaining memory and in causing deletion of reactive lymphocytes were compared for self and for viral foreign antigens. Evidence is reviewed that suggests that B cells essentially recognize antigen patterns, whereas T cells react against antigens newly brought into lymphoid tissues; antigens outside lymphoid tissues are ignored, and antigens always present in, or spreading too fast throughout, lymphoid tissues exhaust and delete T cell responses. Finally, effector mechanisms of antiviral immunity are summarized, as they vary with different viruses. On this basis immunological T and B cell memory against viruses is reviewed. Memory studies suggest that increased precursor frequencies of B and T cells appear to remain in the host independent of antigen persistence. However, in order to protect against cytopathic viruses, memory B cells have to produce antibody to maintain protective elevated levels of antibody: B cell differentiation into plasma cells is driven by persisting antigen. Similarly, to protect against infection with a non-cytopathic virus, cytotoxic T cells have to recirculate through peripheral organs. Activation and capacity to emigrate into solid tissues as well as cytolytic effector function are also dependent upon, and driven by, persisting antigen. Because no convincing evidence is yet available of the existence of identifiable B or T cells with specialized memory characteristics, the phenotype of protective immunological memory correlates best with antigen-driven activation of low frequency effector T cells and plasma cells.

1997 The molecular basis of cellular defence mechanisms. Wiley, Chichester (Ciba Foundation Symposium 204) p 105–129

Immunity is characterized by specificity, tolerance and memory. However, this definition does not usually include: (1) that induction of an immune response to soluble protein antigens is rather difficult, whereas infectious agents trigger immune responses readily; (2) that antigen alone drives an immune response; and (3) that immunological reactivity is defined biologically, e.g. by the discrimination between protective antigens of infectious agents or of their toxins (and not by ELISA assays or T cell proliferation *in vitro*). The discrepancy between *in vitro* and *in vivo* parameters of immunity is the source of much controversy. In this paper, immunology learned from viruses is used to illustrate some key differences relevant to our understanding of disease pathogenesis, of specificity and of memory.

If one compares an immune response to bovine serum albumin or chemically defined haptens coupled to a carrier protein, with an immune response to a virus, the distinction between immunology as a theoretical science and immunity mediating survival advantages is obvious. The response to bovine serum albumin is irrelevant for survival of the host. Consequently, definitions of specificity, memory and tolerance based on model antigens (Paul 1993, Nossal & Ada 1971) may have to be reviewed. Having said this, without the detailed studies of responses to model antigens, immunology would not be where it is today as a measurable science (Mitchison 1964, Rock et al 1990, Chothia et al 1989, Goodnow et al 1990, Rajewsky et al 1987, Eisen & Siskind 1964, Berek & Milstein 1987). Because both huge numbers and short generation times favour viruses (or most parasites), the evolutionary pressure to maintain a balance permitting both host and parasite to survive is probably set by the more flexible viruses adapting to the vertebrate immune system rather than the reverse (Ohno et al 1980). Accordingly, the immune system may outline how far viruses or other infectious agents have been permitted to go in evolution to prevent elimination of the host or virus. Both areas where either immunity, or the lack of it, dominate and also the overlapping fringe areas of this coevolution between the immune system and parasites are of great interest because parameters defining them will reveal the limitations of both sides and thereby will define both sides in a biologically meaningful fashion (Lawrence 1959, Mims 1987, Marrack & Kappler 1994, Janeway 1989, Zinkernagel et al 1985, Zinkernagel 1996).

The role of antigen and the specificity problem

Specificity for antigens is defined by the binding characteristics of antibodies for 3D proteins or carbohydrates and of T cell receptors for a peptide presented by an MHC class I or II antigen (Eisen & Siskind 1964, Dulbecco et al 1956, Smith et al 1993, Fazekas de St Groth 1967, 1981, Langman 1989, Cohn & Langman 1990, Sykulev et al 1994). Although such parameters are measurable *in vitro*, we do not understand which parameters are crucial *in vivo* against infectious agents or toxins. As a consequence, we do not understand what spectrum of differences must be recognizable by the available repertoires of antibodies or T cells for a host to survive when exposed to the universe of infectious agents. One may therefore best define specificity as immunologically discernible structures/peptides that induce a protective immune response relevant for host survival. As a result of the coevolution of infectious agents and their host's defence, non-immunological control mechanisms of infections cannot be separated from immune mechanisms; if these non-specific defence mechanisms are absent, the immunological resistance would usually be too slow (Isaacs & Lindenmann 1957, Wheelock 1965, Gresser et al 1976, Huang et al 1993, Müller et al 1994). For example, elimination of interferon $\alpha\beta$ action by specific antiserum or inactivation of the receptors renders immunocompetent mice susceptible to many viruses which otherwise would be dealt with; in fact, non-specific, non-immunological defence mechanisms are probably quantitatively much more important than precise immunological ones (Blanden 1974). The definition of specificity is limited by the method of measurement. The key question is whether the method used employs a detection threshold similar to that operating *in vivo*. By definition, antiviral neutralizing antibodies against one viral serotype are not protective against another. None the less, if these antibodies are tested in sensitive binding assays, they may yield a positive signal on an irrelevant antigen (Fields 1990, McIntosh 1990, Charan et al 1987). Neutralization of viruses or bacterial toxins *in vitro* seems to depend upon antibody affinities/avidities of $>10^8$ l/mol; this is considerably greater than antibody binding qualities needed for binding in an ELISA. Whether this relatively high binding affinity necessary for virus neutralization is representative of other biological activities has not been tested. For example, for opsonizing antibodies such as phosphocholine-specific antibodies, considerably lower affinities of the order of 10^5 l/mol seem highly effective (Claflin & Berry 1988, Chen et al 1992). A valid definition of specificity for T cells is even more difficult because binding affinities cannot yet be easily measured (Matsui et al 1991, Weber et al 1992, Sykulev et al 1994). The few available studies suggest that functional readouts *in vivo* are considerably more stringent than suggested by several *in vitro* methods (Speiser et al 1992, Castelmur et al 1993, Evavold & Allen 1991). Together, these findings signal that many of the cross-reactive specificities measured *in vitro* are of little relevance *in vivo*. Therefore, an appropriate definition of specificity is offered by direct tests *in vivo* for the presence or absence of cross-protection by either antibodies, the helper T cell (Th)-mediated switch from IgM to IgG, or by protective cytotoxic T lymphocyte (CTL) activity. Because this is usually not possible in humans, it is therefore of paramount importance to explore which of the assay methods *in vitro* adequately reflect the relevant specificity in hosts.

The question of which parameters define the protective efficiency and specificity of an immune response is largely unanswered. For antibodies, is it affinity (binding of one high antibody binding site), avidity (the combined binding strength of a number of molecularly interlinked binding sites), or the concentration of the antibody in serum or in tissue after extravasation? What are the corresponding rules for T cells?

The immune system must be able to cope with the most extreme situations, e.g. infection with an acute highly cytolytic virus, all milder ones then being covered. For example, a virus that is defeated in a normal host exclusively by antibodies must be able to induce this protective antibody response early on in sufficient numbers of hosts to guarantee survival of both itself and the hosts. Antibodies of sufficiently high affinity/avidity and concentration should be generated within the required minimal time period of, for example, 6–7 days, to guarantee survival of the host. If these are not available within this time, subsequent affinity maturation of the antibody is not of much use to the dead host. This condition is thought to be fulfilled if germline genes code for the relevant specificity, or if the specificity is generated by one or very few mutations within a few days (Rajewsky et al 1987, Linton et al 1989, Roost et al 1995). Whether the germline genes have evolved to comprise such specificities, or the infectious agent has adapted by coevolution to the available basic immunological repertoire, is open to debate.

Why and when, then, does affinity maturation by somatic mutation occur? Affinity maturation by somatic mutation has been defined as increase of affinities/avidities of polyclonal or monoclonal IgG antibodies with time; this process correlates inversely with antigen dose and is usually noticed during or after the second week of an antibody response (Eisen & Siskind 1964, Berek & Milstein 1987, Linton et al 1989, Rajewsky et al 1987, Langman 1989, Cohn & Langman 1990). The following explanations have been offered:

(1) Economy in establishing antibody memory. One hundred times fewer memory antibodies are needed to protect, if they exhibit affinities/avidities that are increased by a factor of 100 (Rajewsky 1989).

(2) For some infectious agents, slow maturation of antibody affinity is needed to permit survival and transmission of these poorly or non-cytopathic viruses. For cytopathic viruses, the germline-encoded repertoire must suffice or somatic mutations improving antibody specificity must occur very efficiently, and very early, i.e. between days 1–6 of a primary B cell response (Roost et al 1995).

(3) Somatic mutation of antibodies is necessary for the response to catch up with virus mutations in the host (Rajewsky 1989).

Alternatively, somatic mutation of antibodies may be a general mechanism for increasing the numbers of specificities in a non-directed, random fashion. Accordingly, the limited, germline-coded repertoire of antibodies is increased to comprise a more diverse immunological repertoire (Linton et al 1989, Cohn & Langman 1990). During affinity maturation, non-cytopathic agents (which will not kill the host and therefore are subject to quite different selection pressures than cytopathic agents) may gain sufficient time for their replication and transmission before being eliminated by an immune response (Lehmann-Grube 1971, Hotchin 1971, Battegay et al 1993a, Rosenberg & Fauci 1989, Peters et al 1991, Mondelli & Eddleston 1984). In fact, without an initially poor antibody response, many viruses may be eliminated too quickly for their successful transmission and survival. If such viruses resemble self-antigen to some extent, the neutralizing antibody affinity is initially poor and by subsequent mutation evolves to effective levels. This evolution of neutralizing antibody responses could be postulated for lymphocytic choriomeningitis virus (LCMV), human immunodeficiency virus (HIV) and hepatitis B virus (HBV) infections (Rosenberg & Fauci 1989, Peters et al 1991, Milich et al 1990). By necessity, these agents must be either non- or very poorly cytopathic. The role of somatic mutation in catching up with viral mutants during an ongoing infection in one and the same patient is an interesting proposal, but is without direct evidence. Since rabbits and birds do not exhibit somatic mutation during antibody responses (Bucchini et al 1987, Weill & Reynaud 1987), they should be a source of such viral mutants more frequently than rodents or primates. Whether influenza reassortants and variants often generated in birds support this notion remains to be seen (Webster et al 1982, Klenk & Rott 1988, Palese & Young 1982).

Definition of immunological specificity by viral mutants escaping immune responses

Viral mutants that evolve under the pressure of an ongoing or established immune response help to define the time-dependent specificity of the immunological effector mechanisms. Influenza viruses are the best-studied example and illustrate how a virus population and a wide ranging host population keep each other alive, modulating within defined limits viral antigens and protective responses (Mims & Wainwright 1968, Webster et al 1982, Fazekas de St Groth 1967, 1981, Pircher et al 1990, Phillips et al 1991). The 50–70 year cycle of pandemic influenza strains reveals these dynamics and the limitations of viral variation and of the protective immune response. An interesting aspect of these kinetics, as influenced by the necessary T help, has recently been revealed.

Since T help for anti-influenza antibodies is cross-reactive between many variants and even amongst influenza subtypes, the question arises as to why pre-existing primed T help fails to enhance antibody responses against new variants so as to prevent them from causing disease. Several studies have indicated that it is likely that B cell precursors are limiting and therefore only excessive T help can accelerate the response; such an excess of cross-reactive T help is only available during about 2–3 weeks of the infection (McMichael et al 1983, Roost et al 1990, Liang et al 1994). Thus although primed Th cells are present, they have no direct influence on subsequent anti-influenza (Liang et al 1994) or anti-vesicular stomatitis virus (VSV) (Zinkernagel 1990) responses.

Viruses may also escape CTL responses by mutating the relevant T cell epitope(s) (Pircher et al 1990, Phillips et al 1991, Waters et al 1992). In contrast to the population dynamics relevant for influenza viruses, this process has been observed in individual hosts over time; whether such MHC-restriction-dependent escape mutations can be fixed and can accumulate in viruses over time is unproven, but theoretically possible; this process may help non-cytopathic viruses to adapt increasingly successfully to the host population by avoiding not only direct cytopathogenicity but also immunopathology. The finding that some Pacific populations exhibit HLA-A11 (a specificity rare amongst caucasians) and that Epstein–Barr virus (EBV) isolates from the same area have a mutation in the epitope that is usually presented and recognized by HLA-A11-restricted CTLs, may be such an example (de Campos-Lima et al 1993). Recent evidence suggests that non-cytopathic viruses may evolve not only to escape CTLs, but also to escape and inactivate them by mutating the original T cell epitope to an antagonistic peptide (Klenerman et al 1994, Bertoletti et al 1994, Allen & Zinkernagel 1994). It remains to be established whether selection of such Promethean capabilities influences virus–host equilibria to improve symbiosis of nearly ideal non-cytopathic viruses with their hosts and whether some of the most successful viruses, such as leukaemia and other tumour-associated viruses, use such tricks (Lilly et al 1964, Lilly 1972, Levy & Leclerc 1977).

Because indifference to antigens in privileged sites or exhaustive induction of effector T cells impinge on kinetics of specific responses (see below), specificity cannot be understood in physicochemical terms alone; kinetics dependent upon time, antigen concentrations, and quantity, type and quality of the immune response significantly influence the phenotype of specificity.

Antigen drives the immune responses

Induction of an immune response is usually difficult; similarly, autoimmune diseases are relatively rare. Thus, although overall the immune system is rather inert, it protects vertebrates efficiently against most infectious agents relevant to the species concerned, probably because they induce immune responses most efficiently. This can be illustrated as follows. Experimental animals have been successfully forced to make measurable antibody responses to bovine serum albumin, lysozyme or phenolic haptens, but these responses have always been difficult to induce (Freund et al 1947). Only when Freund converted these antigens into pseudoinfectious antigens by mixing them with oil and tubercle bacilli did such attempts succeed. Before employment of these powerful adjuvants immunology dealt with bacterial toxins, bacterial surface antigens or viral antigens which, by coevolutionary necessity, all readily induced efficient immune responses.

The simple concept that foreign antigen drives the immune response is based on the following findings. If antigen is optimally distributed along a gradient and if antigen is presented on antigen-presenting cells (APCs) and in lymphoid organs, effector T cells are induced efficiently; these T cells are capable of reducing or clearing the antigen. Antigen not on APCs and not in lymphoid organs is ignored by T cells (Ohashi et al 1991, Lafferty & Cunningham 1975, Lafferty & Woolnough 1977), whereas too much antigen on too many APCs in most lymphoid organs may cause exhaustion by induction and peripheral deletion of T cells (Webb et al 1990, Moskophidis et al 1993a, Held et al 1992). Some antigens will persist with distinct kinetics and will continue to drive low levels of T and/or B cell responses in a balanced state, so that antigen is usually not completely eliminated, or cleared only very slowly (Nossal & Ada 1971, Tew et al 1990, Gray & Sprent 1990, Oehen et al 1992, Zinkernagel 1990). The most effective persistence of antigen is guaranteed by low level infection by non-cytopathic viruses in privileged sites, such as kidney, testes, parotis and neurons. This strategy is used by varicella, measles and herpes viruses and, in addition, probably by many more viruses than is commonly suspected (Mims 1987, Fields 1990, Jamieson et al 1991, Stevens & Cook 1971, Planz et al 1993).

For class II-associated antigens, immune complexes stored on follicular dendritic cells seem to be antigen depots of prime importance, particularly for maintaining antibody memory dependent upon both memory B and T helper cells (see below). It is theoretically possible that these complexes may also enter the class I pathway

(Matzinger & Bevan 1977, Bachmann et al 1994, Schirmbeck et al 1992, Rock et al 1990), but this is probably not usually the case. On the other hand, several studies suggest that peptides on class I (and/or class II) may have surprisingly long half-lives that may be involved in supporting low level triggering of T cells (Widmann et al 1991). Taken together, these data indicate that proposed regulatory networks (such as antigen-specific suppressor T cells, idiotypic networks involving antibodies and/or T cell receptors) are probably not necessary and not responsible for regulation of immune responses; antigen suffices and regulates immune responses very efficiently indeed.

Induction of immune responses

Rules for T cell responses

Induction. Successful induction of a T cell response depends on the expression of antigens on cells in lymphoid tissues, usually best accomplished by the so-called professional APCs, particularly dendritic cells (Macatonia et al 1989, Tew et al 1990). Important human viral pathogens (e.g. smallpox, polio) are good examples of this situation; they preferentially infect macrophages and dendritic cells.

Ignorance or indifference. If antigen is not properly presented, T cells ignore it and are not activated (e.g. rabies virus). Papilloma viruses in differentiating keratinocytes are ignored by T cells because these infected cells are not APCs, nor are they located in a lymphoid organ. These viruses only induce immune responses if the relevant antigens, released from destroyed cells, are taken up and brought into presenting cells (Wagner 1987, Tindle & Frazer 1994, Dietzschold & Ertl 1991). Thus T cells are indifferent not only to self antigens but also to foreign antigens out of the reach of T cells. This extends the concept of what were formerly called immunologically privileged sites. These T cells are not deleted and remain potentially iducible; if activated they would destroy host cells.

Exhaustion. There is another extreme situation where there is too much antigen on APCs and in lymphoid organs. If a non-cytolytic virus infects many APCs and cells in lymphoid organs so rapidly that virtually all $CD8^+$ T cells become induced, these seem to die within 2–3 d, thereby depleting the T cell repertoire of this particular specificity. Self antigen widely distributed on APCs and foreign (viral) antigens may induce and exhaust T cells and thereby establish peripheral tolerance not only to self but also potentially to foreign (non-cytopathic) viruses. Of course, if the antigen is present in the thymus, T cells are deleted by negative selection (exhaustion?) most efficiently at an early maturational stage.

The existence of exhaustive induction of effector T cells not only modifies our view of the mechanism for self–non-self discrimination, but also raises important questions about the nature of T cell memory. If, once induced, effector T cells are endstage cells and apparently die off, how can memory $CD8^+$ T cells exist, and how is T cell memory maintained (Sprent et al 1990, Rocha & von Boehmer 1991, Moskophidis et al 1993a)?

Although it is experimentally possible to make tolerant $CD4^+$ T helper cells by using high doses of model protein antigen, this has not yet been found to occur naturally during a viral infection. Thus, the conditions for the induction, indifference, and exhaustion of $CD8^+$ T cells and of $CD4^+$ T helper cells are probably not identical, the former apparently being more readily exhausted than the latter (Oldstone & Dixon 1967, Battegay et al 1993a, Moskophidis et al 1994). Therefore, both CD4 and B cell responses may be stimulated by poorly or non-cytopathic viruses, bacteria or parasites for an extended period because these responses are necessary for the transmission of immunological memory to offspring (see next section) and because, by themselves, they may not be efficient enough to eliminate the infectious agent. This continuous immunostimulation can result in secondary immunopathology as a consequence of the generation of excessive amounts of interleukins, enhanced inflammatory reactions due to constant cell stimulation (Doherty et al 1990) and the accumulation of antigen–antibody immune complexes; the latter also activate complement and recruit inflammatory infiltrates to the site of production or at sites where such complexes are filtered out. This occurs particularly in the kidney, the choroid plexus and generally in most basal membranes, resulting in chronic inflammatory diseases (Theofilopoulos & Dixon 1979).

Definition of T cell responses. Thus, the concept self–non-self discrimination should be replaced by a definition of T cell reactivity: T cells ignore self and non-self (foreign) antigen if not presented in lymphoid organs; they react against temporarily (4–7 d) localized antigen in lymphoid organs and APCs. They are exhausted or negatively selected against either permanently expressed or rapidly and overwhelmingly spreading antigen (both foreign and self-antigens) (Zinkernagel et al 1993). From an evolutionary point of view, the CTL exhaustion phenotype represents another (extreme) example of successful adaptation that is diametrically opposite to that developed by cytopathic viruses. While cytopathic viruses must induce an effective T cell response early to prevent lethal viral pathogenicity, non-cytopathic viruses, by exhausting the antivirally protective $CD8^+$ T cell response, avoid generalized immunopathological destruction of infected lymphohaemopoeitic and many other host cells. In this way, the exhaustion of T cell responses may be evolutionarily advantageous for both non-cytopathic viruses and the host (Osmond 1986, Moskophidis et al 1993b).

Rules for B cell responses

Neutralizing antibody responses against viruses prevent *reinfection* in many instances; against some acute cytopathic and human disease-causing viruses they are key for controlling primary infection. For example, infections with rabies virus in humans or VSV in mice are apparently controlled by neutralizing IgG. Studies of the antibody responses to VSV have revealed that without being a polyclonal activator, the repetitive antigen order plays an important role in triggering the completely T-independent IgM response. These results suggest that B cell receptor cross-linking is a critical step in B cell triggering, and when receptor cross-linking is maximal triggering of B cells can occur in the absence of T help or the so-called signal 2 (Schwartz 1990, Cohn & Langman 1990, Lafferty & Cunningham 1975). Interestingly, the neutralizing determinants in the viral envelope are spaced about 10–15 nm apart, a distance that had been found earlier with bacterial repetitive antigens (flagellin) or artificial repetitive hapten antigens (dinitrophenol) to be optimal in triggering B cells (Nossal & Ada 1971, Dintzis et al 1983, Langman 1972, Feldmann & Nossal 1972). Again, the estimated distance spanned by two antigen binding sites of an antibody is within the same range (10 nm) as the average distance between two Ig receptors on B cells, if one assumes about 50 000 Ig receptors on an average B cell. Also, antigen–antibody complexes trapped on dendritic cells as icosomes have a spacing of about 10–50 nm (Tew et al 1990, Bachmann et al 1993). Rhabdoviruses are amongst the most potent immunogens known, and the particular antigen order appears to represent optimal conditions for B cell triggering. This antigen spacing is crucial since all other forms of VSV-G (derived from purified or recombinant protein materials or produced by vaccinia VSV-G recombinant virus) behave either as T-independent type 2 or conventional T-dependent antigens. Therefore, B cell IgM responses are induced by extensive cross-linking of their receptors in the absence of linked T help; in the absence of direct cross-linking by antigen, T help is mandatory to trigger IgM B cell responses. Switch to IgG almost always depends upon T help. Thus antigen in a strictly repetitive paracrystalline rigid form as defined here is a hallmark of infectious agents and antigen pattern seems to be the crucial parameter determining B cell induction.

Immunological memory

After infection or vaccination of a host, its immune system is primed. This specific immunological memory is usually assessed by increased frequencies of specific T cells or by increased antibody titres, and results in improved immune performance (i.e. enhanced protection against reinfection). From an evolutionary perspective, the most obvious requirement for immunological memory appears to be the protection of young immunoincompetent offspring from otherwise lethal pathogens (Ehrlich 1882, Brambell 1970, Porter 1972, Cerny et al 1986). This may be explained as an evolutionary consequence of MHC-restricted T cell recognition. MHC polymorphism causes histoincompatibility and immunological rejection of histoincompatible cells or organs, and also the mutual rejection between mother and fetus. For this reason, immunological T cell memory cannot be transferred from mother to offspring (Zinkernagel et al 1985, Zinkernagel 1996). From this perspective therefore — and in contrast to antibody and Th cell memory maintaining

B cell memory — CTL memory is probably important for the individual by promoting immune surveillance against immunopathology and tumours.

Natural selection has eliminated immunological low-responders to cytopathic viruses and in addition should have rendered unlikely cytopathic agents that could not be checked efficiently by maternal antibodies during the critical time needed for the full development of immune responses after birth. For this reason, natural selection must have favoured viruses that persist at low levels of infection, or viral determinants that were stored in a stable fashion as IgG-complexes on follicular dendritic cells to maintain B and Th cell memory. One consequence of MHC polymorphism was histoincompatibility between mother and offspring and therefore both the offspring and the mother had to stay immunoincompetent or immunosuppressed at the T cell level to reduce sensitization by and rejection of the fetus. Accordingly, protective antibody memory may also provide compensatory protection of the immunosuppressed mother during this period. The hormone-dependent up-regulation of antibody responses is probably a major reason for the higher susceptibility of females to many autoimmune diseases. Thus MHC-restricted T cell recognition (phylogenetically thought to be the oldest part of the immune system), requiring a highly polymorphic MHC antigen system, necessitated the development of a long-lasting antibody response. This took the form of soluble transmissible polyvalent immunological maternal memory, protecting offspring while they developed T cell competence and gained the capacity to generate Th-dependent antibody responses.

Whether persistence of antigen is also necessary for the maintenance of memory CTLs is currently a subject of intense investigation. Evidence is conflicting (Gray & Sprent 1990, Gray & Skarvall 1988, Gray & Matzinger 1991, Sprent 1994, Oehen et al 1992, Lau et al 1994, Hou et al 1994, Roost et al 1990). Since experimental studies on T cell memory are difficult, these investigations frequently depend upon somewhat artificial adoptive transfer experiments into irradiated hosts. The role of antigen in maintaining memory has recently been reviewed in detail (Kündig et al 1996).

Immunological effector functions

Cellular versus humoral immune responses. Is there an overall regulation of humoral versus cellular immunity (Bretscher & Cohn 1970, Langman 1989, Cohn & Langman 1990, Parish 1972)? There have been several experiments and concepts supporting the notion that humoral and cellular immunity are mutually exclusive. These ideas gained momentum when T helper cells were further subdivided into Th1 and Th2 subsets (Mosmann & Coffman 1989). Experiments on immunity to viruses have so far not readily supported these concepts. Many viruses induce excellent $CD8^+$ and $CD4^+$ Tcell as well as B cell responses, in most cases very early with similar kinetics. Differences in this pattern are apparently genetically fixed in different strains of mice. C57BL/6 mice are γ-interferon (IFN-γ) high responders and therefore predominantly locked into a Th1 response. In contrast, BALB/c mice are low IFN-γ responders and preferentially mount Th2 responses, irrespective of the antigen or parasite used. Nevertheless, IFN-γ induced by many viral infections causes preferential induction of IgG2a rather than IgG1 antibody responses (Coutelier et al 1987, 1988). That is, however, not to deny that differences exist with respect to the relative importance of immunological effector pathways for protective immunity. For example, pox viruses induce CTLs, neutralizing antibodies, $CD4^+$ Th cells and activated macrophages, and there is convincing evidence that IFN-γ, tumour necrosis factor (TNF) and activated macrophages are key for recovery from primary infection (Blanden 1974, Ruby & Ramshaw 1992, Kündig et al 1993). For recovery from LCMV, lytic $CD8^+$ T cells are crucial and against VSV neutralizing IgG is key for recovery and protection, but all other effector pathways are induced efficiently in parallel. It will be interesting to search for viruses that misuse one of the postulated pathways exclusively for their own purpose, i.e. induce 'only' humoral or cellular or only Th1 or Th2 cell responses, as has been found for several classical parasites.

Class I versus class II restricted and lytic versus non-lytic T cells have different functions. Class I antigens have one transmembrane chain; class II have two. Class I antigens are found on all host cells whereas class II antigens occur mostly on lymphohaemopoetic cells. Class I presents largely internally synthesised and processed and precisely trimmed peptides of 8–10 residues, whereas class II presents phagocytosed and processed variable length peptides that are exchanged for the variant chain (Zinkernagel et al 1977, Morrison et al 1986, Germain 1994). Although several arguments have been made that the two peptide presentation structures are mediating similar functions, such as cell killing, evidence *in vivo* is still lacking (Muller et al 1992, Swain 1983, Stalder et al 1994). Accordingly, cytolytic $CD4^+$ T cells are thought to regulate antigen presentation and B cells negatively. However, this notion has not yet been supported by evidence *in vivo*. This does not negate the possibility that $CD4^+$ T cells may release interleukins that cause impairment of cell function or cell destruction. IFN-γ/TNF or a Fas-mediated mechanism may play a regulatory role by destruction of host cells, particularly if the latter are impaired by infections (Rouvier et al 1993, Suda et al 1993). The demonstration that in class I knockout mice or in perforin$^{o/o}$ mice drastic immunopathology may develop after LCMV infection could reflect such mechanisms (Muller et al 1992, Kägi et al 1994a,b). The simple view that antigens synthesized within the cell and transported on pathways of cellular antigens are presented by class I pathways, whereas external antigens that are taken up by cells via phagolysosomes are presented via the class II pathway, is well established and separates the two arms of cell-mediated immune effector mechanisms rather distinctly, but not absolutely. As usual in most biological processes one process is considerably more efficient, but limitations of the other may be overcome by increase of dose and/or local concentration. For example, minute amounts of formaldehyde-inactivated virus will induce a class II-restricted Th-dependent neutralizing antibody response. At least 10^4–10^6 times more is necessary to force non-replicating viral antigen into the class I presentation pathway. Although this does not usually occur in natural infections, for vaccination purposes conditions may be chosen so that protective $CD8^+$ T cell immunity is induced (Bachmann et al 1994, Zazopoulos & Haseltine 1993, Schirmbeck et al 1994).

Cytolytic T cells versus soluble mediator-releasing effector T cells. The debate about the protective role of cytolytic versus interleukin-releasing $CD8^+$ T cells is also coming to a similar salomonic end in that both pathways are induced but are differentially efficient in controlling different viruses (Blanden 1974, Lehmann-Grube et al 1988, Zinkernagel & Althage 1977, Ramsay et al 1993, Martz & Howell 1989, Kägi et al 1994a). For example, perforin-dependent target cell lysis seems essential for the control of LCMV infection in mice, but not against vaccinia or VSV. In contrast, lack of IFN-γ or TNFα or of their respective receptors renders vaccinia virus infection lethal for mice, but does not impair resistance against LCMV or VSV. The absence of $CD4^+$ T cells or of B cells prevents generation of neutralizing IgG which is mandatory for controlling VSV, but this has little influence on resistance against primary LCMV or vaccinia virus infections in mice (Zinkernagel 1996, Lehmann-Grube 1971, Kägi et al 1995, Blanden 1974). From these examples the question arises as to whether the Th1 versus Th2 phenotypes are similar to the distinct B cell Ig class or subclass selections. That is, rather than representing predetermined subclasses of T cells, it appears that specific T cells are being locked into producing a particular spectrum of interleukins similar to the way in which B cells are locked into eventually making one class of Ig. For both processes, pre-existent or dominant interleukin concentrations and patterns crucially determine the direction of differentiation (Mosmann & Coffman 1989, Reiner et al 1993).

Immunoprotection versus immunopathology: a question of balance. Most biological systems are optimized and rarely (if ever) function on an all-or-none basis. For example, lysis of acutely virus-infected host cells during the so-called eclipse phase of infection (i.e. when the virus has ceased to exist as infectious particle), is often necessary to stop virus spread, but at the cost of host cell destruction (Blanden 1974, Zinkernagel & Althage 1977, Zinkernagel & Doherty 1979). Therefore, antiviral $CD8^+$ T cell responses are beneficial for the host infected with a cytopathic virus; if the virus is not stopped early enough, it will cause the death of too many host cells and therefore kill the host. Infections with non-cytopathic viruses are, however, not directly life threatening. In these infections the extent of immunological destruction of infected host cells may determine the extent of pathology and disease. Two examples illustrate this notion. HBV is non-cytopathic (Bianchi 1981, Peters et al 1991, Ando et al 1993, Kohler et al 1974). In most patients it will infect relatively few liver cells, which are then destroyed by immune T cells, thereby eliminating the virus. If the T cell and/or neutralizing antibody responses are delayed, many more liver cells will be infected and the subsequent antiviral CTL response causes partial, substantial or excessive liver cell destruction, causing aggressive hepatitis. If no $CD8^+$ T cell response is generated, infection will result in a virus-carrier state, and the host will be free of liver disease (except for being at risk of developing primary liver cell carcinoma some 20–50 years later).

A second example is virus-induced acquired immunodeficiency caused by infection of mice with LCMV (Mims & Wainwright 1968, Jacobs & Cole 1976, Leist et al 1988, Odermatt et al 1991, Althage et al 1992, Wu-Hsieh et al 1988, Kohler et al 1990) or possibly in humans infected with HIV (Levy 1988, Zheng et al 1991, Zinkernagel & Hengartner 1994). While some cytopathic viruses may cause transient immunosuppression (von Pirquet 1908), they cannot overdo it since the host would die of the infection. In contrast, non-cytopathic viruses may use immunosuppression to establish persistent infections. LCMV has been shown to infect many macrophages and APCs and also a few $CD4^+$ T cells; these LCMV-infected cells are not destroyed by the virus but by the antiviral $CD8^+$ T cell response, depriving the host of efficient antigen-presentation and thereby causing immunosuppression and increased susceptibility to super-infection by intracellular parasites and to tumour growth (Wu-Hsieh et al 1988, Kohler et al 1990, Rosenberg & Fauci 1989). Recent analyses have revealed that anti-LCMV $CD8^+$ effector T cells are at least partially responsible for the slow kinetics of protective neutralizing antibodies to LCMV (Battegay et al 1993a, Moskophidis et al 1992). This example is a fascinating form of virus–host adaptation whereby the virus infects antigen-presenting and phagocytic effector cells of the immune system so that the induced effector T cells destroy that compartment to reduce the specific antiviral immune defence as well. Because there is no direct evidence that HIV is cytolytic *in vivo*, a similar immunopathologically T cell-mediated destruction of infected $CD4^+$ T cells, APCs and macrophages may be invoked to explain pathogenesis and the variable kinetics of AIDS (reviewed in Levy 1988, Rosenberg & Fauci 1989, Zinkernagel & Hengartner 1994).

Changing the balance by vaccination

Vaccination provides protection from disease but not necessarily from infection (Mims 1987, Ada 1990). If a limited infection does not cause lethal disease this is all that is needed. The classical example and one of the most efficient vaccinations of all has been the eradication of smallpox (Behbehani 1983). However, if restricted infection causes limited disease it may additionally (via a distinct pathogenetic mechanism) induce a second type of disease — usually at a much lower frequency than the first. Limitation of the first disease pattern may be insufficient to control the later consequences of the same infection. For example, immunity against EBV may well prevent mononucleosis, a first and direct consequence of infection, but may not always exclude the subsequent development of Burkitt's lymphoma, a much rarer consequence of chromosomal translocation and cellular transformation which is sometimes associated with the escape of the virus from immunological surveillance (Miller 1990). Similar considerations may apply to HIV. Therefore, the assumption that vaccine equals prevention of infection has not been generally true prior to HIV, and certainly cannot apply to HIV infection itself.

It is accepted that vaccinations usually shift the overall balance in favour of an immune response in more than 95% of cases, but this efficiency is, not surprisingly, virtually never 100% (Ada 1990). Such a shift is usually beneficial for the host infected with a cytopathic virus. However, against infections with non-cytopathic virus, induction of T cell responses by vaccination may cause increased immunopathology. Thus, under these special circumstances, vaccination may paradoxically enhance, rather than prevent, immunopathologically mediated disease (Oehen et al 1991, Battegay et al 1992, Ada 1991). This unfortunate vaccination for — rather than against — diseases is probably only going to complicate vaccination strategies against non-cytopathic viruses that cause immunopathology and that may establish a carrier status in immunocompetent hosts (Camp et al 1991, Battegay et al 1992, Ada 1991).

Another interesting conflict may arise if vaccination induces protective neutralizing antibodies but not a concomitant — or only a very short lived — protective T cell response. The problem develops if neutralization eliminates virtually (but not quite) all virus and thereby prevents or at least delays initial induction of $CD8^+$ cells by the residual infecting virus. Because virus may still spread from the local site of infection from cell to cell, the more slowly generated CTL response will eventually cause considerably more extensive immunopathology than would have developed in absence of neutralizing antibodies. Some of the complications of vaccinations with formalin-inactivated respiratory syncytial virus (RSV) (Cannon et al 1988, Alwan et al 1994, Murphy & Chanock 1990), some characteristics of Dengue fever (Halstead 1988), or a model situation described for LCMV (Battegay et al 1993b), may represent such unwanted consequences of unbalanced antibody, $CD4^+$ T cell and $CD8^+$ T cell responses.

Conclusion

This view is obviously incomplete. The interdependence between viruses and immune responses could have been illustrated with many other examples of viral, bacterial or other parasitic infections. The few examples discussed show how viruses and the host immune system are more-or-less well balanced with the host immune system. The analysis of immune responses against viruses has helped to define parameters of immunity *in vivo*. The equilibria between different viruses and the host immune system are wonders of nature worthy of our greatest admiration and respect: viruses and infectious agents in general have 'learned' their immunology well enough to teach us important lessons.

References

Ada G 1990 The immunological principles of vaccination. Lancet 335:523–526
Ada G 1991 Vaccine development: real and imagined dangers. Nature 349:369

Allen PM, Zinkernagel RM 1994 Promethean viruses? Nature 369:355–356

Althage A, Odermatt B, Moskophidis D et al 1992 Immunosuppression by lymphocytic choriomeningitis virus infection: competent effector T and B cells but impaired antigen presentation. Eur J Immunol 22:1803–1812

Alwan WH, Kozlowska WJ, Openshaw PJM 1994 Distinct types of lung disease caused by functional subsets of antiviral T cells. J Exp Med 179:81–89

Ando K, Moriyama T, Guidotti LG et al 1993 Mechanisms of class I restricted immunopathology. A transgenic mouse model of fulminant hepatitis. J Exp Med 178:1541–1554

Bachmann MF, Hoffmann Rohrer U, Kündig TM, Bürki K, Hengartner H, Zinkernagel RM 1993 The influence of antigen organisation on B cell responsiveness. Science 262:1448–1451

Bachmann MF, Kündig TM, Freer G et al 1994 Induction of protective cytotoxic T cells with viral proteins. Eur J Immunol 24:2128–2236

Battegay M, Oehen S, Schulz M, Hengartner H, Zinkernagel RM 1992 Vaccination with a synthetic peptide modulates lymphocytic choriomeningitis virus mediated immunopathology. J Virol 66:1199–1201

Battegay M, Moskophidis D, Waldner H et al 1993a Impairment and delay of neutralizing antiviral antibody responses by virus specific cytotoxic T cells. J Immunol 151:5408–5415

Battegay M, Kyburz D, Hengartner H, Zinkernagel RM 1993b Enhancement of disease by neutralizing antiviral antibodies in the absence of primed antiviral cytotoxic T cells. Eur J Immunol 23:3236–3241

Behbehani AM 1983 The smallpox story: life and death of an old disease. Microbiol Rev 47:455–509

Berek C, Milstein C 1987 Mutation drift and repertoire shift in the maturation of the immune response. Immunol Rev 96:23–41

Bertoletti A, Sette A, Chisari FV et al 1994 Natural variants of cytotoxic epitopes are T cell receptor antagonists for anti-viral cytotoxic T cells. Nature 369:407–410

Bianchi L 1981 The immunopathology of acute type B hepatitis. Springer Semin Immunopathol 3:421–438

Blanden RV 1974 T cell response to viral and bacterial infection. Transplant Rev 19:56–84

Brambell RWR 1970 The transmission of immunity from mother to young. North Holland Publishing, Amsterdam

Bretscher P, Cohn M 1970 A theory of self–nonself discrimination. Science 169:1042–1049

Bucchini D, Reynaud CA, Ripoche MA, Grimal H, Jami J, Weill JC 1987 Rearrangement of a chicken immunoglobulin gene occurs in the lymphoid lineage of transgenic mice. Nature 326:409–411

Camp RL, Kraus TA, Birkeland ML, Puré E 1991 High levels of CD44 expression distinguish virgin from antigen-primed B cells. J Exp Med 173:763–766

Cannon MJ, Openshaw PJ, Askonas BA 1988 Cytotoxic T cells clear virus but augment lung pathology in mice infected with respiratory syncytial virus. J Exp Med 168:1163–1168

Castelmur I, DiPaolo C, Bachmann MF, Hengartner H, Zinkernagel RM, Kündig TM 1993 Comparison of the sensitivity of *in vivo* and *in vitro* assays for detection of antiviral cytotoxic T cell activity. Cell Immunol 151:460–466

Cerny A, Heusser C, Sutter S et al 1986 Generation of agammaglobulinaemic mice by prenatal and postnatal exposure to polyclonal or monoclonal anti-IgM antibodies. Scand J Immunol 24:437–445

Charan S, Hengartner H, Zinkernagel RM 1987 Antibodies against the two serotypes of vesicular stomatitis virus measured by enzyme-linked immunosorbent assay: immunodominance of serotype -specific determinants and induction of asymmetrically cross-reactive antibodies. J Virol 61:2509–2514

Chen C, Roberts VA, Rittenberg MB 1992 Generation and analysis of random point mutations in an antibody CDR2 sequence: many mutated antibodies lose their ability to bind antigen. J Exp Med 176:855–866

Chothia C, Lesk AM, Tramontano A et al 1989 Conformations of immunoglobulin hypervariable regions. Nature 342:877–883

Claflin JL, Berry J 1988 Genetics of the phosphocholine-specific antibody response to Streptococcus pneumoniae. Germ-line but not mutated T15 antibodies are dominantly selected. J Immunol 141:4012–4019

Cohn M, Langman RE 1990 The Protecton: the unit of humoral immunity selected by evolution Immunol Rev 115:7–147

Coutelier J-P, Van der Logt JTM, Heessen FWA, Warnier G, Van Snick JV 1987 IgG2a restriction of murine antibodies elicited by viral infections. J Exp Med 165:64–69

Coutelier J-P, Van der Logt JTM, Heessen FWA, Van Snick J 1988 Virally induced modulation of murine IgG antibody subclasses. J Exp Med 168:2373–2378

de Campos-Lima P-O, Gavioli R, Zhang Q-J et al 1993 HLA-A11 epitope loss isolates of Epstein-Barr virus from a highly $A11^{+}$ population. Science 260:98–100

Dietzschold B, Ertl HC 1991 New developments in the pre- and post-exposure treatment of rabies. CRC Crit Rev Immunol 10:427–439

Dintzis RZ, Middleton MH, Dintzis HM 1983 Studies on the immunogenicity and tolerogenicity of T-independent antigens. J Immunol 131:2196–2203

Doherty PC, Allan JE, Lynch F, Ceredig R 1990 Cellular events in a virus-induced inflammatory process: promotion of delayed type hypersensitivity by $CD8^{+}$ T lymphocytes. Immunol Today 11:55–59

Dulbecco R, Vogt M, Strickland A 1956 A study of the basic aspects of neutralization of two animal viruses, western equine encephalitis virus and poliomyelitis virus. Virology 2:162–205

Ehrlich P 1882 Ueber Immunität durch Vererbung und Säugung. Z Hyg Infektionskr 12:183

Eisen HN, Siskind GW 1964 Variations in affinities of antibodies during the immune response. Biochemistry 3:996–1008

Evavold BD, Allen PM 1991 Separation of IL-4 production from Th cell proliferation by an altered T cell receptor ligand. Science 252:1308–1310

Fazekas de St Groth S 1967 Cross recognition and cross-reactivity. Cold Spring Harbor Symp Quant Biol 32:525–536

Fazekas de St Groth S 1981 The joint evolution of antigens and antibodies. In: Steinberg CM, Lefkovitz I (eds) The immune system, vol 2. Karger, Basel, p155–168

Feldmann M, Nossal GJV 1972 Tolerance, enhancement and the regulation of interactions between T cells, B cells and macrophages. Immunol Rev 13:3–34

Fields BN 1990 Fields virology, 2nd edn. Raven, New York

Freund J, Stern ER, Pisani TM 1947 Isoallergic encephalomyelitis and radiculitis in guinea pigs after one injection of brain and mycobacteria in water-in-oil emulsion. J Immunol 57:179–194

Germain RN 1994 MHC-dependent antigen processing and peptide presentation: providing ligands for T lymphocyte activation. Cell 76:287–299

Goodnow CC, Adelstein S, Basten A 1990 The need for central and peripheral tolerance in the B cell repertoire. Science 248:1373–1379

Gray D, Matzinger P 1991 T cell memory is short-lived in the absence of antigen. J Exp Med 174:969–974

Gray D, Skarvall H 1988 B-cell memory is short-lived in the absence of antigen. Nature 336:70–73

Gray D, Sprent J 1990 Immunological memory. Curr Topics Microbiol Immunol 159:1–138

Gresser I, Tovey MG, Maury C, Bandu M-T 1976 Role of interferon in the pathogenesis of virus diseases as demonstrated by the use of anti-interferon serum. II. Studies with herpes simplex,

Moloney sarcoma, vesicular stomatitis, Newcastle disease and influenza viruses. J Exp Med 144:1316–1323
Halstead SB 1988 Pathogenesis of dengue: challenges to molecular biology. Science 239:476–481
Held W, Shakhov AN, Waanders G et al 1992 An exogenous mouse mammary tumor virus with properties of M1s-1a (Mtv-7). J Exp Med 175:1623–1633
Hotchin J 1971 Persistent and slow virus infections. Monogr Virol 3:1–211
Hou S, Hyland L, Ryan KW, Portner A, Doherty PC 1994 Virus-specific $CD8^+$ T-cell memory determined by clonal burst size. Nature 369:652–654
Huang S, Hendriks W, Althage A et al 1993 Immune response in mice that lack the interferon-γ receptor. Science 259:1742–1745
Isaacs A, Lindenmann J 1957 Virus interference. 1. The interferon. Proc R Soc Lond B Biol Sci 147:258–267
Jacobs RP, Cole GA 1976 Lymphocytic choriomeningitis virus-induced immunosuppression: a virus-induced macrophage defect. J Immunol 117:1004–1009
Jamieson BD, Somasundaram T, Ahmed R 1991 Abrogation of tolerance to a chronic viral infection. J Immunol 147:3521–3529
Janeway CA 1989 Approaching the asymptote? Cold Spring Harbor Symp Quant Biol 54:1–13
Kägi D, Ledermann B, Bürki K et al 1994a Cytotoxicity mediated by T cells and natural killer cells is greatly impaired in perforin-deficient mice. Nature 369:31–37
Kägi D, Vignaux F, Ledermann B et al 1994b Fas and perforin pathways as major mechanisms of T cell-mediated cytotoxicity. Science 265:528–530
Kägi D, Seiter P, Pavlovic J et al 1995 The roles of perforin and Fas-dependent cytotoxicity in protection against cytopathic and noncytopathic viruses. Eur J Immunol 25:3256–3262
Klenerman P, Rowland-Jones S, McAdam S et al 1994 Cytotoxic T cell activity antagonized by naturally-occurring HIV-1 GAG variants. Nature 369:403–407
Klenk H-D, Rott R 1988 Biology of influenza virus pathogenicity. Adv Virus Res 34:247–281
Kohler M, Rüttner B, Cooper S, Hengartner H, Zinkernagel RM 1990 Enhanced tumor susceptibility of immunocompetent mice infected with lymphocytic choriomeningitis virus. Cancer Immunol Immunother 32:117–124
Kohler P, Trembath J, Merrill DA, Singleton JW, Dubois RS 1974 Immunotherapy with antibody, lymphocytes and transfer factor in chronic hepatitis B. Clin Immunol Immunopathol 2:465–471
Kündig TM, Hengartner H, Zinkernagel RM 1993 T cell dependent IFN-γ exerts an antiviral effect in the central nervous system but not in peripheral solid organs. J Immunol 150:2316–2321
Kündig TM, Bachmann MF, Ohashi PS, Pircher HP, Hengartner H, Zinkernagel RM 1996 On T cell memory: arguments for antigen dependence. Immunol Rev 150:63–90
Lafferty K, Woolnough J 1977 The origin and mechanism of allograft reaction. Immunol Rev 35:231–262
Lafferty KJ, Cunningham AJ 1975 A new analysis of allogeneic interactions. Aust J Exp Biol Med Sci 53:27–42
Langman RE 1972 The occurrence of antigenic determinants common to flagella of different salmonella strains. Eur J Immunol 2:582–586
Langman RE 1989 The immune system. Academic Press, New York
Lau LL, Jamieson BD, Somasundaram T, Ahmed R 1994 Cytotoxic T-cell memory without antigen. Nature 369:648–652
Lawrence HW 1959 Homograft sensitivity. An expression of the immunologic origins and consequences of individuality. Physiol Rev 39:811–859
Lehmann-Grube F 1971 Lymphocytic choriomeningitis virus. Virol Monogr 10:1–173

Lehmann-Grube F, Moskophidis D, Löhler J 1988 Recovery from acute virus infection: role of cytotoxic lymphocytes T in the elimination of lymphocytic choriomeningitis virus from spleens of mice. Ann N Y Acad Sci 532:238–256

Leist TP, Rüedi E, Zinkernagel RM 1988 Virus-triggered immune suppression in mice caused by virus-specific cytotoxic T cells. J Exp Med 167:1749–1754

Levy JA 1988 Mysteries of HIV: challenges for therapy and prevention. Nature 333:519–522

Levy JP, Leclerc JC 1977 The murine sarcoma virus-induced tumor: exception or general model in tumor immunology? Adv Cancer Res 24:1–66

Liang S, Mozdzanowska K, Palladino G, Gerhard W 1994 Heterosubtypic immunity to influenza type A virus in mice. J Immunol 152:1653–1661

Lilly F 1972 Mouse leukemia: a model of a multiple-gene disease. J Nat Cancer Inst 49:927

Lilly F, Boyse EA, Old LJ 1964 Genetic basis of susceptibility to viral leukemogenesis. Lancet II:1207–1209

Linton P-J, Decker DJ, Klinman NR 1989 Primary antibody-forming cells and secondary B cells are generated from separate precursor cell subpopulations. Cell 59:1049–1059

Macatonia SE, Taylor PM, Knight SC, Askonas BA 1989 Primary stimulation by dendritic cells induces antiviral proliferative and cytotoxic T cell responses *in vitro*. J Exp Med 169:1255–1264

Marrack P, Kappler J 1994 Subversion of the immune system by pathogens. Cell 76:323–332

Martz E, Howell DM 1989 CTL: virus control cells first and cytolytic cells second? DNA fragmentation, apoptosis and the prelytic halt hypothesis. Immunol Today 10:79–86

Matsui K, Boniface JJ, Reay PA, Schild H, Fazekas de St Groth B, Davis MM 1991 Low affinity interaction of peptide–MHC complexes with T cell receptors. Science 254:1788–1791

Matzinger P, Bevan MJ 1977 Induction of H-2-restricted cytotoxic T cells: *in vivo* induction has the appearance of being unrestricted. Cell Immunol 33:92–100

McIntosh K 1990 Diagnostic virology. In: Fields BN, Knipe DM (eds) Fields virology, 2nd edn. Raven, New York, p 411–440

McMichael AJ, Gotch FM, Dongworth DW, Clark A, Potter C 1983 Declining T cell immunity to influenza 1977–1982. Lancet II:762–764

Milich DR, Jones JE, Hughes JL, Price J, Raney AK, McLachlan A 1990 Is a function of the secreted hepatitis B e antigen to induce immunologic tolerance *in utero*? Proc Natl Acad Sci USA 87:6599–6603

Miller G 1990 Epstein–Barr virus. In: Fields BN, Knipe DM (eds) Fields virology, 2nd edn. Raven, New York, p 1921–1958

Mims CA 1987 Pathogenesis of infectious disease. Academic Press, London

Mims CA, Wainwright S 1968 The immunodepressive action of lymphocytic choriomeningitis virus in mice. J Immunol 101:717–724

Mitchison NA 1964 Induction of immunological paralysis in two zones of dosage. Roy Soc Proc 161:275–292

Mondelli M, Eddleston AL 1984 Mechanisms of liver cell injury in acute and chronic hepatitis. Semin Liver Dis 4:47–58

Morrison LA, Lukacher AE, Braciale VL, Fan DP, Braciale TJ 1986 Differences in antigen presentation to MHC class I- and class II-restricted influenza virus specific cytolytic T lymphocyte clones. J Exp Med 163:903–921

Moskophidis D, Pircher HP, Ciernik I, Odermatt B, Hengartner H, Zinkernagel RM 1992 Suppression of virus specific antibody production by $CD8^+$ class I-restricted antiviral cytotoxic T cells *in vivo*. J Virol 66:3661–3668

Moskophidis D, Lechner F, Pircher HP, Zinkernagel RM 1993a Virus persistence in acutely infected immunocompetent mice by exhaustion of antiviral cytotoxic effector T cells. Nature 362:758–761

Moskophidis D, Laine E, Zinkernagel RM 1993b Peripheral clonal deletion of antiviral memory CD8$^+$ T cells. Eur J Immunol 23:3306–3311

Moskophidis D, Lechner F, Hengartner H, Zinkernagel RM 1994 MHC class I and non MHC-linked capacity for generating an antiviral CTL response determines susceptibility to CTL exhaustion and establishment of virus persistence in mice. J Immunol 152:4976–4983

Mosmann TR, Coffman RL 1989 TH1 and TH2 cells: different patterns of lymphokine secretion lead to different functional properties. Annu Rev Immunol 7:145–173

Muller D, Koller BH, Whitton JL, LaPan KE, Brigman KK, Frelinger JA 1992 LCMV-specific, class II-restricted cytotoxic T cells in β2-microglobulin-deficient mice. Science 255:1576–1578

Müller U, Steinhoff U, Reis LFL et al 1994 Functional role of type I and type II interferons in antiviral defense. Science 264:1918–1921

Murphy BR, Channock RM 1990 Immunisation against viruses. In: Fields BN, Knipe DM (eds) Fields virology, 2nd edn. Raven, New York, p 469–502

Nossal GJV, Ada GL 1971 Antigens, lymphoid cells and the immune response. Academic Press, New York

Odermatt B, Eppler M, Leist TP, Hengartner H, Zinkernagel RM 1991 Virus-triggered acquired immunodeficiency by cytotoxic T-cell dependent destruction of antigen-presenting cells and lymph follicle structure. Proc Natl Acad Sci USA 88:8252–8256

Oehen S, Hengartner H, Zinkernagel RM 1991 Vaccination for disease. Science 251:195–198

Oehen S, Waldner HP, Kündig T, Hengartner H, Zinkernagel RM 1992 Antivirally protective cytotoxic T cell memory to lymphocytic choriomeningitis virus is governed by persisting antigen. J Exp Med 176:1273–1281

Ohashi PS, Oehen S, Bürki K et al 1991 Ablation of "tolerance" and induction of diabetes by virus infection in viral antigen transgenic mice. Cell 65:305–317

Ohno S, Matsunaga T, Epplen JT, Hozumi T 1980 Interaction of viruses and lymphocytes in evolution, differentiation and oncogenesis. Progr Immunol 4: 577–598

Oldstone MBA, Dixon FJ 1967 Lymphocytic choriomeningitis: production of anti-LCM antibody by "tolerant" LCM-infected mice. Science 158:1193–1195

Osmond DG 1986 Population dynamics of bone marrow B lymphocytes. Immunol Rev 93:103–124

Palese P, Young JF 1982 Variation of influenza A,B, and C viruses. Science 215:1468–1474

Parish CR 1972 The relationship between humoral and cell-mediated immunity. Immunol Rev 13:35–66

Paul WE 1993 Fundamental immunology. Raven, New York

Peters M, Vierling J, Gershwin ME, Milich D, Chisari FV, Hoofnagle JH 1991 Immunology and the liver. Hepatology 13:977–994

Phillips RE, Rowland-Jones S, Nixon DF et al 1991 Human immunodeficiency virus genetic variation that can escape cytotoxic T cell recognition. Nature 354:453–459

Pircher HP, Moskophidis D, Rohrer U, Bürki K, Hengartner H, Zinkernagel RM 1990 Viral escape by selection of cytotoxic T cell-resistant virus variants *in vivo*. Nature 346:629–633

Planz O, Bilzer T, Sobbe M, Stitz L 1993 Lysis of major histocompatibility complex class I-bearing cells in borna disease virus-induced degenerative encephalopathy. J Exp Med 178:163–174

Porter P 1972 Immunoglobulins in bovine mammary sections: quantitative changes in early lactation and absorption by the neonatal calf. Immunology 23:225–237

Rajewsky K 1989 Evolutionary and somatic immunological memory. Prog Imm 7:397–403

Rajewsky K, Fürster I, Cumano A 1987 Evolutionary and somatic selection of the antibody repertoire in the mouse. Science 238:1088–1094

Ramsay AJ, Ruby J, Ramshaw IA 1993 The case for cytokines as effector molecules in the resolution of virus infection. Immunol Today 14:155–157

Reiner SL, Wang Z-E, Hatam F, Scott P, Locksley RM 1993 Th1 and Th2 cell antigen receptors in experimental leishmaniasis. Science 259:1457–1460
Rocha B, von Boehmer H 1991 Peripheral selection of the cell repertoire. Science 251:1225–1228
Rock KL, Gamble S, Rothstein L 1990 Presentation of exogenous antigen with class I major histocompatibility complex molecules. Science 249:918–921
Roost HP, Charan S, Zinkernagel RM 1990 Analysis of the kinetics of antiviral memory T help *in vivo*: characterization of short lived cross-reactive T help. Eur J Immunol 20:2547–2554
Roost HP, Bachmann MF, Haag A et al 1995 Early high-affinity neutralizing antiviral IgG responses without further overall improvements of affinity. Proc Natl Acad Sci USA 92:1257–1261
Rosenberg ZF, Fauci AS 1989 The immunopathogenesis of HIV infection. Adv Immunol 47:377–431
Rouvier E, Luciani MF, Golstein P 1993 Fas involvement in Ca^{2+}-independent T cell mediated cytotoxicity. J Exp Med 177:195–200
Ruby J, Ramshaw I 1992 The antiviral activity of immune $CD8^+$ T cells is dependent on interferon-γ. Lymphokine Cytokine Res 10:353–358
Schirmbeck R, Zerrahn J, Kuhrober A, Kury A, Deppert W, Reimann J 1992 Immunization with soluble simian virus 40 large T antigen induces a specific response of CD_3^+ $CD4^-$ $CD8^+$ cytotoxic T lymphocytes in mice. Eur J Immunol 22:759–766
Schirmbeck R, Melber K, Mertens T, Reimann J 1994 Selective stimulation of murine cytotoxic T cell and antibody responses by particulate or monomeric hepatitis B virus surface (S) antigen. Eur J Immunol 24:1088–1096
Schwartz RH 1990 A cell culture model for T lymphocyte clonal anergy. Science 248:1349–1356
Smith TJ, Olson NH, Cheng RH et al 1993 Structure of human rhinovirus complexed with Fab fragments from a neutralizing antibody. J Virol 67:1148–1158
Speiser DE, Kyburz D, Stübi U, Hengartner H, Zinkernagel RM 1992 Discrepancy between *in vitro* measurable and *in vivo* virus neutralizing cytotoxic T cell reactivities. J Immunol 149:972–980
Sprent J 1994 T and B memory cells. Cell 76:315–322
Sprent J, Gao EK, Webb SR 1990 T cell reactivity to MHC molecules: immunity versus tolerance. Science 248:1357–1363
Stalder T, Hahn SH, Erb P 1994 Fas antigen is the major target molecule for $CD4^+$ T cell-mediated cytotoxicity. J Immunol 152:1127–1133
Stevens JG, Cook ML 1971 Latent herpes simplex virus in spinal ganglia of mice. Science 173:843–845
Suda T, Takahashi T, Golstein P, Nagata S 1993 Molecular cloning and expression of the Fas ligand, a novel member of the tumour necrosis factor family. Cell 75:1169–1178
Swain SL 1983 T cell subsets and the recognition of MHC class. Immunol Rev 74:129–142
Sykulev Y, Brunmark A, Jackson M, Cohen RJ, Peterson PA, Eisen HN 1994 Kinetics and affinity of reactions between an antigen-specific T cell receptor and peptide–MHC complexes. Immunity 1:15–22
Tew JG, Kosco MH, Burton GF, Szakal AK 1990 Follicular dendritic cells as accessory cells. Immunol Rev 117:185–212
Theofilopoulos AN, Dixon FJ 1979 The biology and detection of immune complexes. Adv Immunol 28:89–220
Tindle RW, Frazer IH 1994 Immune response to human papilloma viruses and the prospects for human papilloma virus-specific immunisation. Curr Opin Microbiol Immunol 186:217–252
von Pirquet C 1908 Das Verhalten der kutanen Tuberkulinreaktion während den Masern. Deutsche Medizinische Wochenschrift 34:1279–1300
Wagner RR (ed) 1987 The rhabdoviruses. Plenum, New York

Waters JA, Kennedy M, Voet P et al 1992 Loss of the common "A" determinant of hepatitis B surface antigen by a vaccine-induced escape mutant. J Clin Invest 90:2543–2547
Webb S, Morris C, Sprent J 1990 Extrathymic tolerance of mature T cells: clonal elimination as a consequence of immunity. Cell 63:1249–1256
Weber S, Traunecker A, Oliveri F, Gerhard W, Karjalainen K 1992 Specific low-affinity recognition of major histocompatibility complex plus peptide by soluble T-cell receptor. Nature 356:793–796
Webster RG, Laver WG, Air GM, Schild GC 1982 Molecular mechanisms of variation in influenza viruses. Nature 296:115–121
Weill J-C, Reynaud C-A 1987 The chicken B cell compartment. Science 238:1094–1098
Wheelock EF 1965 Interferon-like virus inhibitor produced in human leukocytes by phytohemagglutinin. Science 149:310–313
Widmann C, Maryanski JL, Romero P, Corradin G 1991 Differential stability of antigenic MHC class I-restricted synthetic peptides. J Immunol 147:3745–3751
Wu-Hsieh B, Howard DH, Ahmed R 1988 Virus-induced immunosuppression: a murine model of susceptibility to opportunistic infection. J Infect Dis 158:232–235
Zazopoulos E, Haseltine WA 1993 Effect of nef alleles on replication of human immunodeficiency virus type 1. Virology 194:20–27
Zheng LM, Liu CC, Ojcius DM, Young JD 1991 Expression of lymphocyte perforin in the mouse uterus during pregnancy. J Cell Sci 99:317–323
Zinkernagel RM 1990 Antiviral T-cell memory? Curr Opin Microbiol Immunol 159:65–77
Zinkernagel RM 1996 Immunology taught by viruses. Science 271:173–178
Zinkernagel RM, Althage A 1977 Antiviral protection by virus-immune cytotoxic T cells: infected target cells are lysed before infectious virus progeny is assembled. J Exp Med 145:644–651
Zinkernagel RM, Doherty PC 1979 MHC-restricted cytotoxic T cells: studies on the biological role of polymorphic major transplantation antigens determining T cell restriction-specificity, function and responsiveness. Adv Immunol 27:52–142
Zinkernagel RM, Hengartner H 1994 T cell mediated immunopathology versus direct cytolysis by virus: implications for HIV and AIDS. Immunol Today 15:262–268
Zinkernagel RM, Althage A, Adler B et al 1977 H-2 restriction of cell-mediated immunity to an intracellular bacterium. Effector T cells are specific for *Listeria* antigen in association with H-21 region coded self-markers. J Exp Med 145:1353–1367
Zinkernagel RM, Hengartner H, Stitz L 1985 On the role of viruses in the evolution of immune responses. Br Med Bull 41:92–97
Zinkernagel RM, Haenseler E, Leist TP, Cerny A, Hengartner H, Althage A 1986 T cell mediated hepatitis in mice infected with lymphocytic choriomeningitis virus. J Exp Med 164:1075–1092
Zinkernagel RM, Moskophidis D, Kündig T, Oehen S, Pircher HP, Hengartner H 1993 Effector T cell induction and T cell memory versus peripheral deletion of T cells. Immunol Rev 131:1–25

DISCUSSION

Nossal: I am puzzled by one set of your data. For two or three years you have highlighted the high affinity that is achieved quite early on in this anti-VSV response. You told us that after 6 d the response does not mature further, and the antibodies don't get better, yet at the same time you have shown us that germinal centres, which certainly aren't developed fully by 6 d, develop very nicely afterwards and play a continuing role in the infection, and even lead to some antibody-forming cell response 100 d later. By day 6, unless there is something tremendously different in your system, you will not have any IgG but rather IgM. You will not have any V gene mutations. You must be dealing with a rare B cell that somehow displays surprisingly high affinity and divides quickly over the first six days. There is something here which doesn't resemble most of the studies that have been done with haptens.

Zinkernagel: Part of this work has already been completed by Dr U. Kalinke in our lab and at the moment it looks as follows. You actually start with few V_Hs and V_Ls. These immediately start to vary somatically, but overall you lose some and gain some from another pool and you end up with the same range. None of the data say anything against somatic variation; they simply show that there are some germline-coded V_Hs and V_Ls that give a very high affinity to begin with.

von Boehmer: You have shown quite clearly that T cell activation is necessary for protection, and you say there is no memory in the absence of activated T cells. This may be true for your system and the particular virus you are studying, but we also know that there are cells that become resting cells after exposure to antigen, and upon re-stimulation these populations respond more vigorously. These do not seem to play a part in your system. Can you extrapolate from your system to the rest of the microorganisms we are surrounded by? Is what you have found generally true? Or can you leave room for some other microorganisms that can be dealt with by small resting memory cells that can respond more vigorously?

Zinkernagel: I don't think there is a discrepancy between our data and those of others, it is just that we are asking different questions and measure protection against peripheral rather than i.v. infections. Our *in vitro* read-outs documenting CTL precursor maintenance in the absence of antigen parallel other findings exactly. The delayed-type hypersensitivity (DTH) usually measured in clinical settings involves the assessment of T cell immunity to tuberculosis, Lepra and Boek's sarcoidosis. DTH can be used here for measuring T cell immunity because these infections and their antigens never disappear completely. This situation was regarded to be somewhat exceptional — but it probably isn't.

von Boehmer: You may have picked the wrong microorganism with which to show memory: with another you might see beautiful memory.

Zinkernagel: Most microorganisms would give a more acute response than those I used because they are more acutely life threatening (and actually often controlled by antibodies or soluble mediators). The only ones that this would not be true for would be the immunopathologically damaging microbes such as hepatitis B and HIV.

Rajewsky: I wanted to ask the identical question Harald asked for the T cells for the B cells. You look at a particular virus where antibodies are essential to protect after a certain type of infection. There must be many other organisms where classical memory as localized in resting cells must play a critical role. It is a question of picking the right system. I am not sure that you should call an on-going immune response 'memory': this is not what is classically meant by this term.

Zinkernagel: I would say that there is no protective memory in absence of this ongoing response.

Rajewsky: But you said yourself that these resting cells are present.

Zinkernagel: It all depends on whether you think that measuring CTL precursors or potentially antibody-producing memory B cells *in vitro* is significant. This can be measured, but in terms of protection against reinfection these measurements are not important.

Rajewsky: You can't base your conclusions on results obtained with just one virus.

Zinkernagel: But all acute viruses against which vaccines have been successful belong to the VSV type. There is one interesting exception — smallpox: von Pirquet did an interesting experiment in 1906 in which he vaccinated children and then looked at how long after priming he could re-challenge these children in the skin to get reinfection. He found it took only four weeks. He did not report whether these children developed encephalitis with any greater or lesser frequency. One would predict that although they made the lesion locally, they would not have developed encephalitis, because that is the critical protective mechanism of vaccination. Although I wouldn't argue with either you or Harald that there are certain situations where my argument may be unrepresentative, in general and for epidemiologically important viruses it is applicable.

Melchers: From these experiments, your prediction would have to be that peptide vaccinations are no good at all, because peptides don't hang around for long.

Zinkernagel: We initially tried to prime mice with peptide without any adjuvants, but this does not work.

Melchers: Does this mean that Alexandra Livingstone's experiments are false?

Zinkernagel: She does them differently. She either once or repeatedly injects peptide-loaded dendritic cells into the animal.

Melchers: What is the difference?

Zinkernagel: The difference is that we have no idea of the half-life of these MHC–peptide complexes on cells. If we simply inject peptide, we don't know where most of it lands nor its kinetics.

Melchers: You also seemed to indicate that there is only one type of B cell and two types of T cell. How much are your infectious systems influenced by all the other types of B and T cell that exist?

Zinkernagel: This is what I tried to show for T help in the VSV system. In the VSV model infection we need IgG for protection. Many people have tested CD4s and CD8s in protective/adoptive transfer experiments in the influenza system and have shown protection, but *in vivo* these pathways are not critical, only antibodies are. My argument is that if you do experiments *in vivo* and show that one or the other mechanism is limiting you learn what is important. For example, perforin knockout mice survive influenza virus infections well. If you choose IgM$^{o/o}$ mice you will find that they perform badly.

Mosmann: Concerning the general applicability of this reactivation, isn't it a numbers game? If you give the virus a substantial advantage by putting in a fair amount of virus before you have many reactivated T cells, the system has trouble catching up. The less virus you put in, the less you should need that reactivation process. By titrating down the virus, can you show that these quiescent cells that are at high frequency are in fact protective if you give them more of a chance to catch up? Secondly, if you go the other way, how short can your period of reactivation be? I was impressed with your 24 h reactivation: can you go shorter than that?

Zinkernagel: The shortest we have done is 16 h. In answer to the first question, if you challenge hosts in the big toe or the meninges, what you say doesn't apply. That is, the local amplification of the virus is too quick and the spread to local lymph nodes too slow for the immune system to notice anything, and the response is too delayed to get a substantial enough protective effect.

Mosmann: What about normal infection? Isn't VSV a surface infection?

Zinkernagel: VSV infection is mucosal. LCMV or HIV would enter either via mucosa or skin and would hit peripheral cells. To get rid of these viruses you need neutralizing antibodies or activated T cells (for skin infections). We use injection into the foot pad and you still get a swelling reaction, or in the case of the brain, if you give the virus intracerebrally the mouse will die even if it is primed; only a few PFUs are needed.

Kirberg: You showed that 100 d after immunization, memory B cells are still proliferating. Am I missing something fundamental, or could these not be recent bone marrow emigrants which start cycling?

Zinkernagel: I wouldn't commit myself on that. All I can say is that in this location we found B cells that proliferated.

Kirberg: It would be interesting to see whether B cell memory absolutely requires ongoing proliferation. One could transfer memory B cells into B cell-deficient, antigen-free mice and see what happens.

Paul: Going back to the normal situation that humans encounter after either vaccination or normal infection, we know that immunity to many viruses is extremely durable. Your position is that under those circumstances local infection occurs normally and what you are blocking is dissemination through tissues to distant infection sites. Does this not imply that true protection against disease is based upon reactivation of CTL memory cells or precursors of some other effector cell? That is, the assessment of such cells is not truly unphysiological except when you use a viral challenge in which the disease occurs with a pace in which reactivation cannot account for control.

Zinkernagel: What you have stated is more-or-less true, but one has to look at each example separately. There are two broad distinctions between viruses: cytopathic versus non-cytopathic ones, and those that enter the host via mucosa or blood versus those entering via skin or directly via a peripheral solid organ. Cytopathic viruses are taken care of by soluble factors, mostly antibodies, and all the viruses that have been successfully vaccinated against belong to this category. HIV doesn't fit, because in this case both antibodies and T cells are needed, as in certain other infections, e.g. smallpox. In the case of smallpox, infection can occur both via mucosal surfaces and via skin lacerations. Against the former, antibodies protect; against the latter you need activated T cells which are antigen-dependent.

Answering your question in another way, measles has always been taken as a prime example where it is unreasonable to ask for persistence of antigen for 70 years, because of the epidemiological evidence from the South Pacific or the Faroe Islands. Epidemiology has shown that protective immunity is maintained even with a gap of 60 years between epidemics. People looked for replicating virus in surviving patients and never found it. It turns out that this was the wrong thing to look for. Katayama et al (1995) recently analysed autopsies of brain sections from old people using probes for measles matrix or nuclear protein; PCR revealed that about 20% of 50 autopsies were positive for measles message.

Paul: The key is not that but rather whether they have activated cells.

Zinkernagel: No; the key issue is whether they still have elevated antibody levels, and they do. They may also still have activated T cells but to my knowledge this has never been checked.

Reference

Katayama Y, Hotta H, Nishimura A, Tatsuno Y, Homma M 1995 Detection of measles virus nucleoprotein messenger RNA in autopsied brain tissues. J Gen Virol 76:3201–3204

Dendritic cells and T lymphocytes: developmental and functional interactions

Ken Shortman, Li Wu, Gabriele Süss, Vadim Kronin, Ken Winkel, Dolores Saunders and David Vremec

The Walter and Eliza Hall Institute of Medical Research, Post Office, Royal Melbourne Hospital, Victoria 3050, Australia

Abstract. Dendritic cells (DCs) are specialized for presentation of antigen to T cells and are essential for primary T cell activation. Although DCs are generally considered to be myeloid derived, we now have evidence that a subgroup are of lymphoid origin. In particular, the DCs of the adult mouse thymus appear to be derived from the same early, lymphoid-restricted precursor cells that generate T lymphocytes. Purified early thymic T precursors have the capacity to produce T cells, B cells, NK cells and DCs, but not myeloid cells, on transfer to irradiated recipients. They also produce thymic DCs on culture with a mix of cytokines; this mix does not include GM-CSF, needed to generate myeloid-derived DCs. A subgroup of DCs in other lymphoid organs, which like thymic DCs express CD8 as an $\alpha\alpha$ homodimer, may likewise be of lymphoid origin. These $CD8^+$ DCs in mouse spleen differ functionally from the conventional $CD8^-$ DCs. $CD8^+$ DCs efficiently activate $CD4^+$ T cells but then kill them via Fas ligand on the DC surface. $CD8^+$ DCs efficiently recruit $CD8^+$ T cells into the cell cycle, but their proliferation is then restricted by an inadequate production of interleukin 2. This subgroup of $CD8^+$ DCs therefore appears to have a regulatory role.

1997 The molecular basis of cellular defence mechanisms. Wiley, Chichester (Ciba Foundation Symposium 204) p 130–141

Dendritic cells (DCs) are a minor component of lymphoid tissue but are essential for the functioning of the immune system (Steinman 1991). They are specialized for interaction with T cells and for presentation to them of peptide antigens. DCs in peripheral lymphoid organs initiate immune responses to foreign antigens and are probably the only antigen presenting cells able to activate primary T cells (Knight & Stagg 1993, Epstein et al 1995). In contrast, DCs in the thymus mediate the death of any developing T cells that respond to self-antigens (Fairchild & Austyn 1980). DCs are migratory, bone marrow-derived cells and are generally considered to be related to or derived from monocytes and macrophages. There is evidence for a common progenitor of granulocytes, macrophages and DCs in bone marrow (Inaba et al 1993). Recently a DC-committed precursor has also been identified (Young et al

1995), presumably reflecting a later developmental step. A form of DC can be generated directly from blood monocytes (Rossi et al 1992). Most procedures for the generation of DCs in culture require the myeloid hormone GM-CSF (granulocyte/macrophage colony-stimulating factor) (Caux & Banchereau 1996, Romani et al 1994). In view of this support for a myeloid origin, it is surprising that our own studies have indicated a lymphoid origin for a particular subgroup of DCs. We consider this type of DC to represent a separate lineage and to have a role in regulating T cell responses.

Our initial interest was in the development of DCs and T cells within the thymus. Some years ago this laboratory isolated what appears to be the earliest T-precursor cell in the adult mouse thymus (Wu et al 1991a). This minute population resembled bone marrow multipotent haemopoietic stem cells (BMSCs) in having T cell antigen receptor (TCR) genes in germline state, and in the expression of many surface markers (c-Kit^{+}, Thy-1^{lo}, Sca-1^{+}, CD44^{hi}, negative for most lineage markers). However, it differed from BMSCs in expressing the Ly6 family member Sca-2 (Classon & Coverdale 1994) and in transient expression of low levels of CD4. We termed it the 'low CD4 precursor'. By injecting these precursors intravenously into irradiated recipient mice differing in Ly5 allotype, we were able to track their progeny and so determine their developmental potential. The thymic low CD4 precursors differed from BMSCs in having lost the capacity to form macrophages, granulocytes or erythrocytes. However, they were able to form not only T cells but also B cells (Wu et al 1991b). Others have since confirmed these findings and indicated that these early thymic precursors can also form NK cells (Godfrey & Zlotnik 1993, Matsuzaki et al 1993). Despite the lack of strict clonal evidence, we began to consider these as 'lymphoid-restricted' precursors. However, their potential to form B cells and NK cells would not normally be expressed in the thymic environment, since it was revealed by intravenous but not by intrathymic transfer.

The thymic lymphoid precursor did not produce any detectable macrophages or granulocytes when injected directly into the thymus. This contrasts with BMSCs which do produce such myeloid progeny after intrathymic injection, demonstrating that the thymic environment will support myeloid development. Yet this apparently lymphoid-committed thymic precursor did produce DC progeny when injected into a recipient thymus (Ardavin et al 1993, Wu et al 1995). These DC progeny had all the surface markers and morphological characteristics of normal thymic DCs. The kinetics of DC and T-lineage cell development following reconstitution of an irradiated thymus with the 'low CD4' precursor is shown in Fig. 1. This makes the following points:

(1) Reconstitution was transient, the lifespan of the DC progeny being as short as that of the thymocyte progeny. The low CD4 precursor cell clearly has a restricted self-renewal capacity.
(2) Over 1000 T-lineage thymocytes were produced for each DC. However, this is close to the normal balance in the thymus (1 DC per 2000 thymocytes). Hence it

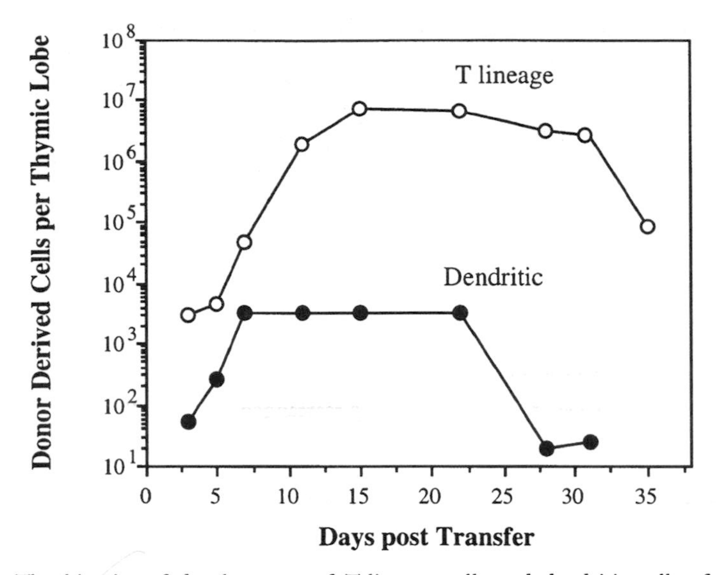

FIG. 1. The kinetics of development of T-lineage cells and dendritic cells after thymic reconstitution with the 'low CD4' precursor. The precursors were purified from the thymus of Ly5.2 mice and 10^4 cells were injected into one thymic lobe of each of 10 irradiated Ly5.1 mice for each time point. At various times after transfer the injected thymic lobes were pooled and cell suspensions prepared by collagenase and EDTA treatment. T lineage progenitors were assays directly on the suspensions as Thy-1^+ Ly5.2^+ cells. DC progenitors were assayed as class II MHC high Ly5.2^+ cells after depletion of T-lineage cells and enrichment of the DCs. The figure is reproduced from Wu et al (1995) which gives full experimental details.

is likely that thymic DCs normally derive from this intrathymic precursor, rather than arriving pre-formed from the periphery.

(3) Each cohort of developing T cells in the thymus was accompanied by a parallel cohort of thymic DCs, newly formed, endogenously produced and short-lived. This seems a good mechanism for ensuring that negative selection of developing T cells by thymic DCs is restricted to endogenously produced self-antigens, rather than to foreign antigens collected in the skin or the gut and then transported into the thymus.

How closely is thymic DC development linked to T cell development? Table 1 summarizes current data on the developmental potential of various precursors on the mainstream leading to T cell development. A precursor later than the low CD4 population is the $CD4^- 8^- 25^+ 44^+$c-Kit^+ 'pro-T' cell, which still has TCR β and α genes in germline state. Even though this population has lost the capacity to form B cells and NK cells (Godfrey & Zlotnik 1993, Zuniga-Pflücker et al 1995), it retains full potential to produce thymic DCs (L. Wu, unpublished results 1995). However, one step later the $CD4^- 8^- 25^+ 44^{lo}$ c-Kit^- 'pre-T' cell, which now has rearranged TCR β

TABLE 1 Developmental potential of T cell precursor populations

Precursor population	*Surface phenotype*	*TCR gene status*	*Developmental potential*
Multipotent stem cell bone marrow	$CD4^-8^-$ $c\text{-}Kit^+Sca\text{-}2^-Thy\text{-}1^{lo}$ $CD44^{hi}25^-$	β germline α germline	Erythroid, myeloid, T, B, NK, DC
Low CD4 precursor thymus	$CD4^{lo}8^-$ $c\text{-}Kit^+Sca\text{-}2^+Thy\text{-}1^{lo}$ $CD44^{hi}25^-$	β germline α germline	T, B, NK, DC
Pro-T thymus	$CD4^-8^-$ $c\text{-}Kit^+Sca\text{-}2^+Thy\text{-}1^{int}$ $CD44^{hi}25^+$	β germline α germline	T, DC
Pre-T thymus	$CD4^-8^-$ $c\text{-}Kit^-Sca\text{-}2^+Thy\text{-}1^{hi}$ $CD44^{lo}25^+$	β rearranged α germline	T
Double-positive cortical thymocyte	$CD4^+8^+$ $c\text{-}Kit^-Sca\text{-}2^+Thy\text{-}1^{hi}$ $CD44^-25^-$	β rearranged α rearranged	T

Data summarized from Wu et al (1991), L. Wu, unpublished work (1995), Godfrey & Zlotnik (1993), Matsuzaki et al (1993) and Zuniga-Pflücker et al (1995).

genes, no longer produces DCs. This all argues for a close linkage and a relatively late choice between thymic DC and T cell development, with TCR β-gene rearrangement being the best indicator of T-lineage commitment.

These studies on DC development, using intrathymic injection of purified precursors followed by tracking the rare DC progeny using the Ly5 marker, are tedious and expensive! A system for DC development in culture would be far more amenable to study. We have now succeeded in growing cells with the surface phenotype and morphology of thymic DCs from purified low CD4 precursors; in fact we have had more success growing DCs from these T precursors than in growing T cells. Although culture with various combinations of one to three cytokines was ineffective, a cocktail of seven cytokines gave after 2–4 d a substantial increase in cell numbers. The product cells, which formed large clusters, had both the surface antigenic phenotype and the morphological appearance of thymic DCs. The optimal mix of cytokines is still being determined (D. Saunders, K. Lucas & K. Shortman, unpublished work 1996). However, it is already clear that in marked contrast to other studies on DC development in culture, this production of DCs does not require GM-CSF, is not enhanced by the addition of GM-CSF, and is not inhibited by the addition of antibodies against GM-CSF. This is further evidence that the type of DC we are producing in culture from this thymic precursor is not the usual, myeloid-derived DC.

We still lack clonal evidence that T cells and DCs can arise from one single lymphoid-restricted precursor cell. Is there other evidence which indicates that these thymic DCs are of lymphoid origin? The presence on the surface of thymic DCs of some markers characteristic of lymphocytes points in that direction, although care must be taken to distinguish molecules that are merely picked up by the DCs from those that are synthesized by the DCs themselves (Table 2). The Thy-1 seen on the surface of thymic DCs is clearly derived from T-lineage thymocytes and is not a true DC product (Shortman et al 1995, Wu et al 1995). However, the early B cell marker BP-1 and the mature T cell marker CD8α are integral components of the thymic DCs, since both protein and mRNA have been demonstrated in the purified DCs. The CD8 on murine DCs is expressed at levels as high as on T cells, but is in the form of an $\alpha\alpha$ homodimer as on $\gamma\delta$-TCR T cells, rather than the $\alpha\beta$ heterodimer characteristic of most conventional $\alpha\beta$-TCR T cells (Wu et al 1995, Vremec et al 1992).

Are DCs with properties similar to these putative lymphoid-derived thymic DCs found outside the thymus? CD8α-bearing DCs are found as a major subpopulation of splenic DCs (Fig. 2) (Vremec et al 1992) and as a minor subpopulation of lymph node DCs. CD8$^+$ DCs had been detected previously by others, but only as a minor subpopulation. We believe our isolation procedure, involving collagenase digestion and EDTA treatment, more efficiently extracts these CD8$^+$ DCs from tissues. These CD8$^+$ splenic DCs are not of thymic origin, since they are present in the spleens of athymic (nude) mice. Nevertheless, they may derive from a type of precursor cell similar to that found in the thymus. If our thymic low CD4 precursor is injected intravenously into irradiated recipients, it produces progeny in the spleen as well as in the thymus. Progeny T cells, B cells and DCs are all found in the spleen, but no progeny myeloid cells (L. Wu & K. Shortman, unpublished work 1995). If the Ly5

TABLE 2 Lymphoid markers on mouse thymic dendritic cells

Marker	*Evidence*
Integral components	
CD8α	Maintained on culture mRNA present
BP-1	Maintained on culture mRNA present
CD25 (IL-2Rα)	Induced on culture of pure DC
CD2 (?)	Maintained on culture
Absorbed on surface	
Thy-1 (moderate)	Reduced on culture Thy 1.1$^+$ and 1.2$^+$ in chimeras
CD8β (moderate)	Reduced on culture mRNA not detected
CD4 (low)	Lost on culture

Data summarized from Vremec et al (1992), Wu et al (1995) and Shortman et al (1995).

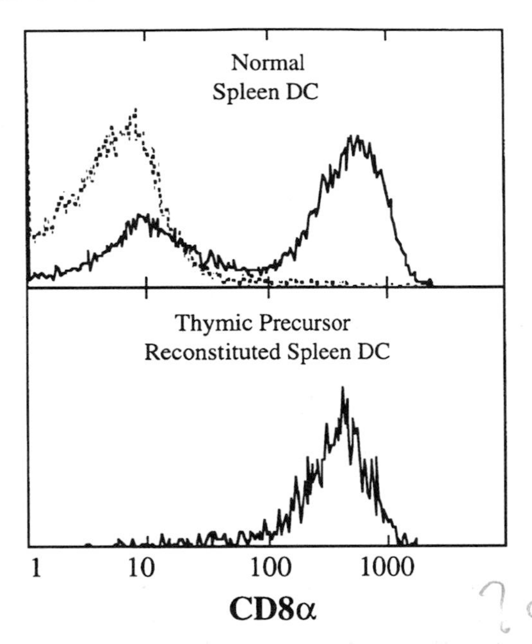

FIG. 2. The expression of CD8α on normal and thymic-precursor reconstituted splenic DCs. The upper panel shows the distribution of CD8α on class II MHC high DCs purified from normal C57BL/6 mouse spleen; the broken line is the background fluorescence. Both $CD8^+$ and $CD8^-$ DCs are present. The lower panel shows the distribution of CD8α on class II MHC high $Ly5.2^+$ DCs purified from the spleens of irradiated Ly5.1 mice that had been reconstituted 4 weeks previously with 2×10^4 low CD4 precursor cells isolated from Ly5.2 mouse thymus. Only $CD8^+$ DCs are detected, at this and other times post-reconstitution.

marker is used to selectively analyse the thymic precursor-derived DCs in the recipient spleen, they are all $CD8^+$, just like thymic DCs (Fig. 2). In contrast, BMSCs produce both $CD8^-$ and $CD8^+$ DCs if injected intravenously. On the basis of these observations it seems possible that CD8α marks a separate, lymphoid-derived DC lineage. The $CD8^-$ DCs may represent the conventional, myeloid-derived DCs.

Do these distinct populations of splenic DCs—$CD8^+$ and $CD8^-$—differ in function? Our preliminary investigation showed they both expressed high levels of major histocompatibility complex (MHC) proteins (both class I and class II) and of co-stimulatory molecules (B7.1 and B7.2). Both proved capable of activating mature peripheral T cells in culture. However, the $CD8^+$ and $CD8^-$ DCs differed markedly in the extent of T cell proliferation they produced. The $CD8^+$ DCs appear to be equipped with mechanisms which regulate or restrict T cell responses, and these mechanisms differ between $CD4^+$ and $CD8^+$ T cells.

The extent of T cell proliferation following culture of pure $CD4^+$ T cells with pure allogeneic splenic $CD8^-$ or $CD8^+$ DCs is shown in Fig. 3. $CD8^+$ DCs induce much less T cell proliferation than $CD8^-$ DCs, and this is evident early in the response. We have

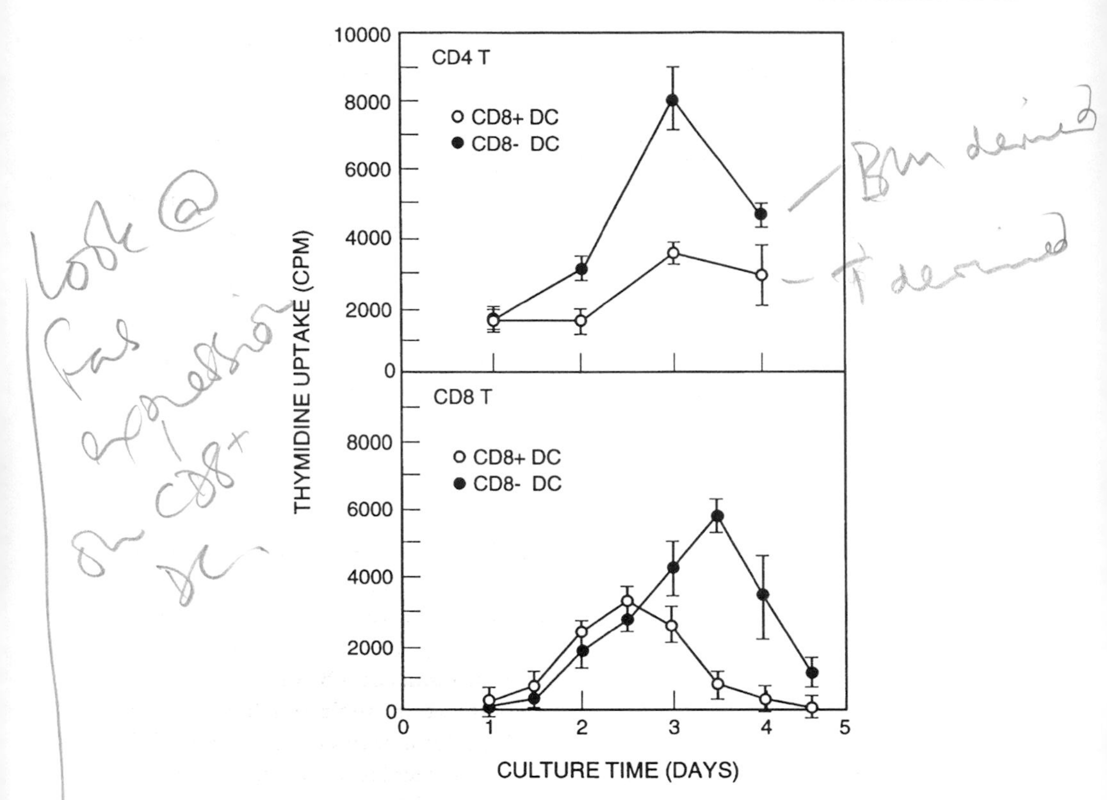

FIG. 3. Kinetics of the proliferative response of $CD4^+$ and $CD8^+$ T cells to $CD8^-$ and $CD8^+$ allogeneic splenic DCs. The responding $CD4^+$ or $CD8^+$ T cells were purified from lymph nodes of CBA mice. The stimulating $CD8^-$ or $CD8^+$ DCs were purified from the spleens of C57BL/6 mice. The T cells (20 000) and DCs (500) were cultured together in 0.2 ml medium in V well trays, then pulsed with $[^3H]$TdR at the times indicated. Incorporation of $[^3H]$TdR was measured by gas-phase scintillation counting. Cultures of T cells or DCs alone gave <100 cpm. The reduced proliferative response to $CD8^+$ DCs throughout the response (for $CD4^+$ T cells) or late in the response (for $CD8^+$ T cells) was obtained consistently with DC numbers ranging from 200 to 2000 per culture.

found this is due to Fas-induced apoptotic death of many of the T cells on interaction with $CD8^+$ DCs. If $CD4^+$ T cells from Fas-deficient *lpr* mutant mice are used as responders, the proliferation induced by $CD8^+$ DCs is then equivalent to that induced by $CD8^-$ DCs. The killing of the T cells is signalled by the Fas ligand present on the $CD8^+$ DCs. We have shown by direct staining that freshly isolated murine splenic $CD8^+$ DCs express Fas ligand at high levels, whereas $CD8^-$ DCs are either negative or express low levels (Süss & Shortman 1996).

The culture of pure $CD8^+$ T cells with pure allogeneic splenic $CD8^-$ and $CD8^+$ DCs is also shown in Fig. 3. Again $CD8^+$ DCs produce much less T cell proliferation than

CD8$^-$DCs, but in contrast to the situation with CD4$^+$ T cells, this difference is only apparent late in the response. T cell death is not evident in these cultures, the CD8$^+$ T cells being apparently resistant to the Fas ligand on the CD8$^+$ DCs. In contrast to the behaviour of CD4$^+$ T cells, CD8$^+$ T cells from the *lpr* mutant mice show the same differences in response to CD8$^+$ and CD8$^-$ DCs as do normal CD8$^+$ T cells. In the case of CD8$^+$ T cells, the difference in proliferation is a consequence of inadequate endogenous interleukin (IL)-2 production by the T cells on stimulation with CD8$^+$ DCs. IL-2 can be detected in the supernatants of cultures stimulated with CD8$^-$ DCs, but not in those stimulated with CD8$^+$ DC. In contrast to results with CD4$^+$ T cells, the differences in proliferation of CD8$^+$ T cells to the two types of DC is totally abolished by the addition of exogenous IL-2 to the cultures (V. Kronin, K. Winkle, G. Süss & K. Shortman, unpublished work 1995). This is an interesting case of a dissociation between the stimulus for cell division and the stimulus for cytokine production.

Conclusions

Overall, these results suggest that splenic CD8$^+$ DCs are equipped with mechanisms to regulate T cell responses. Whether this is in order to control responses to foreign antigens, or to prevent responses to self antigens, is not yet clear; we must now determine whether both of these types of DC collect, process and present foreign antigens to T cells. We have not yet integrated these results on DC function (which have concerned splenic DCs and their interaction with mature, peripheral T cells) with our studies on the lineage origin of DCs (which have focused on thymic DC development). But it now seems likely that there are distinct DC lineages, that as well as the myeloid-derived DCs there are lymphoid-derived DCs, and that these distinct DC types will differ in the signals they transmit while presenting antigenic peptides to T cells.

Acknowledgements

This work was supported by the National Health and Medical Research Council, Australia, by the Human Frontier Science Program Grant Organization and by the Cooperative Research Centre for Vaccine Technology.

References

Ardavin C, Wu L, Li C-L, Shortman K 1993 Thymic dendritic cells and T cells develop simultaneously within the thymus from a common precursor population. Nature 362:761 – 763

Caux C, Banchereau J 1996 *In vitro* regulation of dendritic cell development and function. In: Whetton AD, Gordon J (eds) Blood cell biochemistry. Plenum, New York

Classon BJ, Coverdale L 1994 Mouse stem cell antigen Sca-2 is a member of the Ly-6 family of cell surface proteins. Proc Natl Acad Sci USA 91:5296–5300

Epstein MM, Di Rosa F, Jankovic D, Sher A, Matzinger P 1995 Successful T cell priming in B cell-deficient mice. J Exp Med 182:915–922

Fairchild PJ, Austyn JM 1990 Thymic dendritic cells: phenotype and function. Int Rev Immunol 6:187–196

Godfrey DI, Zlotnik A 1993 Control points in early T-cell development. Immunol Today 14:547–553

Inaba K, Inaba M, Deguchi M et al 1993 Granulocytes, macrophages, and dendritic cells arise from a common major histocompatibility complex class II-negative progenitor in mouse bone marrow. Proc Natl Acad Sci USA 90:3038–3042

Knight SC, Stagg AJ 1993 Antigen-presenting cell types. Curr Opin Immunol 5:374–382

Matsuzaki Y, Gyotoku J, Ogawa M et al 1993 Characterization of c-Kit positive intrathymic stem cells that are restricted to lymphoid differentiation. J Exp Med 178:1283–1291

Romani N, Gruners S, Brang D et al 1994 Proliferating dendritic cell progenitors in human blood. J Exp Med 180:83–93

Rossi G, Heveker N, Thiele B, Gelderblom H, Steinbach F 1992 Development of a Langerhans cell phenotype from peripheral blood monocytes. Immunol Lett 31:189–198

Shortman K, Wu L, Ardavin C et al 1995 Thymic dendritic cells: surface phenotype, developmental origin and function. In: Banchereau J, Schmitt D (eds) Dendritic cells in fundamental and clinical immunology. Plenum, New York, p 21–29

Steinman RM 1991 The dendritic cell system and its role in immunogenicity. Annu Rev Immunol 9:271–296

Süss G, Shortman K 1996 A subclass of dendritic cells kills CD4 T cells via Fas/Fas-ligand induced apoptosis. J Exp Med 183:1789–1796

Vremec D, Zorbas M, Scollay R et al 1992 The surface phenotype of dendritic cells purified from mouse thymus and spleen: investigation of the CD8 expression by a subpopulation of dendritic cells. J Exp Med 176:47–58

Wu L, Scollay R, Egerton M, Pearse M, Spangrude GJ, Shortman K 1991a CD4 expressed on earliest T-lineage precursor cells in the adult murine thymus. Nature 349:71–74

Wu L, Antica M, Johnson GR, Scollay R, Shortman K 1991b Developmental potential of the earliest precursor cells from the adult thymus. J Exp Med 174:1617–1627

Wu L, Vremec D, Ardavin C et al 1995 Mouse thymus dendritic cells: kinetics of development and changes in surface markers during maturation. Eur J Immunol 25:418–425

Young JW, Szabolcs P, Moore MAS 1995 Identification of dendritic cell colony-forming units among normal human $CD34^+$ bone marrow progenitors that are expanded by c-kit ligand and yield pure dendritic cell colonies in the presence of granulocyte–macrophage colony-stimulating factor and tumor necrosis factor. J Exp Med 182:1111–1119

Zuniga-Pflücker JC, Jiang D, Lenardo MJ 1995 Requirement for TNF-α and IL-1α in fetal thymocyte commitment and differentiation. Science 268:1906–1909

DISCUSSION

Miller: If you pulse spleen DCs with an antigen such as ovalbumin and then use both types of DCs to activate T cells from transgenic mice, is the response different when using those that are positive for CD8 and those that are negative?

Shortman: Yes. The results I have shown for alloantigen stimulation of the $CD4^+$ T cells have been duplicated by Gabi Süss using Harald von Boehmer's class II MHC-restricted, haemagluttinin-specific, T cell receptor transgenic T cells. The results were almost identical to those I have shown. The $CD8^+$ DCs always give much lower responses than the $CD8^-$ DCs.

Miller: So you can actually stimulate by antigen-pulsing the DCs of the $CD8^-$ type.

Shortman: Yes; so far it has all been done by pulsing the DCs with peptide, rather than asking the DCs to process the protein antigen.

Miller: If you inject soluble antigens into a mouse, do you find evidence that the antigen is present on the DCs in the thymus?

Shortman: We are trying to conduct those experiments, but we are having difficulty even detecting antigen-labelled DCs in the spleen with conventional fluorescent labelling, let alone the thymus. I don't know whether we are going to get enough sensitivity out of that sort of procedure. We are in the middle of experiments to find out which DCs process which antigen.

Rajewsky: When you tested these peripheral thymus-type DCs for peptide presentation, they had lower activity than the classical DCs. Was this because they simply didn't present or because they killed?

Shortman: We have so far avoided testing their antigen processing ability. We have bypassed that step by loading them with peptide. The levels of class I and II MHC on the two types of DCs are identical, so I am assuming that their peptide loading was equivalent. The difference in the response to the two types of DC was exactly the same for the loaded peptide antigen as for alloantigen.

Carbone: Do you expect these DCs to process protein and present it for class II-restricted presentation?

Shortman: Yes, in the sense that a response to alloantigen is a sort of summation of multiple responses to already processed peptides.

Carbone: None the less, DCs are generally considered to be inefficient at actually taking up and presenting exogenous antigen.

Shortman: The experiments to test this are going to be hard to do, because it is known that during the life history of DCs there is a phase where they pick up and process antigen, followed by a second phase after their migration to, say, lymph nodes, when they are no longer able to process antigen but instead become able to stimulate T cells. Simply taking them out from one lymphoid organ and testing at that stage of their life history would not be adequate; you would somehow have to load up the animal with antigen and then allow time for further DC maturation before taking the DCs out for testing.

Carbone: So are these the mature type of DCs?

Shortman: Yes. I would like it if $CD8^+$ DCs were to be unable to process foreign antigen whereas the $CD8^-$ DCs could. It would explain the difference.

Mathis: Do you know where they're found in the thymus?

Shortman: We haven't looked. They are meant to be in the medulla and at the cortico-medullary junction.

Mathis: Not in the cortex?

Shortman: I'm not sure. When you stain for class II MHC you don't usually see a classical DC stain in the cortex, but they could be present in a less mature or less evident form.

Metcalf: To what degree has the thymus educated the incoming population or restricted its potential? If you start with bone marrow and apply selective types of culture, is it possible to generate both myeloid-derived and lymphoid-derived DCs?

Shortman: We don't know. We are only just finishing the study culturing the thymic precursors. Our next program is to mix and match precursors from different sources and different types of cytokines. We want to find out whether it is the nature of the precursor or the cytokine mix that determines the type of DC we end up with.

Metcalf: This is important. T cell biologists hold that the thymus is the site of elimination of self-reactive cells. One could postulate that the behaviour of DCs in the thymus is the consequence of living in this very special organ.

Shortman: If you inject a multipotent stem cell into the thymus and then look at the type of DCs that develop, they are exactly the same as you get from the intrathymic precursor, the $CD8^+$ DC. Within the thymic environment, the $CD8^-$ cells don't seem to develop to any great extent.

Mosmann: There is a very large literature on $CD8^+$ suppressor cells from peripheral organs. Have you looked at some of these old publications to see whether your cells could account for this phenomenon?

Shortman: Yes. We wondered whether the Miller-Basten type of suppressor cell from yesteryear was in fact not a $CD8^+$ T cell, but rather a $CD8^+$ DC expressing a few T cell antigens. However, in that case, the answer was clearly negative. $CD8^+$ DCs were not the suppressors. One of the reasons I started thinking this way was because of the work of Sambhara & Miller (1991), which suggested that if an antigen-presenting cell expresses CD8, this molecule interacts with a class I on the T cell, gives a negative signal, and turns an antigen-presenting cell into a veto cell. As a result of testing this theory, we found out that there was a difference between the $CD8^+$ and $CD8^-$ DCs. However, this difference has nothing to do with CD8. We have carried out the same experiments with CD8 knockout mice, separating the same two populations using a surrogate marker which detects the same two DC subsets. The results are unchanged; the difference in response to the DCs persists. It is the Fas ligand that matters, not the CD8. The latter appears to be an accidental marker of this lineage.

Strasser: You seem to think that Flk ligand is essential for the formation of at least one type of DC. Do you know whether the Flk ligand knockout mice generated by Immunex are severely deficient in these DCs?

Shortman: I don't know anything about the Flk ligand knockouts. In the Immunex studies, the Flk ligand enhances the level of DCs about 17-fold. In our cultures, Flk ligand is not obligatory for DC development. It enhances the development about three to fourfold, but Flk ligand alone won't drive it.

Tarlinton: Can one subdivide human DCs with Fas ligand in the same way as you have done in the mouse? Do they have the same functional potential that you have seen?

Shortman: That is an important question, and we're trying to answer it. Fas ligand expression should distinguish the DCs. Unfortunately, our Fas–Fc construct involves human Fc, so staining with it has been giving a high background. We have shown heterogeneity in human DCs by several markers, including CD4. We are separating these different DCs to test whether we have a corresponding situation in the human.

Goodnow: Do the DCs that crawl out of skin explants, presumably from Langerhans cells, include these cells? And are these cells affected in the Rel B knockout mice?

Shortman: I have been told that those that crawl out of skin are of the CD8$^-$ type. This may be a nice source of pure myeloid-derived DCs. In the Rel B knockout mice our type of DC may be missing; we are about to study this in collaboration with D. Lo. We have also begun collaborative experiments with K. Georgopoulos on the Ikaros knockout mice. These animals have no lymphoid cells. They do have Langerhans cells in the skin, but so far we have found no DCs in the thymus or the spleen.

Melchers: The fact that you don't find DCs in the Ikaros knockout mouse doesn't prove that they can't develop. I thought that to a certain extent the lymphocytes were stimulating the DCs to develop, so that mutual interaction was necessary for the development of both of these lineages.

Shortman: I agree; this is one of the reasons why Li Wu is repeating these experiments with K. Georgopoulos. She will make bone-marrow chimeras using mice differing at the Ly5 locus, and test whether Ikaros-mutant bone marrow can produce DCs if normal T cell development occurs in the mice.

Miller: Do the CD8$^+$ DCs express molecules of the tumour necrosis factor (TNF) receptor family on their surface?

Shortman: I don't know about the TNF receptor. They may have a whole spectrum of similar molecules to the Fas ligand.

Hodgkin: Then are the T cells from the CD8$^+$ DC-stimulated cultures Fas positive?

Shortman: We haven't looked to see whether there is any selectivity in Fas expression between the T cells that die and those that don't. We don't know the answer, but that is a very good question.

Reference

Sambhara SR, Miller RG 1991 Programmed cell death of T cells signalled by the T cell receptor and the α_3 domain of class I MHC. Science 252:1424–1427

General discussion IV

Immune recognition, control of infections and vaccination strategies

Nossal: So far this morning we have heard from Mark Davis about a recognition phenomenon that depends on T cell–target cell interaction with great specificity. We have had a challenge meted out to us that this specificity is generated by a repertoire which is largely made up of D genes and junctional diversity, in other words CDR3 of the T cell receptor. In this view, CDR1 and 2 are slightly incidental to the main reaction.

We've also had a paper that has genuinely challenged our reductionistic approaches towards the understanding of real life immunity. Rolf Zinkernagel's paper has posed for me quite serious questions about immunization strategies in situations when you are close to elimination of a disease from the globe. He has stressed repeated exposure as a way of boosting immunological memory, but when a disease is eradicated from a country, there can be no such boosts. Recently, *Haemophilus influenzae* B vaccination has been introduced in many industrialized countries to combat meningitis. Part of its spectacular success has been the degree to which it has removed the burden of the reinfecting organism from the susceptible population. If the challenge that Rolf Zinkernagel points out to us is a general one, the world is going to face some real difficulty as we begin to approach disease eradication. Polio and measles will be eliminated from most of the world, but while pockets persist, the risk of imported infections remains. If Rolf is right and memory in the sense of being protective fails at the rate he has described, we are going to face some daunting challenges. Clearly, artificial booster doses of vaccine will be needed until the germ involved is totally eradicated from the globe.

Finally, Ken Shortman has presented us with a beautiful interface between developmental haematology and immunology.

Hodgkin: Concerning the high affinity antibody response Rolf described that occurs within 6 d — how important for this response is the highly crystalline structure of the particle? If you immunize with an uncrystallized structure, presumably the antigen now presents a lot more B cell epitopes: do you then see much delay in the acquisition of a high affinity antibody response?

Zinkernagel: We have not sequenced monoclonal antibodies derived from immunization with monomers. If we look at protection after immunization with monomers coupled to a carrier and complete or incomplete Freund's adjuvant, the protective capacity and the neutralizing capacity are comparable to the quality of the paracrystalline sera after immunization with the highly organized antigen form. It appears that the B cell response induced by the exposed neutralizing epitope has that high affinity quality.

Gus Nossal has raised the question of what will happen when the elimination of measles is achieved, paralleling the elimination of smallpox. I think this is a particularly interesting case, because there is epidemiological evidence that the measles vaccine strain currently being used actually is less attenuated than is generally thought. One would predict that a peptide without adjuvant or special packaging would be a poor vaccine. An attenuated but still reasonably virulent vaccine like the measles vaccine will probably do a reasonable job for a fairly long time.

However, a few years ago, for various reasons measles protection achieved with the vaccine was relatively poor, particularly in Africa. The World Health Organization therefore upped the vaccine doses; as a consequence the vaccinated children became sick or, overall, did less well than before.

Nossal: What actually happened was that people tried to address the question of measles mortality in infants less than nine months old. Between four and six months of age maternal immunity wanes. In industrialized nations children are vaccinated at 12 months, but in the Third World the vaccine is given at nine months. This leaves a window of tremendous vulnerability between four to six months and nine months. As a consequence, people raised the dose of the live attenuated vaccine in the hope of therefore being able to give it earlier. Unexpectedly, this actually *increased* the mortality of those vaccinated children over the next couple of years. There were many peculiar aspects to this. One of them was that it was gender specific—it happened more in the girls. Secondly, the increased mortality was due to death from a total miscellany of causes; certainly not specifically deaths from measles. Whatever the cause, this way of tackling the problem can obviously not be pursued any further.

Much better examples of the dilemma I was describing would be *Haemophilus influenzae* B, a conjugate subunit vaccine, and hepatitis B, which is also a molecular vaccine. With these, if Rolf Zinkernagel is right, effective protection should fade reasonably quickly.

In classical thinking, re-exposure to small amounts of antigen, i.e. encounters with the pathogen leading only to very limited replication, is what gives the boost. In situations where the total disease burden in a community has been so affected by effective vaccination, you can't rely on this re-exposure. That's where Rolf's claim of the failing nature of protection becomes exceedingly important. I would agree with the point made in the discussion after Rolf's paper that he still bears a certain burden of proof. He has to generalize the observations from the vesicular stomatitis virus model that he has presented.

Tarlinton: I have a point concerning Mark Davis' theoretical assertion that most specificity of immunoglobulin for antigen is due to CDR3 of the heavy chain, and the two examples he gave, one from Chris Goodnow's work with a fixed heavy chain (Hartley & Goodnow 1994) and the other from the transgenic immunoglobulin heavy chain mini-locus. It could be that in both of those instances the data actually support the opposite contention: that the combination of heavy and light chain was most important. In one case half of the B cell repertoire is completely fixed, and yet still only 5% of the cells react with the antigen hen egg lysozyme (HEL), even though it's a highly mutated and selected V_H chain gene. In the second case, heavy chain CDR3 and the light chain are random and there is still reactivity to three different antigens. The best test would be to fix everything except for the CDR3 of the heavy chain.

Davis: I agree. Those examples are not proof of my assertion, although they suggest it may be the case. The transgenic experiment is the weakest evidence that the heavy chain is dominant. I believe that the reason why 95% of the B cells don't bind HEL in Chris Goodnow's single chain transgenic mice is probably because the light chains are interfering, but one could also believe that it shows the necessity of a proper light chain with respect to Taylor et al (1994). The authors of that paper concluded that they were getting this unexpectedly high serum titre against all these antigens using the single human V_H because the light chain must be taking up the slack, but again I would suggest that this shows how versitile a V_H can be.

Rajewsky: Diane Mathis, how well do your terminal deoxynucleotidyltransferase (TdT) knockout mice behave in immune responses? After all, they lack this massive diversity in CDR3. According to what Mark Davis has said they should have impaired immune function.

Mathis: They're completely normal in their immune responses to everything that we've looked at, including several of Rolf Zinkernagel's viruses.

Nossal: Rolf, I got the impression that you were happy to accept from us that this early IgM response wouldn't mean a tremendously high serum titre?

Zinkernagel: No, it is the same range.

Nossal: Is Diane Mathis therefore presenting you with a problem?

Zinkernagel: We have looked at the neutralizing antibody response in these mice and they do fine. They produce protective antibodies of sufficient avidity and with other necessary qualities.

Mathis: We also looked at a response which is characterized by restricted V-region usage and has a specific residue in the CDR3 which requires N-region diversity to create it-this was also normal. The mice just switch to another V_β.

Davis: We are intending to study these mice, because this is a crucial experiment where we would expect to see a diminution of diversity. The experiments that have been done so far with them are very limited in terms of the antigens that have been looked at and specificity. However, Mike Bevan has reported that he sees a difference in the repertoire of positive selectives out of the thymus (Gavin & Bevan 1995).

Mathis: It is not that there aren't changes in the repertoire; rather, that there is no change in the efficacy of the response. Bevan's study did not really deal with this issue.

Davis: For a T cell receptor, we've estimated that there are something in the order of 10^{10} different amino acid sequences in the $\alpha\beta$ junctional regions. Of these, about 10^5 are due to N region diversification. This still leaves a lot of diversity in that region because of the way that the D regions and the junctions and so forth are all jumping around.

Nevertheless, I would like to see some difference in terms of the repertoire in TdT-deficient mice.

von Boehmer: These TdT-deficient mice seem to have a more efficient positive selection! Mark Davis, how do you fit positive selection into your model, and what are you actually selecting for in your model? If everything is determined by the CDR3, the CD3 just binds to the peptide and the MHC is more-or-less irrelevant, what is the purpose of positive selection?

Davis: It is hard to separate the variables. The different MHCs are holding collections of peptides differently. The polymorphism creates different pockets, each of which holds different categories of peptides in slightly different and unnatural ways. None of the peptides anyone has looked at is bound to the MHC in the same conformation as the peptide occurs in nature. People used to talk about conformations such as α-helices being important as immunogens for T cells, but in fact the peptide has only the structure that that particular MHC says it can have. Therefore, by definition, the MHC has enormous influence, even if the TCR receptor were not to contact the MHC at all. I happen to believe that the TCR is contacting the MHC: we know this from experiments in which we have made mutations on the surface of the MHC and have found that some can inhibit and some can enhance interaction with the TCR. The contacts exist, but they don't seem to contribute enormously to the specificity (Ehrich et al 1993). The peptide–CDR3 interaction in the thymus is still a key event, but we cannot say to what extent the T cell is actually taking notice of that particular MHC in terms of direct contact and to what extent the affect of restriction as a positive selection is due to the way particular peptides are held by the MHC.

Carbone: I was intrigued by the dominance of the peptide in T cell recognition over the MHC. If one changes the MHC peptide binding site this affects T cell recognition considerably more than changes at the MHC sites that directly interact with the TCR. Moreover, changes in the peptide-binding cleft seem to affect not only the recognition of the mature T cell, but also of the developing T cell. In the K^b-restricted ovalbumin response, if you introduce binding cleft mutations, not only do you abrogate the ovalbumin-specific T cell response, but you also eliminate the thymic selection of those T cells. Presumably there are two different peptides: the antigenic peptide and the selecting peptide. They are both affected by the same changes in the peptide-binding cleft. It seems that the peptide presentation is very dominant.

Davis: Everyone would love to have the natural ligand—presumably a peptide in thymic stromal cells—that is mediating the selection of some T cell receptor that you know would be one of the foreign antigens. The data indicate that some of these antagonist analogues can trigger possible selection. This makes one think of some weak interaction, like some of the antagonist peptides. In general, this suggests that there is a new category of peptides, which includes the foreign peptide as well as the selecting peptide, and that these have similar physical properties with respect to both TCR and MHC binding.

Mathis: We have a system where we have T cells which have been selected on a specific peptide. We can clone the T cells out, and in every case that we've looked at so far none of them are antagonized by the ligand which was used to select them. These results were obtained in *in vivo* systems where we employed an adenovirus vector to target a neo-peptide to the thymus.

Metcalf: We accept that lymphoid and haemopoietic cells have a common ancestor. In some situations, both populations behave in a similar manner but, when antigens are involved, lymphocytes then seem to respond to a bizarrely different regulatory control system. Haemopoietic cells respond to very low concentrations of regulators, involving very few receptors, and there is a subsequent cellular response, such as proliferation or functional activation. To a degree, lymphocytes respond to regulatory growth factors in a very similar manner. However, lymphocytes also respond to foreign antigens—a system in which there are much higher concentrations of antigen molecules and enormous numbers of low affinity membrane receptors. How did these two totally different regulatory control systems develop for lymphocytes? How many things had to change in a common stem cell, which, until then, was essentially behaving like a haemopoietic cell, to allow this bizarre behaviour of lymphocytes? It conjures up the possible necessity for some quite bizarre networks of inductive signalling.

What do people think is the key first step that heads a cell down the lymphoid pathway? For example, is it a commitment to transcribe and express massive numbers of antigen receptor chains? How many cell divisions are required and how many progeny are made before these changes are completed?

Shortman: There are several discrete selective steps that a cell must take along this pathway. Harald von Boehmer identified one of the first, which occurs after the TCRβ gene is rearranged. This is one of the reasons why it so hard to develop T cells in culture, because it involves a series of sequential controlled steps and a progenitor has to pass each gate before it moves on.

Metcalf: But you are describing membrane expression, and that requires transcriptional activation and many earlier events.

Melchers: You shouldn't even be worried about whether these cells make an antigen receptor or not, because they are all as short-lived as granulocytes, although they keep expressing these receptors up to the point when they are being selected either negatively or positively. It is at that point, somewhere in that interphase from an immature to a mature lymphocyte, that cells become different from granulocytes because they enter what is by cell biological standards a resting state, which is quite different from a granulocytic cell just dying, and they assume different circulatory pathways. Whether at that point they are already memory cells is another matter. It is the shape of a resting cell and the extension of the half-life of that cell that makes a difference. This is where the confusion enters. We are worried by the repertoires of lymphocytes being selected in that process, but if you were a haematologist and you looked at 99% of the bone marrow and 99% of the thymus, you would say the lymphoid cells die just as rapidly as granulocytes and are not even useful at that point of development, in contrast to the granulocytes which are.

Metcalf: My question was more concerned with the nature of the inducing signal that causes cells to take the first step towards the lymphoid lineages.

Strasser: Perhaps we can solve this by bringing everything back to cell cycle control. The types of lymphoid cells which respond in what you would call a 'normal' manner are early precursors which are generally not in a quiescent, post-mitotic G0 state. They might be in G1, S, G2 or M phase. IL-7 can push them through the cycle and they proliferate. Lymphocytes are the only cells that I know of that can go through a number of divisions, become post-mitotic and terminally differentiated as a resting T cell or resting B cell, and then be stimulated again later to divide a further 10 times. With all the other haemopoietic cells, once they are in G0 they either function or die.

Metcalf: This is not true for basophils or eosinophils; these apparently mature cells can be restimulated into cell cycle.

Strasser: Perhaps the critical function of antigen receptors is to get cells from the quiescent G0 state into G1. Whilst they are in G1 or S phase, the kind of cytokine receptors that you like to work with can kick in and do the rest of the business.

Paul: There are myeloid cells which behave in a simple way: for example, the mast cell. It has a cell surface receptor that interacts with an immunoglobulin ligand (Fc_{ε} or Fc_{γ}). The signalling pathways are interchangeable with the T cell and B cell pathways.

Goodnow: I don't see that there is any difference amongst the haemopoietic cells at the level of recognition of their targets. I would highlight the recognition of high mannose by macrophages, which is also a relatively low affinity affair buttressed by multi-site binding.

References

Ehrich EW, Devaux B, Rock EP, Jorgensen JL, Davis MM, Chien T-H 1993 T cell receptor interaction with peptide/MHC and superantigen/MHC ligands is dominated by antigen. J Exp Med 178:713–722

Gavin MA, Bevan MJ 1995 Increased peptide promiscuity provides a rationale for the lack of N regions in the neonatal T cell repertoire. Immunity 3:793–800

Hartley SB, Goodnow CC 1994 Censoring of self reactive B cells with a range of receptor affinities in transgenic mice expressing heavy chains for a lysozyme-specific antibody. Int Immunol 6:1417–1425

Taylor LD, Carmack CE, Huszar D et al 1994 Human immunoglobulin transgenes undergo rearrangement, somatic mutation and class switching in mice that lack endogenous IgM. Int Immunol 6:579–591

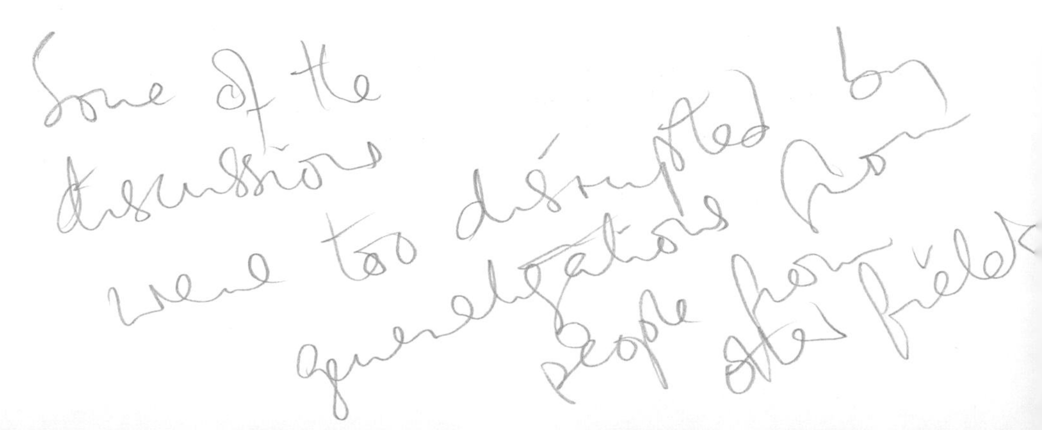

Differentiation and functions of T cell subsets

Tim R. Mosmann, Li Li, *Hans Hengartner, †David Kagi, Wayne Fu and Subash Sad

*Department of Medical Microbiology and Immunology, University of Alberta, Edmonton, †Ontario Cancer Institute, Toronto, Canada and *Institute of Experimental Immunology, Department of Pathology, University of Zurich, Switzerland*

Abstract. The Tc1 and Tc2 subsets of $CD8^+$ T effector cells secrete different patterns of cytokines, but have similar functions, including perforin- and Fas-dependent cytotoxicity, and induction of delayed type hypersensitivity (DTH) reactions involving oedema and granulocytic infiltration. The characteristic cytokines of Tc1 (γ-interferon) and Tc2 (interleukins 4 and 5) are expressed *in vivo* during the DTH reaction. Tc1 cells that are deficient in cytokine synthesis also induce similar levels of DTH, supporting the lack of correlation between $CD8^+$ T cell cytokine patterns and DTH. $CD8^+$ T cells often produce lower cytokine levels than CD4 cells because the CD8 cells kill their antigen-presenting cells before full stimulation can occur. This effect can be counteracted by increasing the frequency of stimulation, or using perforin-deficient T cells. A multiparameter analysis of cytokine effects on $CD8^+$ T cell differentiation has been initiated, on the basis of the principle that normal immune responses involve complex cytokine mixtures. All combinations of seven cytokines were tested. In some combinations, the combined effect could not have been predicted from individual cytokine functions. Conditions were identified in which each of interleukins 4, 10 and 12 could have opposite effects on $CD8^+$ T cell differentiation.

1997 The molecular basis of cellular defence mechanisms. Wiley, Chichester (Ciba Foundation Symposium 204) p 148–158

Subsets of $CD4^+$ T cells secreting different patterns of cytokines play an important role in determining the set of effector functions induced during immune responses against various pathogens. Although the reciprocal responses mediated by the T helper (Th) 1 and Th2 subsets are crucial for mediating resistance against some parasites (Mosmann & Sad 1996), the actual complexity of the immune system is considerably greater. $CD8^+$ T cell subsets have also been identified, and in this paper we summarize the functions and further differentiation of subsets of $CD8^+$ T cells, as well as preliminary results on the multiparameter analysis of cytokine effects on T cell differentiation.

$CD8^+$ T cell subsets — Tc1 and Tc2

Although mouse $CD8^+$ T cells have a strong tendency to differentiate into effector cells secreting the Th1 set of cytokines, stimulation of naïve $CD8^+$ T cells in the presence of interleukin (IL)-4 and antibodies against γ-interferon (IFN-γ) induces differentiation into CD8 effectors secreting Th2 cytokines (Croft et al 1994, Sad et al 1995). Both T cytotoxic (Tc) 1 and Tc2 subsets are relatively stable, and we have not been able to induce redifferentiation of either subset into the other phenotype. Both Tc1 and Tc2 cells are highly cytotoxic for cells bearing the target antigen, and both kill mainly by the perforin-dependent pathway, and to a lesser extent via Fas (Carter & Dutton 1995, S. Sad, L. Krishnan, R.C. Bleackley, D. Kagi, H. Hengartner & T. R. Mosmann, unpublished results 1995).

Although Tc2 cells secrete the same cytokines as Th2 cells, which are excellent B cell helpers, Tc2 cells do not help B cells in a cognate interaction. This is mainly due to their ability to kill target cells, including resting and activated B cells (S. Sad, L. Krishnan, R. C. Bleackley, D. Kagi, H. Hengartner & T. R. Mosmann, unpublished results 1995). Even in the absence of the perforin pathway, Tc2 cells provide only minimal cognate help, possibly due to killing via the Fas pathway. However, Tc2 cells can provide bystander help, probably by expression of CD40 ligand and secretion of copious amounts of Th2 cytokines (Cronin et al 1995, Sad et al 1995).

Delayed-type hypersensitivity induced by Tc1 and Tc2 cells

Th1 cells mediate a delayed inflammatory reaction (Cher & Mosmann 1987), and Th1 but not Th2 cytokines are often associated with delayed-type hypersensitivity (DTH) reactions during infection (Tsicopoulos et al 1992, Yamamura et al 1991). Surprisingly, both Tc1 and Tc2 cells induced a DTH-like reaction when injected into the footpads of mice bearing the target alloantigen (L. Li, S. Sad & T.R. Mosmann, unpublished results 1996). Both swelling reactions peaked at about 20 h, and Tc1 and Tc2 cells induced similar levels of vascular permeability as measured by Evans Blue leakage.

To ensure that the Tc2 DTH reaction was not caused by secretion of Th1 cytokines *in vivo*, we extracted footpads and measured the cytokines by ELISA. Both Tc1 and Tc2 expressed their characteristic cytokine patterns *in vivo*, i.e. IFN-γ was produced during Tc1 but not Tc2 DTH, whereas IL-4 and IL-5 were produced only during Tc2 DTH (L. Li, S. Sad & T. R. Mosmann, unpublished results 1996). Interestingly, substantial levels of tumour necrosis factor (TNF) and IL-6 were produced during both Tc1 and Tc2 DTH. As Tc1 cells do not secrete IL-6 *in vitro*, these cytokines may have been derived from secondary inflammatory cells such as macrophages.

Histology of the inflamed footpads revealed infiltrating polymorphonuclear cells in Tc1 or Tc2 reactions. Immunohistochemistry of footpad sections showed strong infiltration of cells positive for the Gr1 granulocytic marker, and also some staining for the macrophage marker Mac3. As for all other parameters, Tc1 and Tc2 cells induced similar reactions. Finally, the infiltrating cells were examined by extraction from the footpad and staining of cytospin preparations. In addition to large numbers of granulocytes, there were moderate numbers of eosinophils in both Tc1 and Tc2 DTH. The number of eosinophils increased at later times, well after the peak of the swelling reaction. The later increase was slightly more marked in Tc2 inflammatory reactions, possibly because of the activities of IL-4 and IL-5 in eosinophil recruitment and survival.

Although cytokines are often considered to be major determinants of the inflammatory reaction, these results suggest that, at least for CD8 cells, their distinctive cytokine patterns are not the major causes of the inflammatory reaction.

IL-4-induced cytokine deficiency

Although Tc1 and Tc2 cells do not interconvert or revert to a naïve cytokine phenotype, Tc1 cells can undergo an additional differentiation step. In the presence or absence of antigen-presenting cells (APCs), IL-4 induces loss of the ability to synthesize IL-2 (Sad & Mosmann 1995). Although the IL-4-treated cells still produce other cytokines in response to concanavalin A, the synthesis of all cytokines, but particularly IL-2, is impaired during the response to APCs. The cytokine-deficient Tc1 cells kill as effectively as normal Tc1 cells, but they are unable to sustain their own proliferation in the absence of exogenous growth factors. Thus the effect of IL-4 on Tc1 cells is to leave their immediate effector functions intact, while reducing their ability to proliferate in the absence of other components of the immune response.

When injected into footpads of mice bearing target antigens, the cytokine-deficient Tc1 cells induce similar levels of DTH to normal Tc1 cells, although the levels of IFN-γ *in vivo* are considerably reduced (L. Li, S. Sad & T. R. Mosmann, unpublished results 1996). This strengthens the conclusion above, that the ability of CD8 cells to induce DTH does not correlate well with their cytokine synthesis patterns.

Does cytotoxicity limit CD8$^+$ T cell cytokine synthesis?

CD4$^+$ T cells are often considered to be the major source of cytokines during immune responses. However, *in vitro* CD4 and CD8 clones secrete similarly high levels of many cytokines. Recent data suggest that CD8 cells may often not produce their potential cytokine levels because they kill the APC too rapidly.

Non-cellular stimuli such as concanavalin A induce high levels of cytokine production by CD8 effector cells, but APCs often induce much lower levels. This contrasts with CD4 cells, which produce the same levels of cytokines in response to either concanavalin A or optimal antigen stimulation. Perforin-deficient CD8 effector cells produce much higher levels of cytokines (S. Sad, D. Kagi, H. Hengartner & T. R. Mosmann, unpublished results 1995). The Fas killing pathway may also reduce stimulation and cytokine synthesis, as perforin-deficient CD8 cells produce higher levels of cytokines in response to normal targets compared to Fas-transfected targets. The reduced levels of cytokine synthesis can be at least partially compensated by increasing the target : effector ratio to high levels. Under these conditions, the CD8 cells synthesize more cytokines, and the levels made by normal and perforin-deficient CD8 cells converge. Thus our working model is that CD8 cells require sustained stimulation to produce optimal levels of cytokines, but that normal killing is so rapid that stimulation is abrogated before the CD8 cell has been fully activated. The requirement for sustained stimulation can apparently be met by repeated stimulation, e.g. by providing an excess of targets. The significance of these findings *in vivo* could be that CD8 cells normally synthesize minimal levels of cytokines if they encounter only an occasional infected cell. However, at a site with many infected cells, CD8 cells may undergo repeated stimulation and become a major source of cytokines.

hypothesis

Further complexity of T cell cytokine patterns

Although the crucial role of the Th1 and Th2 cytokine patterns has been amply demonstrated in a number of models of parasite infection, immune responses can show considerably more complexity than just the simple Th1 or Th2 patterns. Both sets of effector functions can be important at different times in the same infection (Taylor-Robinson et al 1993). Many rapid responses may not have time to polarize into one of the extreme responses, resulting in a mixed cytokine pattern (Sarawar & Doherty 1994, Baumgarth et al 1994). Different immune responses may occur in different locations, particularly if infections are localized and contained by the immune response. Several additional T cell cytokine patterns can exist, and non-T cells can be major contributors of cytokines of the Th1 or Th2 patterns (reviewed in Mosmann & Sad 1996).

Multiparameter analysis of cytokine effects on T cell differentiation

Considerable complexity also exists in the differentiation process that gives rise to effector T cells. For both $CD4^+$ and $CD8^+$ T cells, IL-4 is the major inducer of differentiation to the Th2 and Tc2 phenotypes, whereas IFN-γ and IL-12 encourage Th1 and Tc1 differentiation. However, these general rules do not take into account the complex interactions that occur between the effects of individual cytokines. We have recently set up a method for multiparameter analysis of T cell differentiation (W. Fu & T. R. Mosmann, unpublished results 1996), on the basis of the principle that normal immune responses include a relatively large number of cytokines, and so studying the effects of one or two cytokines at a time may lead to oversimplification.

Naïve CD8 cells were stimulated with alloantigen for six days, and during this stimulation they were exposed to combinations of seven different cytokines. A simple matrix was designed that allowed rapid setup of all possible combinations (2^8, i.e. 256) of eight different treatments (seven cytokines, duplicate cultures). Where possible, treatment with a cytokine was compared with treatment with a blocking monoclonal antibody against the same cytokine, to minimize contributions of endogenous cytokine synthesis. At the end of the six days, the cultures were thoroughly washed and restimulated to determine the differentiation status of the resulting cells. Proliferation was measured by the MTT (3-[4,5-dimethylthiazol-2-yl]-2,5-diphenyltetrazolium bromide) assay (Mosmann 1983), and synthesis of IL-2, IL-4, IL-5, IL-10 and IFN-γ was evaluated by ELISA.

As the resulting 1536 data points were difficult to evaluate directly, the results were analysed by separating out the contributions of each of the eight factors. For example, to evaluate the effects of IL-4 on differentiation to IL-5-producing cells, a two-dimensional plot was generated in which each point represented the values of IL-5 synthesised by cultures with or without IL-4, holding all of the other seven parameters constant (Fig. 1). This allowed rapid identification of conditions that showed a strong positive or negative effect, e.g. IL-4 was necessary but not sufficient for the generation of IL-5-producing cells (all positive values on the 'with' but not 'without' axis, Fig. 1a), whereas IFN-γ was inhibitory for the generation of IL-5-producing cells (Fig. 1b). Only a few conditions allowed differentiation to IL-5-secreting Tc2 cells, due to requirements for the presence of IL-2, IL-4 and IL-10, and the absence of IFN-γ and transforming growth factor β.

Differentiation to IFN-γ-producing Tc1 cells also showed some simple effects, but in some cases much more complex regulation occurred. IL-2 was required for Tc1 production except for a small number of combinations. IFN-γ was mostly inhibitory, due to inhibition of proliferation. However, IL-4, IL-10 and IL-12 had complex effects: for each of these cytokines, conditions could be identified in which differentiation to IFN-γ-synthesizing cells was enhanced (points only on 'with' axis); inhibited (points

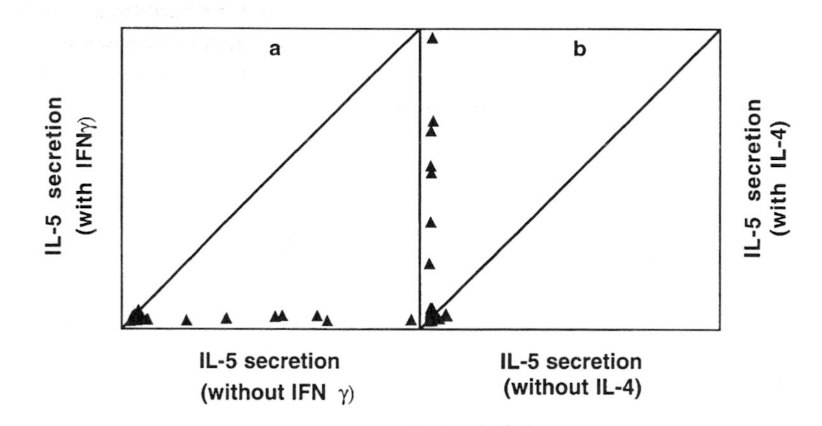

FIG. 1. Naïve CD8^{+} T cells were stimulated with the J774 macrophage cell line for 6 d in all possible combinations (256) of IL-2, IFN-γ, IL-12, IL-4, TGF-β, IL-15 and IL-10 in duplicate. T cells were then thoroughly washed and restimulated with concanavalin A, and the production of IL-5 was measured in the supernatant after 24 h. In the 2D plot in (a), each data point (128 total) represents the pair of IL-5 values obtained with or without IFN-γ treatment, with the other seven parameters held constant. Panel (b) shows a similar analysis of the same IL-5 values, plotted in terms of the presence or absence of IL-4.

Table 1 Effects of IL-4 on CD8⁺ T cell differentiation depend on other cytokines

	IL-2	*IL-10*	*IL-12*	*IFN-γ*	*TGF-β*
IL-4 enhances Tc1 generation	−	+	+	−	−
IL-4 inhibits Tc1 generation	+	+	+ or −	−	+
IL-4 doesn't affect Tc1 generation	+	−	+ or −	−	(−)

IFN-γ, γ-interferon; IL, interleukin; TGF-β, transforming growth factor β.

only on 'without' axis); or unaffected (points near diagonal) by the cytokine. According to the results of this multiparameter analysis, three conditions were chosen in which IL-4 enhanced, inhibited or did not affect Tc1 differentiation (Table 1). Naïve $CD8^+$ T cells were then stimulated in these conditions with titrated amounts of IL-4, and the predictions of the multiple cytokine analysis were confirmed. In condition 1, increasing amounts of IL-4 inhibited subsequent IFN-γ synthesis, whereas the reverse was true for condition 2, and IL-4 had no effect in condition 3. Note that condition 1 requires the specification of five different cytokines in addition to IL-4. Although these results could have been due to effects on either proliferation or differentiation, comparison with MTT values gave similar results on a per cell basis, suggesting that differentiation accounted for most of these effects.

The ability of IL-4, IL-10 and IL-12 to mediate opposite effects on $CD8^+$ T cell differentiation, depending on the other cytokines present, clearly indicates that the actions of cytokines in complex mixtures cannot always be predicted from the properties of the same cytokines in isolation. As normal immune responses involve even more complex mixtures of cytokines, it will be essential in future to assess the effector functions of these complex mixtures.

Acknowledgements

Supported by the Medical Research Council of Canada and the Howard Hughes Medical Institute.

References

Baumgarth N, Brown L, Jackson D, Kelso A 1994 Novel features of the respiratory tract T-cell response to influenza virus infection: lung T cells increase expression of gamma interferon mRNA *in vivo* and maintain high levels of mRNA expression for interleukin-5 (IL-5) and IL-10. J Virol 68:7575–7581

Carter LL, Dutton RW 1995 Relative Perforin- and Fas-mediated lysis in T1 and T2 CD8 effector populations. J Immunol 155:1028–1031

Cher DJ, Mosmann TR 1987 Two types of murine helper T cell clone. II. Delayed-type hypersensitivity is mediated by TH1 clones. J Immunol 138:3688–3694

Croft M, Carter L, Swain SL, Dutton RW 1994 Generation of polarized antigen-specific CD8 effector populations: reciprocal action of interleukin (IL)-4 and IL-12 in promoting Type 2 versus Type 1 cytokine profiles. J Exp Med 180:1715–1728
Cronin DC, Stack R, Fitch FW 1995 IL-4-producing $CD8^+$ T cell clones can provide B cell help. J Immunol 154:3118–3127
Mosmann TR 1983 Rapid colorimetric assay for cellular growth and survival: application to proliferation and cytotoxicity assays. J Immunol Methods 65:55–63
Mosmann TR, Sad S 1996 The expanding universe of T cell subsets: Th1, Th2 and more. Immunol Today 17:138–146
Sad S, Mosmann TR 1995 IL-4, in the absence of antigen stimulation, induces an anergy-like state in differentiated $CD8^+$ Tc1 cells: loss of IL-2 synthesis and autonomous proliferation but retention of cytotoxicity and synthesis of other cytokines. J Exp Med 182:1505–1515
Sad S, Marcotte R, Mosmann TR 1995 Cytokine-induced differentiation of precursor mouse $CD8^+$ T cells into cytotoxic $CD8^+$ cells secreting TH1 or TH2 cytokines. Immunity 2:271–279
Sarawar SR, Doherty PC 1994 Concurrent production of interleukin-2, interleukin-10, and gamma interferon in the regional lymph nodes of mice with influenza pneumonia. J Virol 68:3112–3119
Taylor-Robinson AW, Phillips RS, Severn A, Moncada S, Liew FY 1993 The role of T_H1 and T_H2 cells in a rodent malaria infection. Science 260:1931–1934
Tsicopoulos A, Hamid Q, Varney V et al 1992 Preferential messenger RNA expression of Th1-type cells (IFN-γ^+, IL-2^+) in classical delayed-type (tuberculin) hypersensitivity reactions in human skin. J Immunol 148:2058–2061
Yamamura M, Uyemura K, Deans RJ et al 1991 Defining protective responses to pathogens: cytokine profiles in leprosy lesions. Science 254:277–279

DISCUSSION

Nossal: This fascinating back-killing of the APCs is rather reminiscent of Sasazuki's deconstruction of his suppressor effects, which arose from *in vivo* findings (Sasazuki 1990). Have you discovered the mechanism of $CD8^+$ T cell suppression, for instance in the transplantation situation?

Mosmann: The suppression by CD8 killing would require recognition of the target cell. Which particular kinds of reactions were you thinking of?

Nossal: Both allogeneic and infectious agent-specific T cell responses. In each case, the effector T cell response under consideration would depend on an APC presenting the relevant antigen.

Mosmann: If we move to B cells for a moment, there has been great interest about what the CD8 subsets might do to them. It appears that they simply kill B cells if they recognize them. This doesn't help us with the suppression of an antigen-specific response, because the B cell shouldn't be expressing antigen on MHC class I to be killed. In fact, if it did, we would never get any immune responses. Therefore I don't think this provides an explanation for B cell suppression.

In terms of other kinds of suppression, we have to consider how we get the antigen onto the cell that needs to be suppressed. An effector CD4 cell has to find some way of expressing the correct antigen before this mechanism could operate.

Miller: I thought that there was evidence that CD8 cells played a role in controlling HIV. Would the CD8 cells be killing or controlling HIV-infected cells via the lymphokine pathway?

Mosmann: You are referring to work by Sergio Romagnani, who found human Tc2 cells in a subset of HIV-positive individuals who had Jobs-like syndrome, which is characterized by high IgE levels (Maggi et al 1994). Under those conditions he can extract cells that then become Tc2s. One possibility is that this would contribute to the general suppression of cell-mediated responses that is characteristic of AIDS. How far you want to go down that pathway depends on which camp you belong to — it has become a controversial area. Clearly, an excess of IL-4-secreting Tc2 cells would probably be bad for these patients.

Zinkernagel: Perforin-knockout mice die of lymphocytic choriomeningitis virus infection with a fantastic serum titre of IFN-γ. This is because their $CD8^+$ T cells expand enormously since the stimulating infected cells are not eliminated.

Mosmann: It may make sense that if a CD8 cell is confronted with very few infected targets, it can take care of them simply by killing them and that is the end of it. It is only when it is confronted with a large number of sequential targets that it calls in reinforcements by activating other processes.

Zinkernagel: We would even go further and argue that because of that particular danger, it looks as if the fact that perforin-positive $CD8^+$ T cells get exhausted is necessary to prevent lethal immunopathology, not only caused by killing but also by the release of damaging interleukins.

Tarlinton: How efficiently can you induce peripheral CD8 cells to become one type or the other?

Mosmann: In culture we can induce the population to be pretty homogeneous, but to do this we have to push them pretty hard with large amounts of IL-4 and blocking IFN-γ. We haven't done much *in vivo* yet trying to find these in mice. Sergio Romagnani has found them in human (Maggi et al 1994) and Graham Le Gros has evidence for IL-5-producing CD8 cells during a viral infection (Coyle et al 1995), but my impression is that they would be a minority population and that, especially in a mouse, the majority of cells would be Tc1 (Maggi et al 1994, Coyle et al 1995).

Dexter: I have a great deal of sympathy with you trying to reproduce *in vitro* what is likely to be going in the body, but you might be attempting the impossible. In the general field of myelopoiesis we have learnt that in addition to titrating different combinations of factors, you have to titrate different concentrations. The concentration can well have an effect on the eventual readout. How are you are going to cover all the possible permutations? You have obviously picked the optimal concentration of a cytokine in the readout systems for that cytokine.

Mosmann: Many cytokines that are active on T cells tend to have a saturation curve that doesn't come down again. IL-12 is a cytokine we suspect may have multiple effects at different doses, but this may be because of higher-order effects. In terms of whether the system can be used in this way, we initiated this approach to find out whether there were complex interactive effects, and there are. Now we have that knowledge we have to deal with it somehow. I agree that the concentrations are important. Another complicating variable is the sequence of cytokine addition. All of these factors are significant, and whether we like it or not they are part of normal immune responses.

Dexter: I'm still not sure how you are going to analyse these and then extrapolate to *in vivo* without the ability to measure at any point in time *in vivo* what is going on in terms of combinations and concentrations.

Mosmann: With IL-4, for example, we were able to do quite a lot of *in vivo* predicting on the basis of assays *in vitro* where we used only IL-4. The next layer is to start looking at cytokine combinations. This doesn't mean that times aren't important, but we can only take this one stage at a time. It is certainly not a complete analysis of the system. You are bringing up an important point, and it is one that neurobiologists worry about: is it possible to understand a system with so many interconnections? We are optimists: we keep trying and we keep making progress.

Kirberg: I would like to come back to the target killing. You are saying that CD8 cells do not produce IL-2 because they kill the target and then cannot be sufficiently stimulated any more. Skin graft rejection requires Th cells (Rees et al 1990). When Th cells are absent, CD8 cells normally fail to reject and the target cells persist (van der Vegt & Johnson 1993). But then, according to your model, the CD8 cells should become independent of Th cells as they are further stimulated by the persistent graft, produce IL-2 themselves and reject.

Mosmann: There are two reasons why a CD8 cell might not make IL-2: either because it has seen IL-4 or because it has not received sufficient stimulation. I don't know whether those cells in that graft situation would make IL-2, but exposure to IL-4 would be one possible reason for a failure to make IL-2.

Shortman: When you get the virgin T_0 CD8 cell out of the animal, what is its cytokine production capacity? And secondly, when you persuade such a cell to switch off all its cytokines, can that cell still proliferate in that state, if it is provided with exogenous cytokines?

Mosmann: The population we use gives us a fairly clean differentiation one way or the other. After depletion of B cells, macrophages and $CD4^+$ cells, we sort for $CD8^+$ $CD44^{lo}$. We do a fairly tight cut on $CD44^{lo}$. If we don't get rid of a few of the mature cells, they appear to be able to bias the differentiation very strongly, so the cultures are less reproducible.

Shortman: What do they make if you stimulate them immediately?

Mosmann: Not very much. Certainly, the CD8s are not high IL-2 producers.

Burgess: I don't understand your definition of DTH. You have chosen a set of cytokines to look at, but I would have thought that the chemokines and vascular endothelial growth factors might be factors that are more relevant to DTH responses. What aspect of DTH are you trying to reproduce?

Mosmann: The starting point was our biases from the CD4 system. With the CD4 cells we see quite a strong difference both in terms of what clones can do and in terms of cytokines associated with ongoing DTH reactions. For that reason, when we moved to the CD8 cells and found DTH from both of them, meaning vascular leakage and infiltration, we were surprised. I have been describing mainly the first steps to try to find out why they should be somewhat different from the CD4s.

Burgess: Which cytokines are known to induce vascular leakage?

Mosmann: TNF causes vascular leakage. IFN-γ contributes to a DTH reaction caused by Th1 cells. IL-4 and IL-10 can inhibit DTH.

Paul: Several years ago Bob Schreiber did some experiments in which he showed that anti-IFN-γ blocked elicitation of DTH. This probably isn't so surprising if the bulk of that is CD4 cell dependent.

You described a paradoxical finding where under most circumstances IL-4 is associated with poor development towards IFN-γ or Tc1 production. But there was one set of circumstances in which it induced them. Is this always when IL-2 is absent? The reason I ask is that in our experience IL-2 is essential for any differentiation in the CD4 system to Th1 or Th2. It wouldn't surprise me that under certain circumstances IL-4 might replace this function of IL-2, since in some respects it acts similarly to IL-2, particularly if we are looking at growth rather than differentiation. There are now STAT-6 knockouts in which differentiation function is blocked; it would be interesting to see whether in these mice IL-4 continues to have its function even if it can't signal differentiation.

Mathis: Have you compared the ability of BALB/c and BL/6 mice to make Tc1 and Tc2 cells? Do you see the kinds of differences that have been described for Th1 and Th2 cells?

Mosmann: Although we have used several mouse strains we have not pickled up any strain differences. We can make Th1 and Th2 from any of the mouse strains we have tried.

Melchers: It wasn't clear to me how heterogeneous your test cell populations really were. In other words, how many of the effects do you attribute to one cell as opposed to the cooperation of several cells?

Mosmann: We don't know how many cell interactions we are working with. This analysis gives us a way of quickly looking at where combinations of cytokines are giving us rather unexpected effects. We identified more than we really wanted, because each one now requires painstaking dissection. It's almost a screening tool. The main message from it is that these complex situations exist.

Melchers: Is eight the magic number of cytokines or can you get away with fewer?

Mosmann: Eight gives us 256 combinations, which by the time you have assayed several cytokines, gives about 1500 data points. Beyond nine, my student would probably rebel! Some conditions required at least five cytokines to be set the appropriate way, either on or off—we would like to go further.

Hodgkin: So far in this analysis, have you seen new repertoires that you didn't know about before, in terms of combinations of lymphokines?

Mosmann: Not to date. One we were quite interested in was Tc0 cells, which would make IL-4 and IFN-γ. In fact, IL-4 and IFN-γ are very reciprocal within that population. IL-2 and IFN-γ are fairly strongly associated although there are some interesting differences. IL-4 and IL-5 correlate almost perfectly. Nothing has really come out yet except the Tc1 and Tc2 phenotypes.

Nicola: The idea of the cell as an integrated computer of its environment is an attractive one. The number of cytokine combinations that you would want to test in such a system might be determined by the different kinds of receptors that are expressed by the cell system under test. Are you beginning to see any sort of biological sense in the findings?

Mosmann: I'm currently overwhelmed by the number of interesting and paradoxical combinations. The short answer is therefore that we don't see any biological sense in it.

Nossal: I have a question for Frank Carbone, relating to the first part of Tim Mosmann's talk and the question of the killing of the APCs. Are there now enough exceptions to the general rule that class I presents internal cytosolic antigens and class II external pinocytosed antigens for us not to worry about them? In other words, could the $CD8^+$ T cells be interacting with foreign antigens on the APCs in the way that has been described?

Carbone: This is a paradox. The early *in vitro* assays demonstrated that the rule seemed to be fairly strict: something on the outside of a cell does not easily get into the class I processing pathway. There are exceptions, although you have to extensively manipulate the system to find them. *In vivo* there are many more exceptions and the rule breaks down more readily. The challenge is for us to determine what is going on and why.

Mosmann: Do any of those *in vivo* examples involve B cells? The B cell is rather special in that if you allow a B cell that internalizes an antigen by antibody to present on class I then you run the risk of terminating the response by cytotoxicity. The distinction ought to be particularly apparent for a B cell.

Carbone: To my knowledge no one has actually looked at this. There is some evidence that if you deplete a system of macrophages you don't see this phenomenon. However, there are potential problems with these experiments. In most cases investigators focus on T cell priming where eliminations of macrophages may well make it harder to prime the CD4 cells which are necessary to help CD8 cells in their response.

References

Coyle AJ, Erard F, Bertrand C, Walti S, Pircher H, Le Gros G 1995 Virus-specific CD8+ cells can switch to interleukin 5 production and induce airway eosinophilia. J Exp Med 181:1229–1233

Maggi E, Giudizi MG, Biagiotti R et al 1994 Th2-like CD8+ T cells showing B cell helper function and reduced cytolytic activity in human immunodeficiency virus type 1 infection. J Exp Med 180:489–495

Rees MA, Rosenberg AS, Munitz TI, Singer A 1990 *In vivo* induction of antigen-specific transplantation tolerance to Qa1a by exposure to alloantigen in the absence of T cell help. Proc Natl Acad Sci 87:2765–2769

van der Vegt FP, Johnson LL 1993 Induction of long-term H-Y-specific tolerance in female mice given male lymphoid cells while transiently depleted of $CD4^+$ or $CD8^+$ T cells. J Exp Med 177:1587–1592

Sasazuki T 1990 HLA-linked immune suppression genes. Jpn J Hum Genet 35:1–13

T cell tolerance and autoimmunity

J. F. A. P. Miller, W. R. Heath, J. Allison, G. Morahan, M. Hoffmann*, C. Kurts and H. Kosaka†

The Walter and Eliza Hall Institute of Medical Research, Post Office, Royal Melbourne Hospital, Victoria 3050, Australia

Abstract. Many T cells with auto-aggressive potential are deleted in the thymus. Although some of these escape to the general circulation, they do not usually damage organs such as the pancreas. To investigate the mechanisms preventing autoimmunity, we generated transgenic mice expressing known genes under the control of various promoters. We found that the occurrence of autoaggression depended on factors such as the precursor frequency of responding T cells, their state of activation, their accessibility to the autoantigen, the physicochemical properties of the autoantigen, the possibility of priming by environmental antigens which mimic the target antigen, and some inflammatory reaction in the target site.

1997 The molecular basis of cellular defence mechanisms. Wiley, Chichester (Ciba Foundation Symposium 204) p 159–171

Self-tolerance is mediated by both central and peripheral mechanisms. The dominant tolerogenic mechanism for T cells is generally thought to involve the intrathymic negative selection of differentiating cells with reactivity to self peptides presented in association with molecules encoded by the major histocompatibility complex (MHC). This has been shown to occur when there is a high frequency of T cells specific for a given self antigen, such as a superantigen or some defined cell surface antigen in the case of T cell receptor (TCR) transgenic mice. Nevertheless, the existence of potentially autoreactive T cells in the peripheral T cell pool of healthy individuals and of experimental animals has been well documented (see review by Miller & Flavell 1994). These cells presumably escape thymus censorship by expressing antigen-specific TCRs of too low an affinity for the self antigen. Alternatively, the self epitope may not exist in the thymus as a stable complex with MHC molecules. Since autoimmunity is not the norm, a variety of factors must operate to hold in check the autoaggressive potential of self-reactive T cell escapees. For example, these cells may

Present addresses: *Medizinische Hochschule Hannover Klinik fur Abdominal- und Transplantationschirurgie, Postfach 61 01 80, D-3000 Hannover 61, Germany and †Department of Dermatology, Osaka University Medical School, 2-2 Yamada-Oka, Suita, Osaka 565, Japan.

ignore the target antigen, either because they cannot penetrate endothelial barriers, or because they cannot be fully activated after recognizing antigen on tissue cells that fail to express co-stimulator molecules. The autoimmune T cells may be anergized, deleted peripherally by the autoantigen or silenced by immunoregulatory cells.

To investigate these possibilities, my colleagues and I have examined various models in which genes coding for known molecules were introduced by transgenic technology.

Ignorance in RIP-K^b mice

We previously showed that mice expressing the allogeneic class I protein, H-2K^b (K^b), in pancreatic islet β cells (RIP-K^b mice) did not develop autoimmunity, although a completely non-immune form of diabetes ensued owing to the over-expression of transgene product (Allison et al 1988). These mice were tolerant of K^b-bearing skin, but this was due to the intrathymic expression of a few molecules of transgenic K^b (Heath et al 1992). To determine whether such few molecules would delete all K^b-specific T cells, we produced double transgenic mice, by mating RIP-K^b mice to mice transgenic for a K^b-specific TCR (Des-TCR mice). Of great importance was the finding that those $CD8^+$ T cells expressing the highest density of the TCR identifiable by the clonotypic Désiré (Des) monoclonal antibody (which is specific for the transgenic TCR) were greatly reduced compared to $CD8^+$ Des^+ T cells in the single Des-TCR mice. This could be seen after gating on $CD8^+$ T cells and looking at levels of clonotype expression. A biphasic peak was seen in single transgenic mice. On the other hand, the $CD8^+$ T cells of the double transgenic mice expressed only the lower level of clonotype seen in the single transgenic mice (Heath et al 1992). These 'low density' T cells were found not to have down-regulated their transgenic TCR but to be bispecific, expressing a second TCR composed of the transgenic β chain associated with a rearranged TCR α chain (Heath & Miller 1993). We interpreted all these findings in terms of the intrathymic deletion of T cells with the highest density of clonotype and presumably the highest avidity for K^b. The immunocompetence of the $CD8^+$ T cells expressing the lower density of clonotype in the double transgenic mice was tested by confronting them with skin grafts from two different lines of B10.BR mice transgenic for K^b, one expressing high levels (150% of normal values), the other low levels (33%). Control single and young double transgenic mice were able to reject $K^{b\text{-}hi}$ skin at the same rate. On the other hand, only a few of the double transgenic mice could reject the $K^{b\text{-}lo}$ skin (Heath et al 1995). Hence, the low density T cells circulating in the double transgenic mice were not anergic but differed from the high density cells in their ability to respond to tissues bearing low antigen dose.

Since interleukin 2 (IL-2) had previously been shown to enhance the *invitro* response of K^b-specific cells from RIP-K^b mice (Morahan et al 1989), we examined whether local IL-2 production might activate K^b-specific cells in double transgenic Des-TCR × RIP-K^b mice (Heath et al 1992). To this end, transgenic mice were generated that expressed IL-2 in the β cells (RIP-IL-2 mice). In earlier work we had shown that IL-2 in these

TABLE 1 Early onset of autoimmune diabetes in RIP-K^b models

	Islet β cell transgene			
T cell repertoire	*K^b*	*K^b+IL-2*	*K^b+TNFα*	*K^b+B7.1*
K^b-specific, clonotype-low[a]	–	+	+	+
K^b-specific, clonotype-low, primed[b]	–	ND	ND	ND
K^b-specific, clonotype-high[c]	–	ND	ND	ND
K^b-specific, clonotype-high, primed	+	ND	ND	ND
Modified[d] normal repertoire	–	–	–	–
Modified[d] normal repertoire, primed	–	–	ND	ND

+, diabetes occurred; –, diabetes did not occur; ND, not done.
[a] Low K^b expression in the thymus has deleted K^b-specific T cells with high density of clonotype$^+$ TCR, allowing only low clonotype T cells bearing a second TCR to peripheralize.
[b] Priming was by intraperitoneal injection of 10^7 C57BL/6 spleen cells and an additional subcutaneous injection of the same amount.
[c] K^b-specific T cells with high density of clonotype$^+$ TCR peripheralized only when the hosts were thymectomized, irradiated, protected with bone marrow from Des-TCR transgenic mice and given a non-transgenic syngeneic thymus graft.
[d] Only low affinity K^b-specific T cells peripheralized since the thymus aberrantly expressed low levels of K^b.

mice attracted mononuclear cells around the islets, a consequence of which was the up-regulation of the antigen-presenting machinery on islet cells but not autoimmune diabetes (Allison et al 1992). Triple transgenic mice, with K^b and IL-2 in the islets and clonotype$^+$ T cells, developed autoimmune diabetes as early as one week after birth. This depended on the high precursor frequency of clonotype$^+$ cells because it did not occur in RIP-K^b × RIP-IL-2 mice with a normal T cell repertoire. Local expression of tumour necrosis factor (TNF)α in the islets also attracted mononuclear cells. TNFα was able to substitute for IL-2 in our triple transgenic model of autoimmune diabetes, which suggested that the inflammatory consequences of local cytokine production were to potentiate the immune response to islets, possibly because of the up-regulation of the antigen-presenting capacity of islet cells. This was up-regulated in another way when the co-stimulator molecule B7.1 was introduced into β cells. Alone or in combination with K^b, it did not result in autoimmunity. If, however, clonotype$^+$ T cells were also present, spontaneous diabetes occurred before 10 days of age (Table 1). It therefore appears that despite the presence of strong inflammatory stimuli or co-stimulator molecules, one of the major requirements for autoimmunity is a high precursor frequency of specific T cells.

What would happen if the high density T cells were allowed to escape thymus censorship? To determine this, we thymectomized RIP-K^b mice and irradiated them with 900 R. They were then protected with bone marrow from Des-TCR transgenic mice and grafted with an irradiated (1000 R) thymus from non-transgenic syngeneic newborn donors (Heath et al 1995). RIP-K^b mice manipulated in this way had circulating high density cells but only two out of 22 mice showed very few $CD8^+$ T cells in some of the 20 islets scored. In the absence of intentional priming, therefore, high density cells ignored islet antigens. These mice were not tolerant of K^b as they rejected K^b-bearing skin grafts and most islets were infiltrated two weeks after such priming. Analysis revealed that β cells were specifically destroyed in this response. Many manipulated mice rapidly lost weight and died. Hence, high density T cells, presumably those with high avidity for K^b (in contrast to low density cells), were capable of inducing lethal diabetes after priming and in the absence of constitutive local IL-2 production (Table 1).

Deletion in MET-K^b mice

We generated B10.BR transgenic mice expressing K^b under the control of the metallothionein promoter (MET-K^b mice) and crossed them to Des-TCR mice (Bertolino et al 1995). The bone marrow-derived cells in the double transgenic progeny expressed K^b and induced intrathymic deletion of T cells with the highest density of clonotype. To prevent thymic expression of K^b, we manipulated these mice in the same way as the RIP-K^b mice. The thymus grafts had comparable numbers of cells and similar proportions of $CD4^+$ and $CD8^+$ T cells and did not show loss of $CD8^+$ T cells when examined 4–11 weeks after grafting. The lymph nodes also had comparable numbers of cells but the percentage of $CD8^+$ T cells was less in the manipulated MET-K^b mice. Some lymphocytes were present in the livers of manipulated control B10.BR mice but, in contrast to these, most of the cells in the profuse liver infiltrates of the MET-K^b mice were $CD8^+$ Des^+ T cells, and had the phenotype of activated cells expressing CD44. The numbers of $CD8^+$ Des^+ T cells in the liver of manipulated MET-K^b mice increased for the first 5 weeks after thymus grafting and subsequently decreased. In the lymph nodes and spleen of the controls, there was a steady increase in the number of $CD8^+$ Des^+ T cells, but in the spleen of the MET-K^b mice this population increased for the first 5 weeks and then decreased, and only a few of these cells were found in the lymph nodes. The livers were heavily infiltrated with lymphocytes by 4 weeks after thymus grafting and to a lesser extent by 20 weeks. Thus, when the thymus did not express the transgene, K^b-specific cells were generated intrathymically, infiltrated the liver and many disappeared, as their numbers did not increase with time after thymus grafting.

In another set of experiments we irradiated thymectomized, marrow-protected MET-K^b mice and injected them 7–8 weeks later with 20 million lymph node cells from Des-TCR transgenic mice (Bertolino et al 1995). Massive infiltration was seen within 3–9 d, but this had largely disappeared by 30 d. The infiltrating K^b-specific T cells showed signs of activation, proliferated *in situ*, caused damage to the hepatocytes and were eventually deleted. Liver damage was also evident from the serum transaminase levels which increased dramatically in correlation with the number of infiltrating cells. As the liver has great regenerative ability, continuous antigenic stimulation must have confronted the infiltrating T cells and this may be the reason

TABLE 2 Incidence of diabetes in RIP-OVA transgenic mice

Strain of mice	*Number of mice*	*Number of mice with diabetes at age (days)*			
		0–20	*21–40*	*41–60*	*61–80*
Littermates	28	0	0	0	0
RIP-OVA	28	0	0	0	0
OT-1	19	0	0	0	0
RIP-OVA × OT-1	9	0	5	3	1
$Rag^{-/-}$ RIP-OVA × OT-1	9	0	8	1	0
$Rag^{+/-}$ RIP-OVA × OT-1	9	0	9	0	0
RIP-OVA × OT-1 (germfree)	8	0	7	1	0

for their deletion. This also took place in mice crossed to *lpr* and *gld* mutant strains suggesting that Fas–Fas ligand interactions are not involved in $CD8^+$ T cell apoptosis (H. Kosaka, unpublished data).

Autoimmunity in RIP-OVA mice

Transgenic mice expressing ovalbumin (OVA) in a soluble form under the control of the rat insulin promoter (RIP-OVA mice) were produced. OVA has an advantage over K^b in that it has both class I- and class II-restricted epitopes. RIP-OVA mice did not show signs of autoimmunity and were tolerant to OVA as measured by antibody production. When the mice were examined for cytotoxic T lymphocyte generation using different priming protocols, only under conditions where cytotoxic T lymphocytes could be induced in a $CD4^+$ T-cell-independent manner were responses detected. Thus $CD4^+$ but not $CD8^+$ T cells were tolerant of islet-expressed OVA.

When RIP-OVA mice were crossed to transgenic mice which produced class I-restricted OVA-specific $CD8^+$ T cells specific for $OVA_{257-264}$ peptide (OT-1 mice), the double transgenic mice showed no evidence of thymic deletion of transgenic T cells and many such cells were found in the peripheral lymphoid system. All the mice developed massive infiltration of the pancreas at an early age and were diabetic by 4–5 weeks. This rapid onset and high incidence occurred both in mice that had been kept germ-free and in mice derived from OT-1 transgenic parents that had been crossed onto a Rag-1-deficient background (Table 2). Despite the rapid onset of disease in RIP-OVA × OT-1 mice, diabetes was rarely observed after reconstituting lethally irradiated adult RIP-OVA mice with bone marrow from B6 or OT-1 mice. When analysed for islet infiltration, all chimeric mice reconstituted with OT-1 bone marrow alone showed mild islet infiltration, whereas only four of eight chimeric mice receiving mixed bone marrow had infiltration (Table 3). We are investigating

TABLE 3 Diabetes incidence and islet infiltration in lethally irradiated RIP-OVA mice reconstituted with OT-1 bone marrow alone or in combination with normal B6 bone marrow

Bone marrow donor	*No. of diabetic mice/total (>100 days)*	*No. of mice with infiltrated islets/total*	*% of islets infiltrated (mean)*
OT-1	2/23	8/8	39
OT-1+B6[a]	0/13	4/8	22

Normal B6 mice were also lethally irradiated and reconstituted with bone marrow as for the RIP-OVA mice. None of these mice became diabetic or showed islet infiltration.
[a]These mice were reconstituted with a 1 : 4 ratio of OT-1 : B6 bone marrow.

the possibility that an immunoregulatory network may account for the lower onset of insulitis and disease in these chimeras.

We used a second transgenic OVA line, the RIP-mOVA mice, which expressed OVA as a membrane-bound molecule in the β cells and in the kidney proximal tubular cells. OVA-specific CD8$^+$ T cells from OT-1 transgenic mice were injected into non-irradiated RIP-mOVA mice. T cell infiltration in the kidneys was not detected, but activated OVA-specific T cells accumulated exclusively in the draining lymph nodes of the kidneys and pancreas (Fig. 1). They expressed CD69 and proliferated, as evidenced by the incorporation of BrdU. Their lifespan was curtailed as shown by a second transfer to non-transgenic syngeneic Rag$^{-/-}$ mice. Controls were OT-1 cells obtained from the inguinal lymph node of the RIP-mOVA mice and from the renal node of a non-transgenic recipient of OT-1 cells (Fig. 2).

Unilateral nephrectomy 7–14 d prior to inoculation of OT-1 T cells into RIP-mOVA mice allowed the injected T cells to home only to the regional lymph node of the remaining kidney (and also to the pancreas), but when the operation was performed 4 h before injecting the T cells, homing to the regional nodes on both sides was evident (Fig. 1). These results indicate that specific CD8$^+$ T cells are activated when they encounter membrane-bound OVA on short-lived antigen-presenting cells (APCs). These must either have migrated from the kidney to the regional lymph nodes or have been resident in the nodes and taken up antigen shed from the kidney. As T cell homing to any other lymph nodes did not occur, these APCs could not have aberrantly expressed the transgenic OVA, but must have taken up membrane-bound OVA molecules from the renal tubular cells and introduced them into the class I pathway for presentation to CD8$^+$ T cells.

Discussion

Although we have obtained different results on the fate of potentially autoreactive T cells in our various transgenic mice, in none have we found evidence of T cell anergy. In the RIP-K^b model, in which there was thymic expression of K^b, only CD8$^+$ T cells

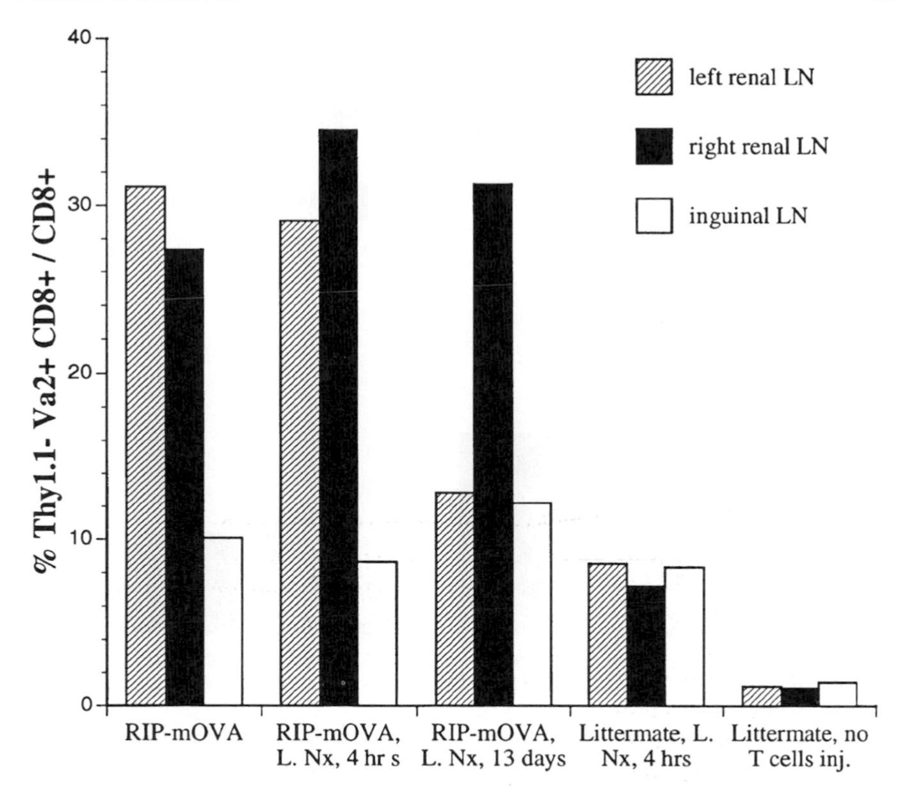

FIG. 1. Homing of OVA-specific $CD8^+$ T cells to the renal lymph node (LN) of nephrectomized RIP-mOVA mice. 8×10^6 Thy-1.1$^-$, Vα2$^+$ and $CD8^+$ T cells from OT-1 donors were injected intravenously into B6 RIP-mOVA × B6.Kathy recipients. *RIP-mOVA*, transgenic recipient, no nephrectomy; *RIP-mOVA, L. Nx, 4 hrs*, transgenic recipient, left nephrectomy 4 h before injection of T cells; *RIP-mOVA, L. Nx, 13 days*, transgenic recipient, left nephrectomy 13 d before injection of T cells; *Littermate, L. Nx, 4 hrs*, littermate recipient, left nephrectomy 4 h before injection of T cells; *Littermate, no T cells inj.*, littermate, no T cells injected.

with TCRs of low clonotype density and presumably with low avidity for the target molecule, K^b, escaped thymus censorship. They circulated but remained ignorant of K^b expressed by the β cells of the pancreas. Autoimmunity occurred only if the T cells were activated by antigen and a local inflammatory reaction was set up by the transgenic production of IL-2. The supply of IL-2 may therefore allow disease to be caused by low avidity cells that would otherwise be incapable of being autoaggressive. Like low avidity $CD8^+$ T cells, high avidity $CD8^+$ T cells in manipulated RIP-K^b mice generally ignored islet antigens, although the increase in avidity enabled them to cause autoimmunity after priming. This suggests that in order to induce disease, low avidity cells may be dependent on $CD4^+$ T cell help, whereas high avidity cells may be effective in its absence.

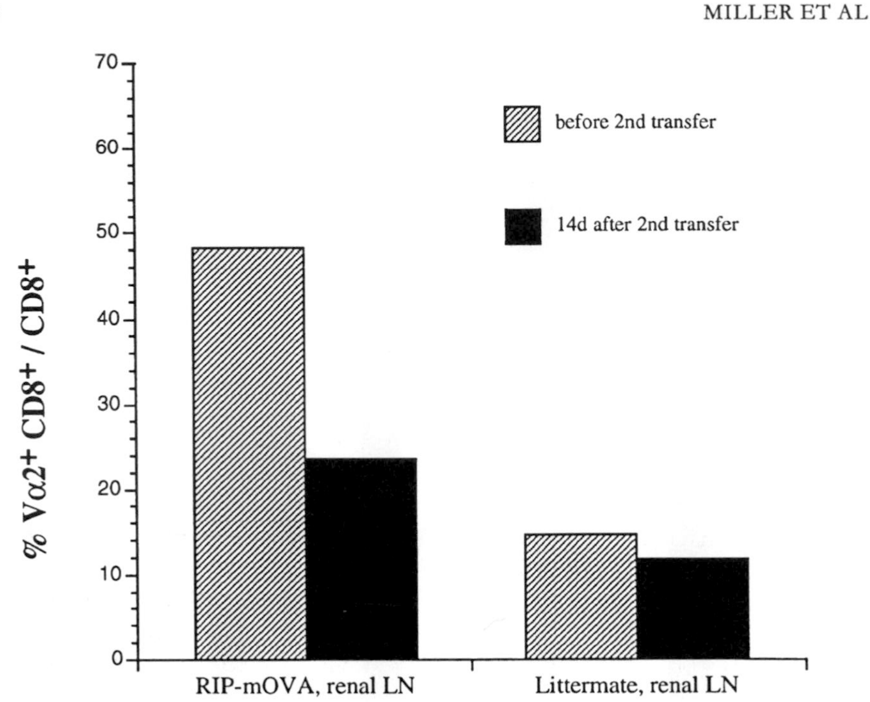

FIG. 2. RIP-mOVA mice and littermate controls were injected intravenously with 8×10^6 OT-1 cells. After 3 d, T cells from the lymph nodes (LN) draining the kidneys were prepared. The proportion of OVA-specific T cells ($V\alpha2^+$ $CD8^+$ T cells) was determined and the T cells were transferred to $Rag^{-/-}$ mice as second hosts. After 2 weeks, the proportion of surviving OVA-specific T cells (black bars) in the lymphoid tissues of the second hosts was compared to the proportion before transfer (hatched bars).

In the Met-K^b model in which high density T cells were allowed to migrate out of the thymus or were injected into thymectomized irradiated recipients, massive infiltration was observed in the liver and was followed by deletion of the invading cells. We have argued that it is unlikely that bone marrow-derived cells were responsible for this deletion because most of these cells were eliminated by the lethal irradiation and purified hepatocytes were able to stimulate very efficiently unprimed K^b-specific T cells *in vitro*. Yet co-stimulator molecules such as B7 have not been detected on the surface of hepatocytes (P. Bertolino, unpublished data), but whether other such as yet undefined molecules exist is not known. It may also be that K^b expression on hepatocytes is so intense that a threshold is reached at which co-stimulation is not necessary. This is in line with the view of Sprent & Schaefer (1990) that virtually any cell type expressing a high density of class I MHC molecules can provide APC function for $CD8^+$ T cells, even in the absence of added IL-2. The reason why unprimed T cells infiltrated the target organ in this model and not in the

RIP-K^b mice may be because lymphocytes can normally enter the liver (Huang et al 1994).

Infiltration and severe diabetes occurred early in life and in the absence of intentional priming in the RIP-OVA × OT-1 double transgenic mice and did so in germ-free mice and in mice crossed onto a Rag-1-deficient background. This raises the question as to why, in contrast to the RIP-K^b model, CD8$^+$ T cells did not ignore islet antigens in the RIP-OVA mice. This may relate to several factors such as the precursor frequency of the cells expressing the OVA-specific transgenic TCR, the avidity of the interaction between the OVA-expressing islet cells and the T cells, the peculiarities of the neonatal environment and the fact that OVA is not an MHC antigen but one that may be transported by lymph or APCs to regional lymph nodes. Cogent evidence for such transport was indeed obtained in the RIP-mOVA mice, but the identity of the transporter remains to be elucidated. Some APCs, either coming from the islets or kidney, or present in regional nodes, must be able to process exogenously acquired antigen (from the islets and the renal tubules) and present it to class I restricted CD8$^+$ T cells. This has been shown to occur in several other systems, many of these being *in vitro* (Gooding & Edwards 1980, Staerz et al 1987, Carbone & Bevan 1990, Rock et al 1990, Suto & Srivastava 1995, Bachmann et al 1996). The curtailed life span of the activated CD8$^+$ T cells in our model begs the question of whether such presentation constitutes a physiological mechanism for removing autoreactive CD8$^+$ T cells.

Acknowledgements

We thank Bernd Arnold and Gunther Hammerling for the Des-TCR transgenic mice, Anne-Marie Schmitt-Verhulst for the Désiré-1 mAb, Pierre Vassalli for the TNFα transgenic mice and Frank Carbone for the constructs used to produce RIP-OVA and OT-1 transgenic mice. Work from Professor Miller's laboratory was supported by the National Health and Medical Research Council of Australia and by NIH grant AI-29385.

References

Allison J, Campbell IL, Morahan G, Mandel TE, Harrison L, Miller JFAP 1988 Diabetes in transgenic mice resulting from over-expression of class I histocompatability molecules in pancreatic β cells. Nature 333:529–533

Allison J, Malcolm L,Chosich N, Miller JFAP 1992 Inflammation but not autoimmunity occurs in transgenic mice expressing constitutive levels of interleukin 2 in islet β cells. Eur J Immunol 22:1115–1121

Bachmann MF, Lutz MB, Layton GT et al 1996 Dendritic cells process exogenous viral proteins and virus-like particles for MHC class I presentation to CD8$^+$ cytotoxic T lymphocytes. Immunity, in press

Bertolino P, Heath WR, Hardy C, Morahan G, Miller JFAP 1995 Peripheral deletion of autoreactive CD8$^+$ T cells in transgenic mice expressing H-2K^b in the liver. Eur J Immunol 25:1932–1942

Carbone FR, Bevan M 1990 Class I restricted processing and presentation of exogenous cell-associated antigen *in vivo*. J Exp Med 171:377–387

Gooding LR, Edwards CB 1980 H-2 antigen requirements in the *in vitro* introduction of SV40-specific cytotoxic lymphocytes T. J Immunol 124:1258–1262

Heath WR, Miller JFAP 1993 Expression of 2 α chains on the surface of T cells in T cell receptor transgenic mice. J Exp Med 178:1807–1811

Heath WR, Allison J, Hoffmann MW et al 1992 Autoimmune diabetes as a consequence of locally produced interleukin 2. Nature 359:547–549

Heath WR, Karamalis F, Donoghue J, Miller JFAP 1995 Autoimmunity caused by ignorant $CD8^+$ T cells is transient and depends on avidity. J Immunol 155:2339–2349

Huang L, Soldevilla G, Leeker M, Flavell R, Crispe IN 1994 The liver eliminates T cells undergoing antigen-triggered apoptosis *in vivo.* Immunity 1:741–749

Miller JFAP, Flavell R 1994 T cell tolerance and autoimmunity in transgenic models of central and peripheral tolerance. Curr Opin Immunol 6:892–899

Morahan G, Allison J, Miller JFAP 1989 Tolerance of class I histocompatability antigens expressed extrathymically. Nature 339:622–624

Rock KL, Gamble S, Rothstein S 1990 Presentation of exogenous antigen with class I major histocompatability complex molecules. Science 249:918–921

Sprent J, Schaefer M 1990 Antigen-presenting cells for $CD8^+$ T cells. Immunol Rev 117:213–234

Staerz UD, Karasuyama H, Garner AM 1987 Cytotoxic lymphocytes T against a soluble protein. Nature 329:449–451

Suto R, Srivastava PK 1995 A mechanism for the specific immunogenicity of heat-shock protein-chaperoned peptides. Science 269:1585–1588

DISCUSSION

Nossal: This is not the first time that you've presented a model in which a tolerance phenomenon involves prior division of the cells. Is it reasonable to postulate that such a tremendously vital thing as purging the repertoire of self-reactive T cells has to proceed *via* a mechanism which obligately creates cells, at least for a time, that are damaging? It does not seem to be a very elegant way to go about things. I know that the liver is such a big organ that it hardly matters if you kill a few hepatocytes, but as a biologist do you believe such mechanisms, first creating autoimmune immunocytes to do damage, then killing them, have plausibility for self tolerance?

Miller: I don't see any problem. It makes things much easier in that there is one rule: a cell becomes activated and eventually will die. If the antigen persists, the cells which have not died become activated again and eventually the entire clone will disappear. Other factors come into this such as geographical boundaries, the possibility of inflammation, and whether you have co-stimulation. In our case we don't believe that we have inflammation in our normal healthy tissue and we have never found evidence of non-lymphoid tissue cell death. We probably don't have any co-stimulation, so the T cells which have been activated in the regional lymph node cannot go to the kidney to kill the cells, and they are probably being activated all the time. Eventually the whole clone is doomed to die.

von Boehmer: We have never found a situation where we have confronted mature T cells with antigens in whatever form, where the cell didn't become transiently activated and then was eliminated. Even when we induce anergy at fairly high antigen doses there is proliferation, because we find more anergic cells than the number of resting cells we started off with: during the transient activation a few of the antigen presenting cells have probably been knocked off, but then it's all over.

Goodnow: We have done a parallel experiment to the one Jacques Miller has just described. When CD4 cells are transferred from a TCR transgenic mouse into one that expresses lysozyme as a membrane bound form on the pancreatic islet or on the thyroid gland, we see no T cell activation or decrease in their numbers. There is no induction of CD69 or CD44 or any change in TCR expression, but after a number of days in these hosts, the T cells proliferate poorly *in vitro* and are unable to help B cells. If we transfer them into mice expressing significant amounts of circulating lysozyme rather than these organ-specific forms, we see CD69 and CD44 go up. The T cells never proliferate, but gradually decrease in numbers by a factor of two. Even after two days in those hosts the cells are anergic and poor helper cells. Suzanne Hartley has just done preliminary experiments with another TCR transgenic to lysozyme, transferring it into a mouse that expresses boatloads of antigen in the periphery. The preliminary evidence is that it is doing exactly what you are seeing: the T cells undergo a wave of expansion initially. We're wondering whether this is simply a quantitative thing: that low level chronic tickling, which in these organ-specific mice is so low that we don't even get good CD69 induction, gradually desensitizes the receptor signalling apparatus.

Zinkernagel: I agree that that dose has something to do with this, because in our RIP-LCMV-GP transgenics, you don't see down-modulation or anergy or tolerance of T cells. It is possible to vary the antigen dose: if we infect with LCMV we induce diabetes, and if we infect with a vaccinia recombinant expressing the same LCMV-GP we get $CD8^+$ T cell priming, but not enough for the effectors to emigrate in sufficient numbers to cause diabetes. For this to happen after infection with vaccinia-LCMV-GP, one needs a tremendously increased precursor frequency of T cells, a situation occurring in RIP-LCMV-GP plus TCR double transgenic mice. In your case the point is that something soluble that is presumably secreted gets fished out locally at a concentration high enough to do that tickling at the level where danger is minimal because the effector function isn't heated up enough to cause destruction.

There is evidence that viruses that avoid APCs and are either non- or poorly cytopathic, such as the papilloma virus, are ignored by T cells. The problem with warts is that the virus differentiates or matures only in keratinocytes, which are out of reach of Langerhans cells. That's why these viruses can cause tumours and not be immunologically bothered. The other case is rabies, where the virus initially infects only neurons. As long as the virus is in the neuronal axon and by a retroaxonal flux gets to the neuronal body, no immune response is induced during this phase. This also tells us that the efficiency of these types of cells to transmit their antigenic load to APCs and induce cytotoxic T lymphocytes seems to be very poor.

Miller: It may depend on the site.

Zinkernagel: Absolutely. It will be interesting to find out how exceptional the behaviour of ovalbumin is in these types of situations.

von Boehmer: Perhaps it is not so exceptional. All minor histocompatibility antigens do this regularly. I think this is a very general mechanism.

Zinkernagel: Minor histocompatibility antigens are rather interesting and specialized. We don't really know what they are, and most of those defined so far have been found to be endogenous viruses. We have tried to induce cross-priming with the LCMV glycoprotein: even by using large numbers of LCMV-GP-transfected fibroblastic cells we cannot demonstrate that this happens.

von Boehmer: Minor histocompatibility antigens are mostly represented by intracellular proteins, not retroviral proteins, for instance in the case of HY antigen.

Miller: I would like to emphasize that our model is not like cross-priming in the sense that we had a completely healthy tissue that has not been manipulated at all. We are not introducing any immunogenic particles from outside.

Tarlinton: I have a comment relating to the relationship between priming in the local lymph node and the observation that Ken Shortman made where dendritic cells have picked up Thy-1. He said this was not an endogenous synthesis but was due to their ability to pick up membranes from adjacent cells. Perhaps a dendritic cell in the kidney has absorbed a bit of membrane from one of the kidney cells and then migrated to the lymph node where it is acting to prime T cells. Are the dendritic cells just getting some membrane from the kidney cells rather than actually having to process antigen through an alternative pathway?

Nossal: It is true that this kind of dendritic cell can keep things on its outside for a very long time, so such a mechanism wouldn't shock me.

Shortman: Thy-1 may be exceptional in that it is lipid-anchored. Usually the attachment of proteins to follicular dendritic cells is indirect, via antibody attached to the follicular dendritic cell.

Carbone: I think David Tarlinton's suggestion was that the complex of MHC and peptide is shed from the kidney cells and then picked up by the APCs. The chimeras address this point. The bone marrow-derived APCs must express the correct MHC for there to be T cell activation. This rules out the transfer of MHC–peptide complexes from the kidney cells to the APCs.

Nossal: So is this great audience of T cell experts prepared to accept that in real life situations the dual pathway rules don't work? Are they an *in vitro* artefact?

Zinkernagel: We have looked at some of these aspects with viral systems. We quantified how efficient loading of class I took place from the outside versus from the inside. In fact, if you push the external pathway you can do it. Usually, soluble protein is very inefficient whereas aggregated or corpuscule-associated proteins are reasonably efficient. If one wants to quantify the difference of efficiency of live virus versus dead viral particles to induce class I-restricted cytotoxic T lymphocytes, it is probably of the order of greater than a 10^6-fold difference.

Miller: The cross-priming experiment and the experiment we have described today favour the idea that there exists an exogenous cross-over presentation to the class I pathway of antigen presentation.

von Boehmer: There is not only the cross priming experiment, but also the cross-tolerance experiments. In H-2^b male plus H-2^k female chimeras there is always tolerance to H-2^k male. Therefore the male antigen from H-2^b cells must be presented by H-2^k cells without introducing anything from the outside. It happens all of the time and is sufficient to induce tolerance!

Control of the sizes and contents of precursor B cell repertoires in bone marrow

Fritz Melchers

Basel Institute for Immunology, Grenzacherstrasse 487, CH-4005 Basel, Switzerland

Abstract. Ordered rearrangements of immunoglobulin (Ig) gene loci, first as D_H to J_H, then as V_H to D_HJ_H, and finally as V_L to J_L segment-specific recombinations occur 'in-frame' and 'out-of-frame'. 'In-frame' rearrangements lead to the expression of truncated D_HJ_H-μC proteins and to μH chains. These H chain proteins have two major effects on precursor B cells. They suppress (as DJCμ proteins) or enhance (as full μH chain) the proliferation of precursor cells at the point where these precursors express these proteins. At the same time, they signal allelic exclusion of the μH chain alleles, so that V_H to D_HJ_H rearrangement at the second allele is suppressed. Regulation of precursor B cell proliferation and H chain allelic exclusion is mediated by a pre-B cell receptor that is composed of the μH chains and a surrogate L chain. This surrogate L chain is made up of two proteins encoded by the Vpre-B and λ_5 genes that are expressed only at the early precursor cell stages just before and when H chain genes are first expressed. They are not found in later B cell development, when L chains are expressed, nor in any other cell of the body tested so far. The physiological roles of surrogate L chain and of the pre-B receptor have been clarified by generating mutant mice in which the λ_5 gene has been inactivated by targeted disruption. Molecular mechanisms and cellular developments, by which the pre-B receptor controls proliferation and allelic exclusion, are discussed.

1997 The molecular basis of cellular defence mechanisms. Wiley, Chichester (Ciba Foundation Symposium 204) p 172–186

Early B cell development is ordered by successive rearrangements of immunoglobulin (Ig) gene segments. First, D_H segments of the Ig heavy (H) chain gene locus are rearranged to J_H segments. Next, V_H segments are rearranged to D_HJ_H-rearranged segments. Finally, the Ig light (L) chain gene loci are rearranged (Tonegawa 1983). In the mouse, V_κ to J_κ rearrangements appear to be 10 to 20-fold more frequent than V_λ to J_λ rearrangements (for review, see Selsing & Daitch 1995).

IgH chain gene products influence the order of Ig gene rearrangements

In principle, both alleles of the Ig gene loci appear accessible when a particular rearrangement is carried out at a given stage of B cell differentiation. However, the results of the rearrangement process at one allele of the H chain gene locus influences the subsequent rearrangement at the other allele, at least whenever a productive rearrangement with subsequent expression of the locus as a μH chain protein occurs. In mouse but not in human B cell development, the first D_H to J_H rearrangements can lead to a productive, i.e. D_HJ_H-μC protein-expressing H chain gene locus whenever these D_HJ_H rearrangements occur in reading frame (rf) II (Fig. 1A) (Reth & Alt 1984, Tsubata et al 1991). Hence, the majority of pre-B cells at this stage of development, called pre-BI cells, have both H chain alleles D_HJ_H rearranged, but the representation of cells with rf II is suppressed (Gu et al 1991, Rolink & Melchers 1991, 1993). This suppression of rf II is not observed when the μH chain cannot be deposited in the surface membrane, i.e. of IgH chain gene loci where the transmembrane portion of the H chain gene has been deleted by targeted mutation. We have seen that the establishment of rf II suppression depends on the proliferation of pre-BI cells (with other rfs) and the concomitant suppression of pre-BI cells with rf II (Haasner et al 1994). In conclusion, a membrane-bound form of a truncated D_HJ_H-μH chain influences the repertoire of rfs in D_HJ_H-rearranged pre-BI cells in the mouse.

The expression of a truncated D_HJ_H-μH chain has a second consequence for the corresponding pre-B cells: they are inhibited from further V_H to D_HJ_H rearrangements (Ehlich et al 1994). Hence, a membrane-bound form of this truncated μH chain suppresses further rearrangements at both H chain alleles.

The rf I and/or rf III D_HJ_H-rearranged pre-BI cells then enter the stage of V_H to D_HJ_H rearrangements. The mechanisms controlling these V_H to D_HJ_H rearrangements have yet to be clarified in detail. It is, however, evident from studies with pre-BI cells, isolated '*ex vivo*', or grown in the presence of interleukin (IL)-7 and stromal cells *in vitro*, that the cytokine IL-7 plays a major role in the induction of the cellular development to a stage where these rearrangements can occur (Namen et al 1988, Nishikawa et al 1988). Pre-BI cells express c-Kit and the IL-7 receptor to control their continuous proliferation on stromal cells. Stromal cells express stem cell factor (also known as Steel factor; the ligand for c-Kit) and secrete IL-7 as a consequence of the pre-BI/stromal cell interaction. *In vitro* removal of IL-7 from the pre-BI cells induces their differentiation into $V_HD_HJ_H$-rearranged pre-BII cells, which thereby are rendered no longer reactive to stromal cells and IL-7, and thus cease to proliferate even in their presence (Rolink & Melchers 1991). *In vitro* such 'removal' of IL-7 may, in fact, occur when pre-BI cells have occupied all stromal cell

FIG. 1. (*see over*) B cell development from B-lineage-committed pro-B cells (with all Ig loci in germline configuration, i.e. the H chain loci in $V_H \neq D_H \neq J_H$, the L chain loci in $V_L \neq J_L$), to pre-BI cells (with H chain gene loci in germline or $V_H \neq DJ$ configuration) to large pre-BII cells (with H chain gene loci in $V_H \neq DJ$ or $V_HD_HJ_H$-rearranged configuration) to small pre-BII and to immature B cells (with H chain loci in configuration of large pre-BII cells, and L chain loci in V_LJ_L-rearranged, or germline configuration). p = productively rearranged; up = non-productively rearranged. (A) genetic constitutions of Ig loci; (B) with expansion of surrogate L chain (-+-); DJc μ proteins, mH chains and L chains.

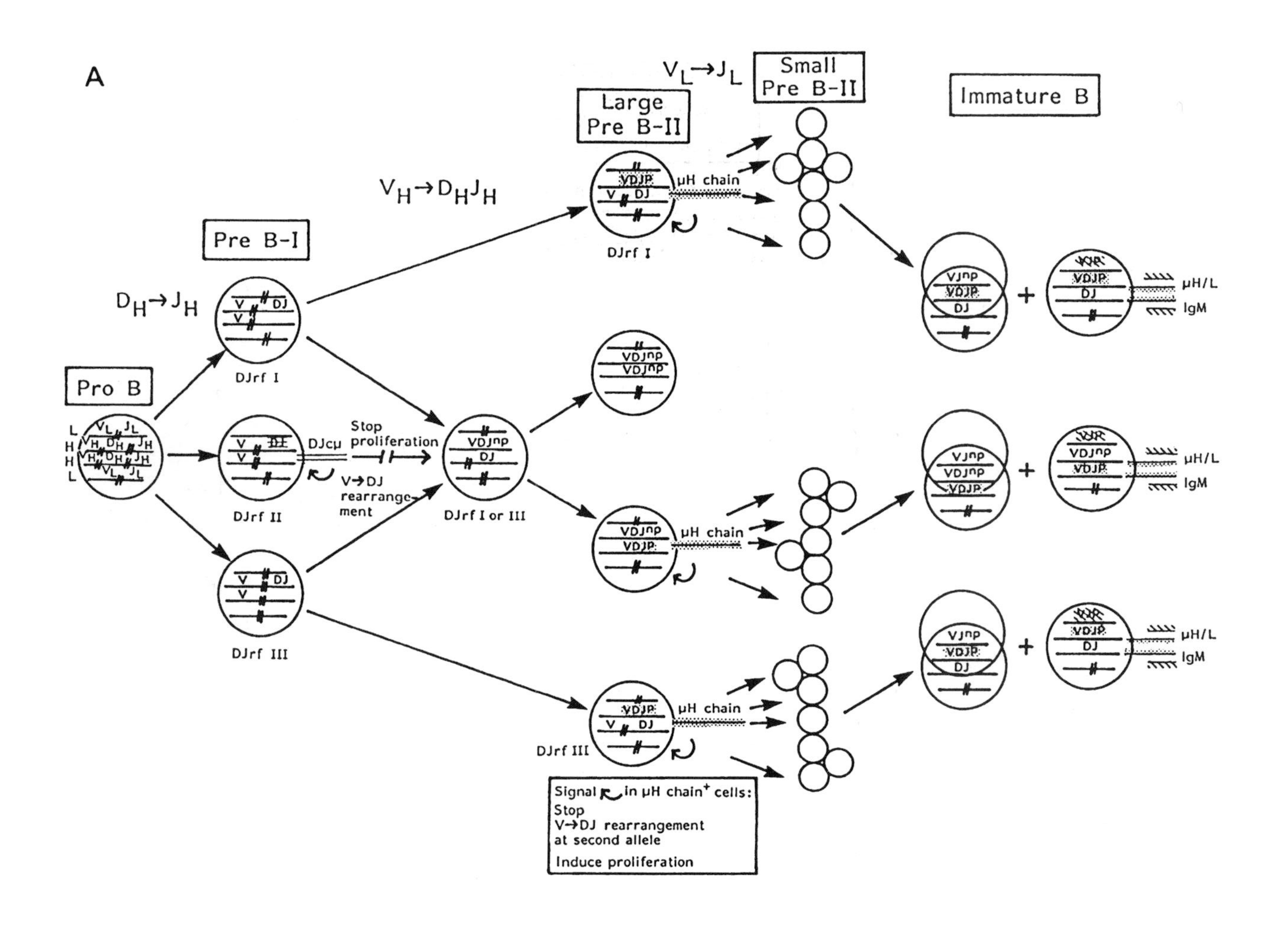

A
Pro B
Pre B-I
Large Pre B-II
Small Pre B-II
Immature B
$D_H \rightarrow J_H$
$V_H \rightarrow D_H J_H$
$V_L \rightarrow J_L$
DJrf I
DJrf II
DJrf III
DJrf I or III
DJcμ
Stop proliferation
V→DJ rearrangement
μH chain
μH/L
IgM
Signal in μH chain+ cells:
Stop
V→DJ rearrangement at second allele
Induce proliferation

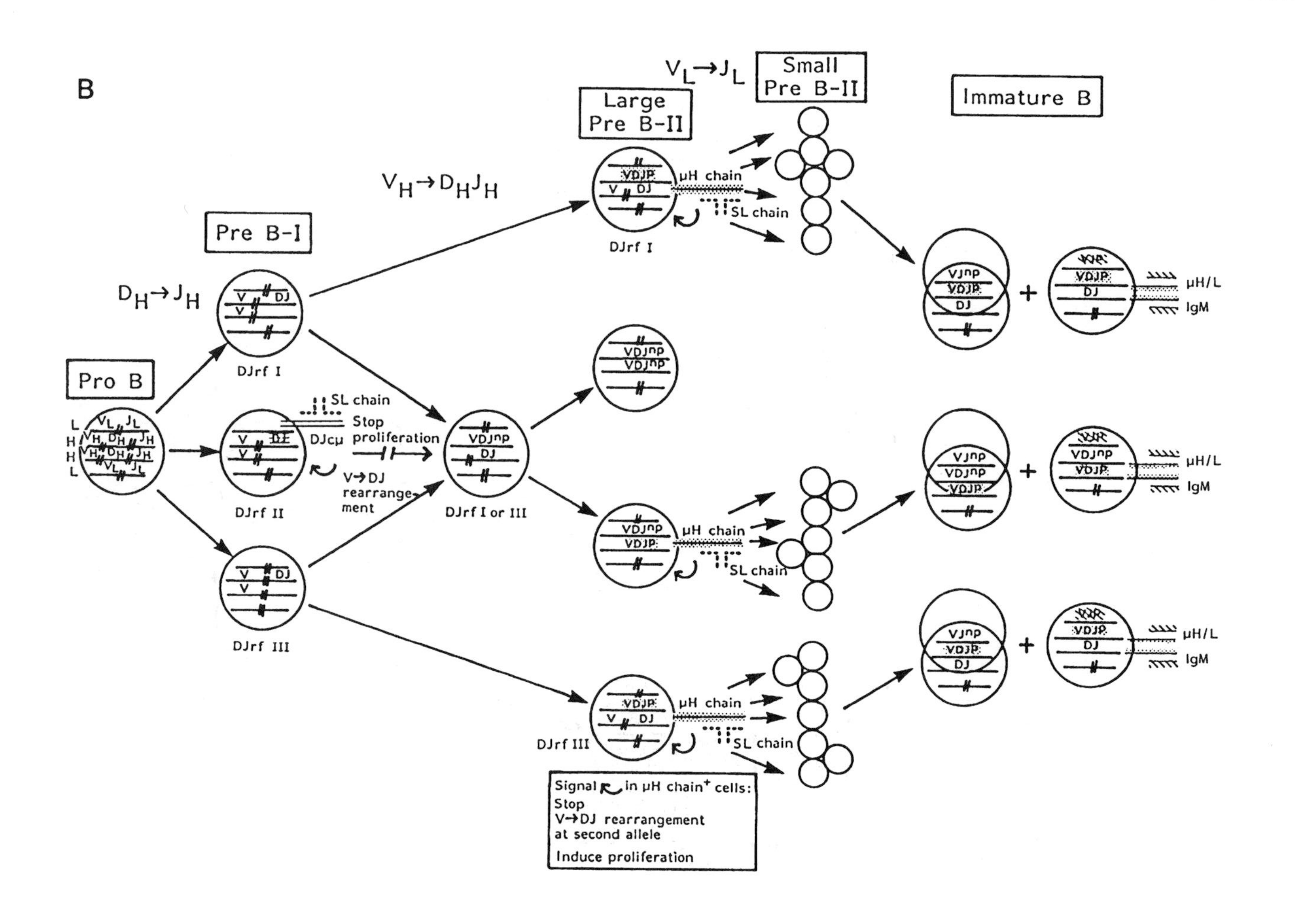
B
Pro B
Pre B-I
Large Pre B-II
Small Pre B-II
Immature B
$D_H \rightarrow J_H$
$V_H \rightarrow D_H J_H$
$V_L \rightarrow J_L$
DJrf I
DJrf II
DJrf III
DJrf I or III
SL chain
Stop proliferation
DJcμ
V→DJ rearrangement
μH chain
μH/L
IgM
Signal in μH chain+ cells:
Stop
V→DJ rearrangement at second allele
Induce proliferation

sites available for contact. The next division of a pre-BI cell will remove one daughter cell from this contact and thus remove it from the source of IL-7. Hence, induction of differentiation to the next step of B cell development is seen as a *removal* of a ligand, rather than its addition, from the cell which is to be induced.

Differentiation of B-lineage cells which do not express Ig chain

It is also worth emphasizing at this point of B cell development that c-Kit$^+$ CD19$^+$ CD45(B220)$^+$ (pre-BI-like) pro-B cells develop in V(D)J recombinase-deficient Rag-1T or Rag-2T mice in normal numbers. Furthermore, removal of IL-7 induces these pro-B cells *in vitro* to further differentiation, as monitored by the down-regulation of c-Kit expression, the loss of clonability on stromal cells in the presence of IL-7, and, most importantly, the induction of (germline, sterile) transcription of the κ and λL chain gene loci (Grawunder et al 1995). Hence, D_HJ_H- and $V_HD_HJ_H$-rearranged Ig loci, and the μH chain products translated from these productively rearranged Ig loci, are not mandatory for further B cell differentiation to pre-BII-like cells.

However, *in vitro* as well as *in vivo*, this differentiation occurs without proliferative expansion (in contrast to μH chain-expressing pre-BII cells—see below). Since the differentiation to pre-BII and immature B-like cells is accompanied by extensive apoptosis, these pre-BII-like cells are scarce and difficult to find *in vivo* in the bone marrow of the V(D)J recombinase-deficient mice.

The roles of μH chains in positive selection, expansion and allelic exclusion of precursor B cells

In vitro, D_HJ_H-rearranged pre-BI cells from bone marrow of normal mice rearrange V_H segments in the D_HJ_H-rearranged H chain loci as expected from a random 'in-frame' and 'out-of-frame' rearrangement process, so that only a fraction of all cells produce a μH chain, first as intracytoplasmic protein, later (on immature B cells with L chains) as surface Ig (sIg). This, again, occurs without any detectable cell proliferation.

In striking contrast, the pre-BII compartment in the bone marrow of normal mice is filled to over 98% with μH chain-expressing (i.e. productively $V_HD_HJ_H$-rearranged) cells. 10–20% of them are in cell cycle, i.e. proliferating, while 80% are resting. We have shown that the proliferating pre-BII cells are the precursors of the resting pre-BII cells, since single-cell PCR analyses of the status of rearrangement of the Ig loci has indicated that the majority of the $V_HD_HJ_H$-rearranged large pre-BII cells have all Ig L chain loci in germline configuration, while the resting pre-BII cells have a majority of L chain gene loci V_LJ_L rearranged (Ten Boekel et al 1995) (Fig. 1).

In mutant mice in which the transmembrane portion-encoding region of the μH chain gene has been deleted by targeted mutation (Kitamura & Rajewsky 1992), B cell development is arrested at the transition from pre-BI to pre-BII cells in bone marrow (Rajewsky 1992, Rolink & Melchers 1993). Over 95% of the pre-BII compartment is missing, as are all subsequent B cell differentiation stages, and large, cycling pre-BII cells are absent. Hence, μH chains have to be deposited in the (surface) membranes of pre-B cells at the transition from pre-BI to pre-BII cells to generate large, cycling cells and a 10–20-fold larger number of resting pre-BII cells. The proliferative expansion of pre-BII cells *in vivo* appears to be the mechanism by which productively $V_HD_HJ_H$-rearranged μH chain-producing cells are positively selected over non-productively-rearranged, μH chain-negative cells.

Again, the membrane-bound expression of μH chains has a second consequence for the emerging pre-BII cells. In all those cells, in which the second allele is still only D_HJ_H-rearranged, further V_H to D_HJ_H rearrangements are suppressed (Fig. 1A). Single-cell PCR analyses of the Ig H chain gene loci in pre-BII cells (cycling as well as resting) have shown that half of all pre-BII cells retain the second H chain allele in D_HJ_H-rearranged configuration (Ehlich et al 1994, Ten Boekel et al 1995). As a consequence, this suppression of V_H to D_HJ_H rearrangements at the second allele upon a productive $V_HD_HJ_H$ rearrangement in these cells at the first allele ensures allelic exclusion at the Ig H chain gene locus. The other half of the pre-BII cell compartment has one allele productively, and the other allele non-productively $V_HD_HJ_H$ rearranged. Here allelic exclusion has been achieved in all likelihood by a first non-productive and a second productive rearrangement. Hence, over 99% of all B cells produce only one H chain.

In summary, the expression of truncated or normal size μH chain proteins before the rearrangements of L chain gene loci appears to have two major effects on precursor B cells. (1) They suppress (as truncated $V_HD_HJ_H$ proteins) or enhance (as full size μH chains) the proliferation of cells at the transition from pre-BI to pre-BII cells; and (2) they signal allelic exclusion of the H chain alleles at the level of V_H to D_HJ_H rearrangements.

Surrogate L chain

How can H chains perform these signalling functions? In principle, they could do so either inside the cell or on its surface, as long as they are membrane-bound. It is likely that they do so from the cell surface, because despite earlier reports (see below), μH chains can be detected on the surface. However, μH chains alone cannot be transported to the cell surface. In immature and mature sIg^+ B cells, L chains associate with μH chains and, thus, enable the surface deposition. However, in precursor cells, i.e. in pre-BI and large pre-BII cells (Fig. 1), L chain gene loci remain in germline configuration, hence L chains cannot be produced. How, then can μH chains be transported to the surface and deposited there?

Some reports had indicated that μH chains could be deposited on the surface of precursor B cells, apparently with other polypeptide chains which had the approximate size of L chains — and which, erroneously and without proof of their structure, were assumed to be normal L chains (Melchers 1977, Rosenberg & Parish 1977). This became untenable when it was found that Ig gene rearrangements occurred in the stepwise fashion described above, and that precursor B cells had Ig L chain gene loci in germline, non-rearranged conformations. Subsequently, μH chains could not be detected by other laboratories on the surface of the majority of the precursor B cells although they expressed the μH chains in easily detectable quantities in the cytoplasm (Levitt & Cooper 1980, Siden et al 1981).

The discovery of two genes, Vpre-B and λ_5, has changed this picture of B cell development again. The two genes are transcribed and expressed as proteins at two early stages of precursor B cell development, namely in pro/pre-BI and in large, cycling pre-BII cells (Fig. 1B). They are not expressed in later stages of B cell development nor in any other cell of the body so far tested (reviewed in Melchers et al 1993).

Their sequence homology with Ig domains provided clues as to their possible functions in B cell development. The Vpre-B genes have weak, but significant homologies with Ig and T cell receptor variable regions. The N-terminal region of the λ_5 gene, again, has weak homology to the variable regions, while the C-terminal region has strong homology to the J- and C-encoded segments of λ L chains. The Vpre-B and λ_5 genes are not rearranged during development. The proteins they encode associate with each other to form an L chain-like structure which has been termed 'pseudo' or 'surrogate' L chain. This structural homology with conventional L chains made the surrogate L chain a good candidate for a partner molecule of μH chains in precursor B cells which have not yet rearranged L chain gene loci. Surrogate L chain, in fact, was found to be disulfide-bonded via its λ_5 C-terminal cλL-like domain to μH chains as well as to truncated D_HJ_H-μC proteins, to form a precursor B cell receptor (pre-B receptor). This pre-B receptor was found on the surface of normal and transformed pre-B cells which produced μH chains, but had their L chain gene loci in non-rearranged conformations.

Surrogate L chain, therefore, provides the missing link in B cell development which clarifies how μH chains can signal proliferative expansion of productively $V_HD_HJ_H$-rearranged cells (and, respectively, suppression of reading frame II D_HJ_H-rearranged cells) at the pre-BI to pre-BII transition, and how they can signal allelic exclusion at the level of V_H to D_HJ_H rearrangements of H chain loci.

Surrogate L chain has been found in other species. Development of mouse and human B lineage cells in bone marrow is strikingly similar. The sizes of the precursor compartments are comparable, they both follow a stepwise order of rearrangements of the Ig gene loci, and both express surrogate L chain at comparable stages of cellular development.

Functions of surrogate L chain in pre-B receptors

The physiological role of surrogate L chain, in particular its functions in proliferative expansion and allelic exclusion of precursor B cells, has been clarified by the generation of mutant mice in which the λ_5 gene has been inactivated by targeted mutation (Kitamura et al 1992). The analysis of bone marrow cells by flow cytometry revealed that the number of small pre-BII cells and of $sIgM^+$ immature and mature B cells was drastically reduced, whereas that of earlier precursor B cells, i.e. of pro/pre-BI cells was not normal, if not increased two- to threefold compared with control littermates (Kitamura et al 1992). An analysis using c-Kit, CD25 and surrogate L chain as distinguishing markers showed that c-Kit$^+$ CD25$^-$ surrogate L chain$^+$ pro-B/pre-BI cells were produced in normal numbers, whereas c-Kit$^-$ CD25$^+$ surrogate L chain$^+$ large pre-BII cells and c-Kit$^-$ CD25$^+$ surrogate L chain$^-$ large and small pre-BII cells, as well as immature B cells were at least 40-fold reduced (Rolink et al 1993). These results indicate that in λ_5-deficient mice B cell differentiation is impaired at the transition from the pro-B/pre-BI to the pre-BII cell stage. Peripheral BI and conventional B cells accumulate more slowly as compared with littermate controls. While the small compartment of BI cells is filled one to two weeks later after birth, it takes more than six months to fill even half of the large compartment of conventional B cells. The retarded accumulation of mature B cells in the periphery can be explained by the severely reduced pool size of immature B cells from which these mature cells are formed. In contrast, the number of T cells appears to be unchanged in λ_5-deficient mice. Thus, surrogate L chain-deficient and μH chain transmembrane segment-deficient mice are severely affected in their B cell development at the same stage, i.e. where they enter the pre-BII compartment by proliferative expansion. This suggests that the μH chain/surrogate L chain pre-B receptor, in fact, signals this proliferative expansion. As a consequence, only cells with productively $V_HD_HJ_H$-rearranged H chain alleles, producing μH chains, are expanded, overgrowing the non-productively rearranged cells.

Two additional sets of experiments illuminate the proliferation-inducing role of the pre-B cell receptor. In one, the transgenic expression of μH chains in Rag-deficient mice fills up the bone marrow with normal number of large and small pre-BII cells (Karasuyama et al 1994, Rolink et al 1994, Spanopoulou et al 1994, Young et al 1994). This result is in accord with the proposed model. Several μH transgenes with different V_H regions, and one without a V_H region, all fulfil this function, indicating that the presence and/or specificity of V_H is not of primary importance for the proliferation-controlling function of the pre-B receptor.

In the other set of experiments, the precocious expression of a conventional λ_2 or κL chain as a transgene in λ_5-deficient mice restores the sIgM$^+$ immature B cell compartment in bone marrow cells at normal rates (Rolink et al 1996). This restoration can be observed only when these L chain transgenes are expressed precociously, i.e. under the control of Eμ enhancer, but not when expressed normally, i.e. under the control of their own enhancers. This indicates that the surrogate L chain can be replaced by conventional L chains if the L chains are expressed abnormally early. Altogether, these results suggest that the pre-B receptor has to be deposited in the (surface) membrane of pre-B cells, but that the composition of its variable and variable-like regions can vary. This makes it less likely that a defined ligand, possibly provided by environmental stromal cells, functions in this proliferative expansion. Since the Vpre-B protein is expressed in all mutant cases it, together with μH chains, is a candidate for such ligand recognition.

When one of the two μH chain alleles has achieved a productive $V_HD_HJ_H$ rearrangement and produces μH chain, the V_H to D_HJ_H rearrangement at the other allele is inhibited to avoid the generation of B cells with double specificity. The surrogate L chain appears also involved in the molecular mechanisms which achieve this allelic exclusion. The feedback inhibition of V_H to D_HJ_H rearrangements at the second allele takes place as soon as a μH chain becomes expressed, i.e. at the earliest phase of the large pre-BII cell stage, in which a membrane-bound form of μH chain becomes associated with surrogate L chain (Karasuyama et al 1994, Winkler et al 1995) (Fig. 1B). Therefore, the pre-B cell receptor appears to participate in signalling for allelic exclusion. In contrast to this prediction, the allelic exclusion seemed to be maintained in λ_5-deficient mice as long as splenic B cells were examined (Kitamura et al 1992). However, a recent analysis of bone marrow cells in these mice revealed that the μH chain genes were not allelically excluded at the precursor B cell stage (Löffert et al 1996). Though it remains uncertain how the cells producing two μH chains are eliminated along the differentiation from pre-B to B cells, this result supports the idea that both μH chain and surrogate L chain (i. e. the pre-B cell receptor) are involved in the process of allelic exclusion at the pre-B cell stage.

Regulated *rag* expression as one mechanism controlling allelic exclusion?

The gene products of two recombination activating genes, *rag-1* and *rag-2*, are crucial for the rearrangements of Ig genes. In fact, mice deficient for either of the two have all Ig heavy and light chain gene loci in germline configuration (Mombaerts et al 1992, Shinkai et al 1992). Hence, one of the ways to stop the rearrangement at the second allele could be to down-regulate the expression of the *rag* genes as soon as a μH chain has been expressed. Therefore, the pattern of *rag* expression along B cell development was examined on mRNA level by reverse transcriptase (RT)-PCR. Earlier work had shown that the expression of both *rag* genes was maintained at high levels through the pro-B and pre-B cell stages until it dropped at the immature B cell stage (Li et al 1993). However, a recent study on *rag* mRNA level by RT-PCR and on protein level by Rag-specific antibodies together with more detailed fractionation of pre-B cell populations has illuminated a sharp and transient drop of Rag-1 and -2 expression at the earliest phase of the large pre-BII cell stage, coincident with the expression of the pre-B cell receptor (Grawunder et al 1995). As expected, high levels of *rag* transcripts and Rag-2 proteins were detected in the pro-B/pre-BI and the small pre-BII stages where the rearrangements of H chain and L chain genes took place, respectively. In contrast, the large pre-B cells between those stages expressed at least 50-fold less *rag* transcripts and Rag-2 protein. Among them, especially, the cells expressing pre-B cell receptors did not express any detectable levels of Rag transcripts or protein. The coincidence of the transient down-regulation of Rag and the expression of the pre-B cell receptor strongly suggests an involvement of the receptor in allelic exclusion of the heavy chain locus through this down-regulation of the *rag* genes. Taken together, the shut down of recombination activity at the large pre-BII cell stage appears to be guaranteed by the down-regulation of Rag expression at both the transcriptional and the post-transcriptional levels. Thus, the pre-B cell receptor might induce allelic exclusion in several different modes; one to shut down the expression of *rag* genes, another to drive cells into cell cycle. Once large pre-BII cells have shut down Rag activity, they are expected to close the second μH chain allele in order to prevent further rearrangements before they open the L chain gene loci, and the *rag* genes become re-expressed.

Finally, it is both striking and rewarding to see many similarities in B cell development in bone marrow and T cell development in thymus. Similar to the pre-B receptor in B cell development, a pre-T receptor composed of VDJ-rearranged TCR β chains and the pre-T α chain appears to control proliferative expansion and allelic exclusion in α/β TCR–T cell development (Groettrup & von Boehmer 1993).

Acknowledgements

The Basel Institute for Immunology was founded and is supported by Hoffmann-La Roche Ltd.

References

Ehlich A, Matin V, Müller W, Rajewsky K 1994 Analysis of the B-cell progenitor compartment at the level of single cells. Curr Biol 4:573–583

Grawunder U, Rolink A, Melchers F 1995 Induction of sterile transcription from the κL chain gene locus in V(D)J-recombinase deficient progenitor B cells. Int Immunol 7:1915–1925

Groettrup M, von Boehmer H 1993 A role for a preT-cell receptor in T-cell development. Immunol Today 14:610–614

Gu H, Kitamura D, Rajewsky K 1991 B cell development regulated by gene rearrangement: arrest of maturation by membrane-bound D_μ protein and selection of D_H element reading frames. Cell 65:47–55

Haasner D, Rolink A, Melchers F 1994 Influence of surrogate L chain on D_HJ_H-reading frame 2 suppression in mouse precursor B cells. Int Immunol 6:21–30

Karasuyama H, Rolink A, Shinkai Y, Young F, Alt FW, Melchers F 1994 The expression of V_{preB}/λ_5 surrogate light chain in early bone marrow precursor B cells of normal and B-cell deficient mutant mice. Cell 77:133–143

Kitamura D, Rajewsky K 1992 Targeted disruption of μ chain membrane exon causes loss of heavy-chain allelic exclusion. Nature 356:154–156

Kitamura D, Kudo A, Schaal S, Müller W, Melchers F, Rajewsky K 1992 A critical role of λ_5 in B cell development. Cell 69:823–831

Levitt D, Cooper MD 1980 Mouse preB cells synthesize and secrete μ heavy chains but not light chains. Cell 19:617–625

Li YS, Hayakawa K, Hardy RR 1993 The regulated expression of B lineage associated genes during B cell differentiation in bone marrow and fetal liver. J Exp Med 178:951–960

Löffert D, Ehlich A, Müller W, Rajewsky K 1996 Surrogate light chain expression is required to establish immunoglobulin heavy chain allelic exclusion during early B cell development. Immunity 4:133–144

Melchers F 1977 Immunoglobulin synthesis and mitogen reactivity: markers for B lymphocyte differentiation. In: Cooper MD, Dayton DH (eds) Development of host defenses. Raven, New York, p 11–29

Melchers F, Karasuyama H, Haasner D et al 1993 The surrogate light chain in B-cell development. Immunol Today 14:60–68

Mombaerts P, Iacomini J, Johnson RS, Herrup K, Tonegawa S, Papaioannou VE 1992 RAG-1-deficient mice have no mature B and T lymphocytes. Cell 68:869–877
Namen AE, Lupton S, Hjerrild K et al 1988 Stimulation of B-cell progenitors by cloned murine interleukin-7. Nature 333:571–573
Nishikawa S, Ogawa M, Nishikawa S, Kunisada T, Kodama H 1988 B lymphopoiesis on stromal cell clone: stromal cell clones acting on different stages of B cell differentiation. Eur J Immunol 18:1767–1771
Rajewsky K 1992 Early and late B-cell development in the mouse. Curr Opin Immunol 4:171–176
Reth MG, Alt FW 1984 Novel immunoglobulin heavy chains are produced from DJH gene segment rearrangements in lymphoid cells. Nature 312:418–423
Rolink A, Melchers F 1991 Molecular and cellular origins of B lymphocyte diversity. Cell 66:1081–1094
Rolink A, Melchers F 1993 Generation and regeneration of cells of the B-lymphocyte lineage. Curr Opin Immunol 5:207–217
Rolink A, Karasuyama H, Grawunder U, Haasner D, Kudo A, Melchers F 1993 B cell development in mice with a defective λ_5 gene. Eur J Immunol 23:1284–1288
Rolink A, Grawunder U, Winkler TH, Karasuyama HFM 1994 IL-2 receptor α chain (CD25, TAC) expression defines a crucial stage in preB cell development. Int Immunol 6:1257–1264
Rolink A, Haasner D, Melchers F, Andersson J 1996 The surrogate light chain in mouse B cell development. Int Rev Immunol, in press
Rosenberg YJ, Parish CR 1977 Ontogeny of the antibody-forming cell line in mice. IV. Appearance of cells bearing Fc receptors, complement receptors, and surface immunoglobulin. J Immunol 118:612–617
Selsing E, Daitch LE 1995 Immunoglobulin λ genes. In: Honjo T, Alt FW (eds) Immunoglobulin genes, 2nd edn. Academic Press, New York, p 193–203
Shinkai Y, Rathbun G, Lam KP et al 1992 RAG-2-deficient mice lack mature lymphocytes owing to inability to initiate V(D)J rearrangement. Cell 68:855–867
Siden E, Alt F, Shinefeld L, Sato V, Baltimore D 1981 Synthesis of immunoglobulin μ chain gene products precedes synthesis of light chains during B-lymphocyte development. Proc Natl Acad Sci USA 78:1823–1827
Spanopoulou E, Roman CAJ, Corcoran LM et al 1994 Functional immunoglobulin transgenes guide ordered B-cell differentiation in RAG1-deficient mice. Genes & Dev 8:1030–1042
Ten Boekel E, Melchers F, Rolink A 1995 The status of Ig loci rearrangements in single cells from different stages of B cell development. Int Immunol 7:1013–1019
Tonegawa S 1983 Somatic generation of antibody diversity. Nature 302:575–581
Tsubata T, Tsubata R, Reth M 1991 Cell surface expression of the short immunoglobulin μ chain (Dμ protein) in murine pre-B cells is differently regulated from that of the intact μ chain. Eur J Immunol 21:1359–1363
Winkler TH, Rolink AG, Melchers F, Karasuyama H 1995 Precursor B cells of mouse bone marrow express two different complexes with the surrogate light chain on the surface. Eur J Immunol 25:446–450
Young F, Ardman B, Shinkai Y et al 1994 Influence of immunoglobulin heavy- and light-chain expression on B-cell differentiation. Genes & Dev 8:1043–1057

DISCUSSION

Nossal: I'm sure the haematologists would be very interested in your speculation, given the enormous importance of the surrogate receptor, as to what you think it might be recognizing.

Melchers: The short answer is that we don't know. The more lengthy answer is that it is disturbing how many heavy chain transgenes work in filling up the pre-BII compartment on a Rag-2 knockout genetic background. Of course, these heavy chains are in a sense positively selected in that they all come from peripheral B cells. What hasn't been tested yet is a truly random heavy chain repertoire before the exit into the periphery, but this is difficult to do.

The other disturbing thing is what happened when we tried to repair the defect of the λ_5 knockout by using transgenic light chains. We found that a λ_2-light chain and a κ-light chain, expressed under Eμ-control, could replace normal λ_5 light chain in the capacity of expanding the pre-B cells and would re-establish the peripheral B cell pools at normal rates.

Nossal: You are therefore retreating somewhat from the previous working hypothesis that this might be a recognition event that held the pre-Bs in a particular geographic compartment in association with stromal cells, thereby preventing some of the later maturation.

Melchers: All I am saying is that it is not the specificity of the variable regions of the heavy chains, and not the variable-like regions of the surrogate light chains. I am still saying that it is the pre-B receptor, but I don't know what part of it.

Metcalf: You said there is normally a 10-fold population increase which does not happen in a variety of knockouts, the common feature of which is the lack of expression of the pre-B receptor. Is the IL-7 responsiveness normal in these mice?

Melchers: Yes, because pre-B cell lines can be established from λ_5-knockout, μH chain transmembrane knockout or Rag-knockout bone marrow, which respond normally to stromal cells and IL-7.

Metcalf: To me the pre-B receptor appears to be a candidate receptor for a ligand that is capable of delivering a proliferative signal. Since it is a surrogate light chain, this is not so difficult to imagine. This may force you to postulate that the IL-7 pathway is a default pathway in normal development, but this does not agree well with the IL-7 knockout data. Is it possible that there is some sort of complexing between the IL-7 receptor and this surrogate light chain?

Melchers: There is no evidence for this. It is dangerous to compare this proliferative expansion with haemopoietic cell proliferation in the marrow. When we look for B cell development in fetal liver, we see the same surrogate light chain expression but a pre-BII compartment is never formed. Hence the pre-BII-type cells might leave the fetal liver as soon as they develop, in contrast to bone marrow where they stay and accumulate.

Nossal: Couldn't that be explained by simply saying you want a richer substrate for immunoglobulin V gene hypermutation and selection in order to cover every possible antigen?

Rajewsky: The extent of clonal expansion of these cells is controversial because the size of the pre-B compartment of the resting cells is determined by both the lifespan and the expansion of cells going into it.

Strasser: I'm particularly interested in the timing of the cell cycle status and the timing of rearrangement of antigen receptor genes, not only in B cell but also T cell development. It makes a lot of sense for a cell not to play around with its genes while it is in S phase. Data generated from T cells on the timing of rearrangement and the cell cycle status correlate beautifully: expansion occurs first to generate a large number of cells, then the cells stop and rearrange TCR β genes. During this time there is no cell division; afterwards cells lacking TCR β die and the ones expressing TCR β proliferate. When the second receptor locus (TCR α) is rearranged, the cells cease to proliferate again. This makes a lot of sense to me. However, on your scheme, the timing of rearrangement of the Ig heavy chain genes seems to occur in a compartment where the cells are rapidly cycling.

Melchers: It depends which cells you're looking at. The DJ-rearranged and rearranging pre-BI cells are clearly in cycle. The pre-B receptor-expressing, large, cycling pre-BII cells are also in cycle. For the latter cells, the hypothesis is that the pre-B receptor could signal the shut-off of VDJ-rearrangements at the heavy chain allele which has not yet been VDJ-rearranged. It is also pretty clear that the pre-B receptor does not induce VJ-L chain gene rearrangements.

Burgess: Is CD19 one of the first of the B cell markers, and can you generate $CD19^+$ cells *in vitro* from $CD19^-$ cells?

Melchers: We have looked at CD19 expression using a monoclonal antibody developed by Doug Fearon. All $CD19^+$ cells express B220, but there's a population of cells that expresses B220 that is $CD19^-$. Ton Rolink has spent the last few months analysing the latter population and has found that about a third of them are NK precursors. There's a second population that expresses CD4 and a third that is split into two sub-populations, one of which expresses MHC class II molecules. Amongst the triple negative sub-population, which is only $B220^+$, there is a high frequency of cells that will become $CD19^+$, and amongst those are also cells that express surrogate light chain. It is interesting that CD19 is not expressed by all cells that could eventually become B lineage cells.

Burgess: So is B220 the earliest-expressed marker?

Melchers: B220 seems to be the earliest, but it isn't a B lineage-specific marker, because of the $B220^+$ cells that can become NK cells.

Tarlinton: You found no bias in the usage of V_H gene families in bone marrow B cells from animals which expressed the surrogate light chains. Do you have any thoughts on how such a situation of random V_H gene usage could arise?

Melchers: The simplest interpretation is that it is somehow connected to the proliferative expansion of pre-B cells at the interphase of pre-BI to pre-BII. Whenever this doesn't happen, as in fetal liver, the randomization doesn't happen. So far all we have done to study this is to measure ratios between on the one hand V_H7183 and V_HQ52 gene usage, and on the other V_HJ558 family gene usage.

Tarlinton: Is that in the pre-BII compartment?

Melchers: In the first pre-B receptor expressing compartment, i.e. in large pre-BII cells, the genes are already randomized in their V_H usage.

Tarlinton: A number of other groups have examined V_H gene usage in bone marrow B cells and have consistently found preferential usage of the families located at the 3′

end of the V_H gene cluster, specifically the 7183 and Q52 families (Freitas et al 1990, Malynn et al 1990). These results would appear to be the opposite of those presented by you.

Melchers: I know.

Nossal: Is anything known about the mechanism whereby, for example, a failure to make productive rearrangement of the Ig chains would signal the apoptotic pathway? How does the cell know that it needs to die?

Melchers: All of the cells are bound to die unless rescued; I don't think apoptosis needs to be induced because it is the default pathway. Perhaps negative selection is accelerated death.

von Boehmer: It is clear in the double positive thymocytes that the cells die whether or not they have a receptor. If they have the receptor of the right specificity they don't die. There is no signal to die: death is preprogrammed. The receptor signals rescue from programmed cell death.

Strasser: I don't agree with Fritz Melchers' notion that negative selection might just be accelerated death. There is good evidence from various labs including our own that there are two fundamental forms of cell death in lymphoid cells. One is the death by neglect of cells with useless or no antigen receptors, and the other involves cells that make perfect receptors but have the misfortune of interacting with an antigen with extremely high avidity, and which therefore get activated into the cell death programme. Bcl-2 blocks the former almost entirely but the latter not at all.

Metcalf: There is such a thing as death from old age after having had a fulfilled life. Red cells, for instance, have a finite lifespan. In principle there is no reason why you couldn't have a granulocyte or a B cell in the marrow with a finite lifespan.

Williams: A good example is provided by the gut: once a progenitor cell in the crypt starts to differentiate the lifespan of the cell may be as short as 24 h.

Melchers: I have not said that these cells aren't useful; I don't think anybody knows what they're used for. Apparently the lymphocytes don't do anything as immature cells except to get selected.

Metcalf: There was a small but active field where lymphocytes were for a time regarded as trophocytes—cells that feed other cells. One of my earlier experiments involved labelling lymphocytes with tritiated thymidine and then measuring the reutilization of radiolabelled DNA. I don't think anyone seriously believes in this hypothesis today.

Melchers: Ian McLennan is a strong proponent of the idea that most of the immature cells do not actually die in the bone marrow, but reach the extrafollicular regions of the spleen, and that is where this selection or death happens. No one would exclude the possibility that the B cells have some function that we haven't discovered yet in these extrafollicular regions. For the thymocytes the problem is different. They never get out of the thymus. The medulla of the thymus may, in fact, be the site corresponding to the extra follicular regions of spleen when we compare immature T and B cell development.

Rajewsky: The function of these cells could be that they simply form a reservoir, from which cells can be drawn in, if there is a need in the periphery.

Metcalf: Along those lines you could argue that myeloblasts don't have a function because they never leave the marrow. It is a little like debating the value of a new-born baby.

References

Freitas AA, Andrade L, Lembezat MP, Coutinho A 1990 Selection of V_H gene repertoires: differentiating B cells of adult bone marrow mimic fetal development. Int Immunol 2:15–23

Malynn BA, Yancopoulos GD, Barth JE, Bona CA, Alt FW 1990 Biased expression of J_H-proximal V_H genes occurs in the newly generated repertoire of neonatal and adult mice. J Exp Med 171:843–859

General discussion V

Cytokine regulation of immune responses and further reflections on self–non-self discrimination

Nossal: In the last three papers, we have been thinking about subsets and how cells come to belong to them. We are coming to grips with the fact that 'choices' depend largely on cytokine regulation. We also have the vast field of self and non-self to discuss, re-visiting perhaps the question of whether in tolerance induction in the periphery, for both B and T cells, the final common pathway is always going to be some form of activation which precedes death. And we have the exciting double life of the lymphocyte to address. What is special about lymphocytes is that their complex development leads to end-stage G0 cells, but they serve only one purpose, and that's either to be selected if they meet their antigen, or not if they don't. Then they go through the whole process of division and differentiation once more, requiring a whole new set of feedback loops, again involving activation-induced cell death in some circumstances, to make sure that this second wave of antigen-induced proliferation doesn't turn the whole body into a jellified crystal of pure γ-globulin.

Mosmann: Perhaps I could address the point you just made about differentiation and whether it is instructional. We showed a while ago that a single T cell can give rise to either Th1 or Th2 cells. It is pretty clear that this decision is made after exposure to antigen, but to get good differentiation in these cultures we normally have to add antibodies against some of the cytokines. There's a strong suspicion that the cell spontaneously starts to produce some of the other cytokines, and if it's allowed to continue making those there's a positive feedback and off it goes in that direction. The regulation of that 'initial testing of the waters' may not be instructed. Instead it could be spontaneously happening at low frequency.

Paul: I'm not convinced that's the case. There may be specialized cells that produce the cytokines required to induce naïve T cells to differentiate into Th1 or Th2 cells.

Burgess: Jacques Miller, you said ovalbumin is expressed in the kidney and the antigen appears in the draining lymph node. Do you envisage that there's some sort of surveillance cell washing through these organs looking for antigens and then trafficking with them to the lymph node?

Miller: I think antigen-presenting cells exist in non-lymphoid organs and traffic to regional lymph nodes.

Burgess: What are the candidate cells?

Miller: We don't know their identity, but among the candidates are dendritic cells, macrophages and perhaps B cells.

Burgess: When these cells see the antigen in organs, do they cleave it off?

Miller: I think the antigen is shed and is picked up exogenously by antigen-presenting cells.

Burgess: Can you see it in the urine and the serum?

Miller: You can actually find it in the serum at very low concentrations.

Melchers: An interesting question is how that antigen presenting cell would induce non-reactiveness in the end rather than reactiveness in a response. An answer might be that the steady-state production of a constant level of antigen doesn't stimulate a response. Instead, changes in the concentration of an antigen appear to induce an immune response.

Goodnow: We feel strongly that this is the case for B cells. One of the key cues that mature cells have to distinguish self from non-self is the kinetics of antigens. It comes into the same realm as the endorphin receptors, which when chronically exposed to their ligand will desensitize.

Hodgkin: Situations must occur where self molecules fluctuate in concentration. How then would you distinguish foreign and self?

Melchers: There is still elimination of self-reactive cells in the primary organs so, with lymphocytes, antigen receptor must play a role.

Nossal: It is possible to remove the thyroid in very early life and later to transplant a syngeneic thyroid, which will then be destroyed by a non-tolerant lymphocyte population.

Mosmann: There is also the possibility in the T cell system that the lymph node environment may prime cells up for a while, but they then need to go into a tissue and react under rather different circumstances to keep the response going. In other words the lymph node environment on its own cannot sustain a response indefinitely. If I understand correctly, in your system you have a lymph node reaction, but it is possible that there is not enough inflammation in the target tissue to bring the cells in so that they can actually react in the tissue as well. Perhaps that tissue reactivity is required to sustain the response in the way an infection model does.

Miller: Absolutely. That's why I brought up the idea that if a tissue is infected by a virus or pathogen that causes an inflammatory response, the CD8 T cells that have been activated in the way I mentioned can now enter the organ and get rid of the infected cells. In contrast, in healthy tissues there is no inflammation nor any up-regulation of B7 or other co-stimulatory molecules. In that case, the antigen presenting cells will stimulate the CD8 cell to the pathway of activation and apoptosis, and there will be no tissue infiltration, unless of course the precursor frequency is very high.

Zinkernagel: Concerning the question of concentration changes and activation, at least in the case we have studied where the transgenic VSV-G is expressed either in the membrane or as a soluble form, we found that when we immunize with a poorly or non-organized antigen form of G, we cannot induce the B cells unless we bring in linked T help in the classical fashion. But if we immunize with highly organized paracrystalline antigen forms such as G on VSV envelopes, we immediately induce the specific B cells, despite a stable steady-state of self-antigen expressed on transgenic host cells or in the serum. There is an additional dimension, in addition to T help, to what is important to trigger unresponsive (anergic?) B cells — i.e. the degree of cross-linking.

Paul: I wanted to come back to a point that Rolf Zinkernagel makes often, *vis à vis* the issue of infection inducing a set of inflammatory activities that are particularly favourable for allowing an immune response. Rolf makes the point that there is a set of viruses that don't induce anything other than immunopathology. Do such viruses induce the inflammation required to elicit a response?

Zinkernagel: Lymphocytic choriomeningitis virus, for example, initially induces a fantastic amount of interferon. Interferon $\alpha\beta$ would be one of the factors that kicks off the whole circus of factors. But one should keep in mind that in biological systems it is all a question of balance. If antigen and inflammation is localized in the lymphocyte then induction happens efficiently, as we usually want to see it. However, if virus inundates the immune system and the periphery, then the same thing happens so overwhelmingly that within a few days all available T cells are induced to become short-lived effector cells and disappear. In particular, non-cytopathic viruses can cause this exhaustion, whereas cytopathic viruses cannot because the host would die.

Glimpses into the balance between immunity and self-tolerance

Christopher C. Goodnow

Howard Hughes Medical Institute and Department of Microbiology and Immunology, Stanford University School of Medicine, Stanford, CA 94305–5428, USA

Abstract. The need to maintain self-tolerance is at odds with the need to draw upon antibody and T cell receptor diversity to fight infection. Advances in genetic manipulation of the mouse have at last brought into view the clonal selection mechanisms that underpin self-tolerance, confirming in general terms the notion of clonal deletion and clonal anergy put forward by Burnet and Nossal. The image that has emerged, however, is much more sophisticated than could have been imagined, revealing that self-reactive clones are deleted or held back in a remarkable series of culling checkpoints placed at many steps along the pathway to antibody production. These checkpoints act in concert to balance the nature and size of the holes in the repertoire generated by self-tolerance against the need to draw upon as many clones as possible for immunity to infection. Spontaneous and induced mutations in the mouse, such as Fas, PTP1C and CD45 mutations, have just begun to yield a few glimpses into the molecular circuitry underpinning these cellular checkpoints. Much more extensive genetic analysis, made possible by the genome project, will be needed to illuminate the details of those circuits and the factors that lead them to fail in autoimmune disease.

1997 The molecular basis of cellular defence mechanisms. Wiley, Chichester (Ciba Foundation Symposium 204) p 190–207

Burnet's (1959) concept of deleting self-reactive clones from the preimmune repertoire, leaving only non-self reactive clones to fight infection, poses a central problem of balancing the opposing needs of self-tolerance and immunity. As pointed out by Nossal (1983), clonal deletion must logically have an affinity cut-off to avoid deleting most of the potential repertoire of B or T cells (Fig. 1). The fact that self-reactive B and T cells can in fact be detected in the preimmune repertoire of healthy individuals and induced into autoaggression by particular immunization regimes has at various times been interpreted either as the footprints of self-reactive clones that fall below the deletion cut-off, or as evidence that self-tolerance is not acquired by clonal deletion at all.

Resolving how self-tolerance is actually acquired and balanced against the need to fight infection has depended on developing ways to track the fate of individual B or T lymphocyte clones within an inpenetrably diverse preimmune repertoire. Elucidation

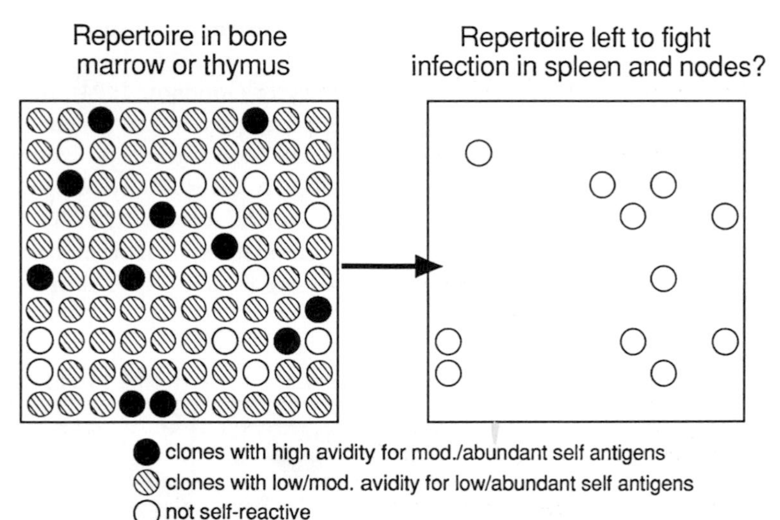

FIG. 1. A schematic illustrating Burnet's model for acquisition of self-tolerance and the problem it poses for balancing tolerance and immunity.

of the process for generating antibody and T cell receptor diversity in the 1980s finally provided two ways: (1) by following superantigen-reactive T cells bearing particular $V\beta$ elements; or (2) by following antigen-specific cells after elevating their frequency by fixing half or all of the diversity-generating machinery through introduction of pre-rearranged antibody H and L or TCR α and β transgenes. Examples of the B cell repertoires that result from fixing either H chain diversity or both H and L chain diversity in antibody-transgenic mice are shown in Fig. 2 (e.g. panels B and D). The latter approach has turned out to have tremendous scope, particularly when combined with the ability to engineer antigens by transgenesis that are expressed as part of the genetic self in different forms, concentrations and tissue locations. Collectively, this new-found ability to see inside the black box of repertoire selection *in vivo* has, firstly, confirmed the hypothesis that self-tolerance is acquired in part by clonal deletion. Secondly, these transgenic windows have provided glimpses into a much more sophisticated process of clonal deletion than could have been imagined, designed to balance self-tolerance by clonal deletion against the need to draw upon clones to fight infection.

Clonal deletion thresholds

Figure 2 (D–I) shows several examples of self-tolerance by clonal deletion, visualized in mice where H chain diversity has been fixed by introducing a pre-rearranged H chain transgene encoding an anti-lysozyme antibody. In H chain transgenic mice where hen egg lysozyme (HEL) is genetically foreign (Fig. 2D and H), subpopulations of B cells can be readily visualized in the spleen that bind lysozyme with distinct apparent affinities ranging from 10^{-9} M to less than 10^{-5} M, presumably due to combining the one H chain with different light chains (Hartley & Goodnow 1994). In H chain transgenic mice that have inherited an HEL gene as part of the genetic self and express 1 nM HEL in soluble form (sHEL) in the bloodstream, by contrast, the highest affinity self-reactive B cells are deleted whereas lower affinity cells are spared (Fig. 2E; Cyster et al 1994). Inheritance of a different HEL construct, with a different promoter and 10-fold higher circulating HEL, results in deletion of the highest affinity B cells and the lower affinity cells (Fig. 2F; Cyster et al 1994). The majority of

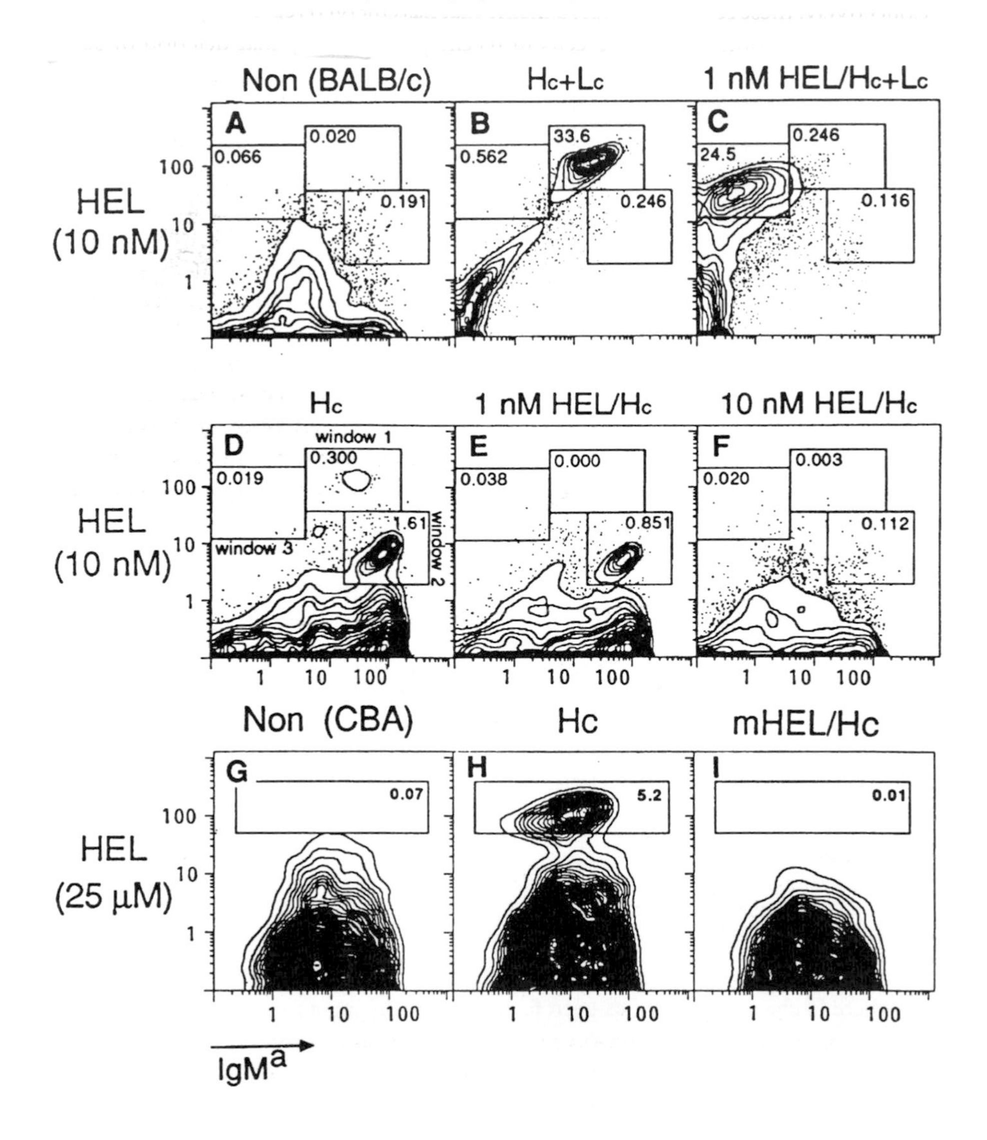

HEL-binding B cells in these H chain Tg mice nevertheless bind HEL with even lower affinity, so that they are only detected when much higher concentrations of HEL are used to visualize them (Fig. 2H). These very low affinity cells escape deletion by sHEL, but are all eliminated in mice that inherit a third type of HEL gene encoding the protein in a membrane-bound form (mHEL; Hartley et al 1991) on the surface of cells exposed to the bloodstream (Fig. 2I; Hartley et al 1993). The membrane-bound lattice of self-antigen presumably cross-links HEL-binding surface immunoglobulins on B cells much more efficiently than the soluble form, so that this increased avidity compensates for low intrinsic affinity.

Collectively, these results illustrate a theme that has emerged repeatedly from studies of repertoire selection in B and T cells in recent years, namely that deletion of self-reactive clones is limited by thresholds that are determined by the amount of self-antigen and by the affinity and avidity with which a clone binds it. These thresholds have probably been set over evolutionary time by opposing forces selecting for rapid antibody responses against infectious pathogens on the one hand, and for minimal destruction of autologous tissues on the other. For example, stringent deletion of low affinity self-reactive B cells that recognize blood-cell surface antigens is almost certainly essential to survival, since secretion even of low affinity but high avidity IgM antibodies to ABO carbohydrates would have devastating consequences in individuals whose erythrocytes and endothelium bore those antigens. By contrast, transient production of low or moderate avidity autoantibodies to oligovalent antigens dissolved in blood plasma, such as low-affinity rheumatoid factors, appears tolerable in the short-term due to the capacity to clear such immune complexes without serious immune pathology. Preservation of such clones in the preimmune repertoire may in this case be advantageous, by allowing them to be drawn upon immediately after infection if no other clone is available.

FIG. 2. Visualization of self-reactive B cell clonal deletion and clonal anergy. Each panel shows spleen cells from different mice, stained by two-colour immunofluorescence for cell surface IgM of the a allotype (IgM^a) on the x-axis and for binding of hen egg lysozyme (HEL) antigen on the y-axis. The concentration of HEL used to stain the cells is indicated. The upper row shows the rarity of high-affinity HEL-binding B cells in the repertoire of normal mice (A), and the high frequency and homogeneity of HEL binding B cells that results when both H and L chain diversity is fixed by H and L chain transgenes (B). In a littermate mouse that has inherited soluble HEL as part of its genetic make-up (C), this homogeneous HEL-specific population is not deleted but down-regulates cell surface IgM as part of the desensitization process. The middle row shows the small subsets of high and medium affinity HEL-binding B cells present in an otherwise diverse repertoire when only the H chain diversity is fixed by transgenesis (D). In a littermate mouse that has inherited the 1 nM serum HEL gene (E), the high affinity HEL-binding subset is deleted (window 1) but the medium affinity subset (window 2) is spared. The latter cells are nevertheless deleted when they develop in a mouse that expresses 10-fold higher serum HEL (F). The bottom row shows the entire pool of low, medium and high affinity HEL-binding cells present in an H chain transgenic mouse, revealed by staining with a very high concentration of HEL antigen (H). All of these cells are deleted when they develop in a mouse that expresses HEL on its cell surfaces (mHEL), where it is recognized with very high avidity.

Clonal anergy: antigen receptor down-regulation and desensitization

Figure 2 also shows an example of clonal anergy, an alternative strategy for acquiring self-tolerance originally championed by Nossal for clones that bound to self at intermediate levels. In the example shown, the entire B cell repertoire has been fixed to have homogenous HEL reactivity by the introduction of pre-rearranged H and L chain transgenes (Fig. 2B). In a mouse that has inherited the sHEL gene as part of its genetic self, the HEL-reactive B cells are not deleted but instead have decreased their complement of surface IgM antigen receptors by some 20-fold (Fig. 2C; Goodnow et al 1988). Interestingly, this is achieved not by decreasing IgM synthesis but by blocking its egress from the endoplasmic reticulum (Bell & Goodnow 1994). The majority of HEL-binding receptors remaining on the self-reactive cells are of the IgD class. These are not blocked in their intracellular transport and are expressed on the surface in normal numbers (Goodnow et al 1988, Bell & Goodnow 1994), but they have a greatly reduced capacity to activate tyrosine phosphorylation of their intracellular ITAMs (immunoreceptor tyrosine-based activator motifs) or of the pp72^{syk} tyrosine kinase that together play essential roles in B cell antigen receptor signalling and activation (Cooke et al 1994).

As a result of these two processes, anergic self-reactive B cells cannot be triggered into mitosis through their antigen receptors by T-cell-independent routes of activation, and they require much more avid receptor cross-linking to induce CD86/B7.2 on their surface and to cooperate successfully with helper T cells (Goodnow et al 1988, Cooke et al 1994). Self-reactive B cells continually exposed to the moderately soluble HEL self antigen thus become functionally tolerant to that ligand in much the same way that cells become tolerant when chronically exposed to drugs or hormones, by down-regulating and desensitizing their receptors.

Preserving clones in this desensitized state provides another way to balance tolerance and immunity, because these cells can be called into clonal expansion and antibody secretion if the need is great enough (Fig. 3A). For that to happen, they must suddenly bind much more avidly to a foreign antigen and immediately collaborate with an anti-foreign helper T cell (Cooke et al 1994). Moreover, under those particular conditions clonal expansion of desensitized self-reactive clones will be overshadowed by any non-self-reactive clones of comparable avidity for the inciting foreign antigen, because the receptors on the latter are not desensitized and consequently they proliferate approximately 10-fold more (Cooke et al 1994).

Clonal deletion takes place at several distinct checkpoints

Deletion of self-reactive B cells in response to high avidity self-antigen binding, as occurs in mice that inherit the mHEL form of antigen (Fig. 2I), occurs in the bone marrow by developmental arrest and apoptosis so as to render these clones completely unavailable for antibody responses (Hartley et al 1993). By contrast, self-reactive cells that recognize the moderate avidity sHEL antigen in the H chain Tg mice

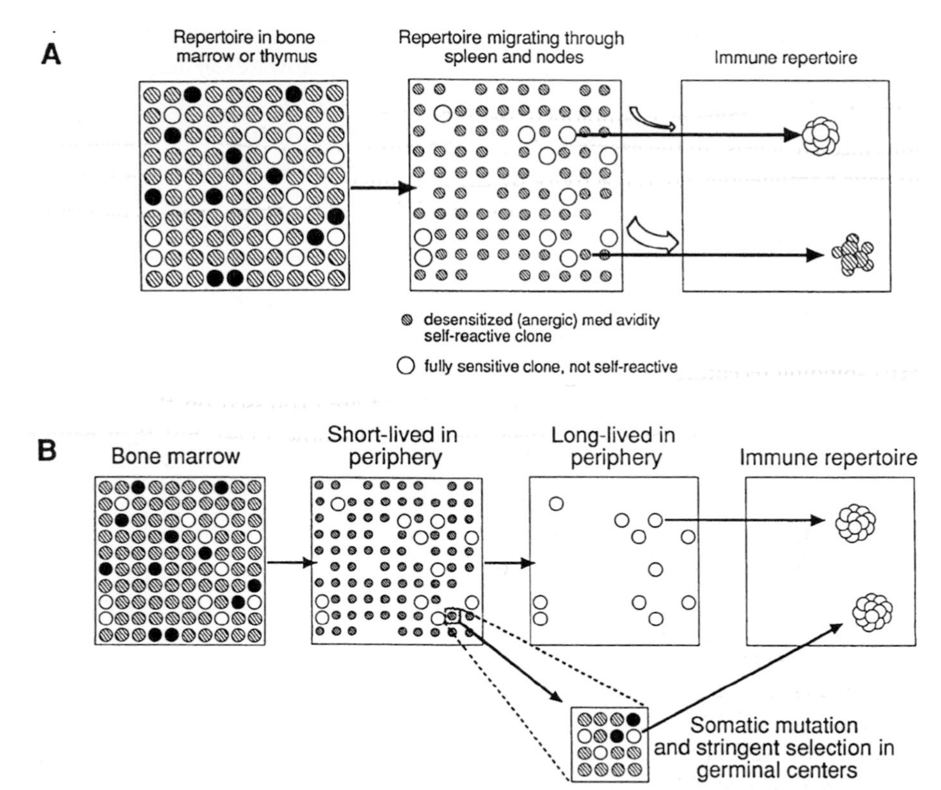

FIG. 3. Schematic illustrating how B cell desensitization and peripheral deletion help to balance tolerance with autoimmunity. In (A), high avidity self-reactive clones (dark grey) are deleted in the bone marrow, while moderate avidity self-reactive clones (light grey) are exported to the spleen and lymph nodes in a desensitized or anergic state. These cells can nevertheless be drawn into a primary antibody response, although they require a much stronger stimulus and their net clonal expansion is reduced. In (B), moderate avidity self-reactive clones are also filtered out in the periphery by competition for primary follicle niches, so that the long-lived repertoire that accumulates becomes dominated by non-self-reactive clones. Moderate avidity self-reactive clones can also potentially be recruited into germinal centres in secondary follicles, where somatic hypermutation and stringent negative and positive selection can give rise to variants that have lost their self-reactivity.

in Fig. 2E and F are deleted in a very different way (Cyster et al 1994). In this case, deletion of self-reactive B cells occurs after the cells have matured, become desensitized to sHEL, and emigrated from the bone marrow to the white pulp cords of the spleen or the paracortex of lymph nodes. After entering these peripheral lymphoid sites, the self-reactive B cells fail to continue their normal migration route into primary follicles and become stuck in the outer T cell zones, where they die prematurely by apoptosis within 1–3 d. Follicular exclusion depends on competition among B cells of different specificities, and appears to be due to an effect of continued antigen receptor signalling on the relative tropism of B cells towards an unknown set of migration cues (Cyster et al 1994, Cyster & Goodnow 1995a).

The primary follicles in spleen and lymph node thus appear to function as filters in the preimmune repertoire, preferentially excluding self-reactive B cells so that the accumulating long-lived repertoire becomes dominated by the least self-reactive clones available, even if these are initially produced at low frequency in the bone marrow (Fig. 3B). Because the short-lived self-reactive cells that are excluded by these filters are nevertheless transiently available to be drawn upon in the T zones, deletion at this stage and site strikes a further balance between tolerance and immunity. This type of peripheral filtering appears to make a large impact on the repertoire of normal animals, since a large fraction of B cells that are exported from the bone marrow are eliminated within 1–3 d in the same site and pattern (Chan & MacLennan 1993) as found for moderate avidity self-reactive B cells above. Moreover, the repertoire that is retained as long-lived recirculating cells is markedly skewed in V region use, except when there are insufficient B cells to support such a competitive filtering process (Malynn et al 1990, Gu et al 1991).

On the occasions when self-reactive clones are drawn into clonal expansion by T cells in the T zone, they may enter the germinal centres of secondary follicles where further receptor diversification by somatic hypermutation occurs. HEL-binding B cells that begin to expand within the germinal centres of mice that express soluble HEL as genetic self are very efficiently eliminated within these structures (K. Shokat & C. C. Goodnow, unpublished results). Deletion of self-reactive B cells occurs within germinal centres by a process of rapid cell death that can be mimicked by the administration of exogenous HEL antigen in soluble form (Shokat & Goodnow 1995). Since the germinal centre also positively selects emerging clones that bind more avidly to foreign antigens displayed by follicular dendritic cells, the germinal centre output of mutated memory B cells and long-lived plasma cells is thus simultaneously enriched for anti-foreign and stringently depleted of any self-reactivity (Fig. 3B). These clones come to dominate antibody responses, accounting for the exquisite fine specificity for the foreign immunizing antigen and lack of self-reactivity that is a hallmark of memory-type antibodies.

The net effect of these filters in primary and secondary follicles is that self-reactive B cells are stringently deleted from the immune repertoire, ending up with essentially the result predicted by Burnet (1959). Stepwise deletion at multiple stages in the periphery, rather than all at one step in the bone marrow as envisaged originally, has a great impact on balancing tolerance and immunity, however, because moderate avidity self-reactive clones are available: (1) to be drawn into primary antibody responses in the T zones, where their immediate benefit may outweigh any transient risk of immunopathology; and (2) to be drawn into germinal centres where hypermutation and stringent selection can chisel away unwanted self-reactivity (Fig. 3).

Deciding B cell fate at each checkpoint

The fact that self-reactive B cells are deleted in the T zones and in the germinal centres, perhaps more than in the bone marrow, raises the central question of how clones passing through each checkpoint decide whether to adopt a tolerant fate and die or adopt an immune fate and proliferate.

Cues deciding B cell fate in the bone marrow

In the bone marrow, B cells appear to use timing as a cue, much as originally envisaged by Burnet (1959) and Lederberg (1959; Fig. 4A): if they are already binding an antigen at this immature stage in their development they are probably self-reactive. Indeed, immature B cells appear almost hard-wired to respond to antigen by arresting their development, inhibiting their proliferative potential, and dying if antigenic stimulation is sustained (Nossal 1983, Hartley et al 1993). The other chief factor determining B cell fate at this stage is the amount and avidity of the antigen bound. More antigen and greater cross-linking results in a progressively increased magnitude of intracellular signalling to nuclear pathways such as ERK, NF-κB, and JNK in B cells (J. Healy & C. C. Goodnow, unpublished results), and differences in the amount of these or other signals to the nucleus probably determine the extent to which B cells are terminated in the bone marrow.

Communication from the antigen receptor to nuclear signalling pathways is normally tuned so that only high avidity self-reactive B cells achieve a sufficiently high pitch of signalling that they are arrested and deleted in the bone marrow, while the rest have only a low rumble within them and are retained for export to the periphery. The importance of inheriting the right tuning set-point is illustrated by mouse mutations in two protein tyrosine phosphatases, PTP1C and CD45, that regulate the intracellular signalling response to antigen in B cells. In HEL-specific B cells carrying a loss-of-function mutation in PTP1C, from the moth-eaten viable strain, the intracellular Ca^{2+} response to antigen is exaggerated demonstrating that PTP1C normally tunes down this response (Cyster & Goodnow 1995b). Self-reactive cells with this signal-exaggerating mutation are developmentally arrested and deleted in the bone marrow when they develop in a mouse that has inherited the soluble HEL gene, encoding a moderate avidity autoantigen that normally deletes only at later checkpoints in the periphery. Moreover, in the absence of any HEL ligand, the PTP1C-deficient B cells down-regulate their receptors as they mature, suggesting that spontaneous ligand-independent signalling by the antigen–receptor complex occurs when PTP1C is not present to oppose intrinsically active tyrosine kinases such as $pp72^{syk}$. Deletion of low and moderate avidity self-reactive B cells in the bone marrow as a result of an exaggerated signalling response to antigen is likely to explain the marked deficiency of mature recirculating B cells that occurs in moth-eaten viable mice (Cyster & Goodnow 1995b).

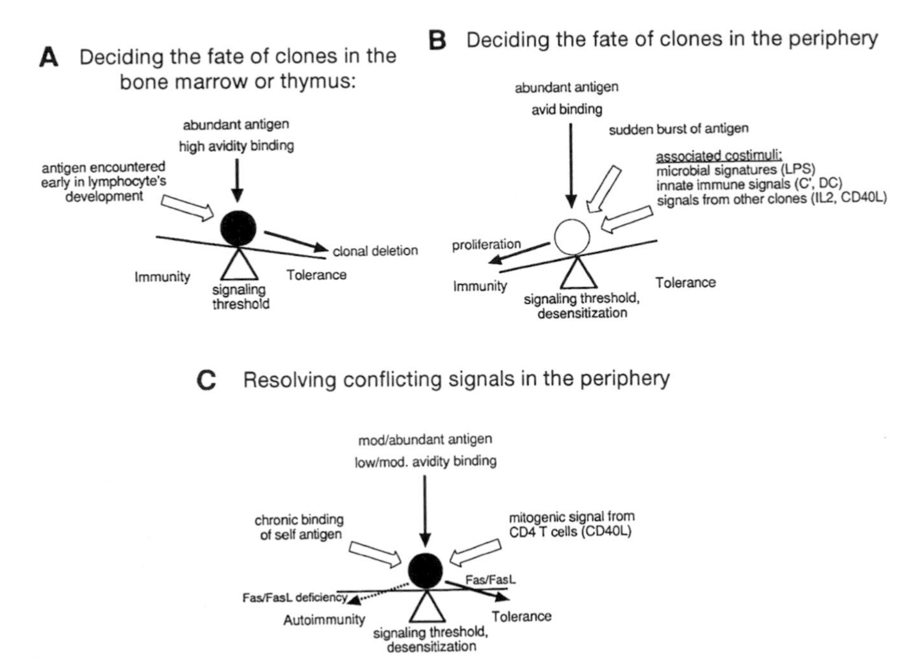

FIG. 4. Schematic illustrating the integration of different factors that shift the balance between tolerance and immunity.

By contrast with PTP1C deficiency, HEL-specific B cells lacking the CD45 tyrosine phosphatase have diminished intracellular Ca^{2+} and ERK responses to antigen, confirming that this phosphatase tunes-up signalling by antigen receptors (Cyster et al 1996). Selection of B cells with this signal-depressing mutation is altered in the opposite direction. While CD45-deficient B cells that develop in mice carrying the high avidity mHEL self antigen are still deleted in the bone marrow, they are positively selected into the long-lived peripheral pool instead of being filtered out when they develop in mice carrying the moderate avidity sHEL autoantigen. In the absence of an exogenous ligand, CD45-deficient HEL-specific B cells fail to persist after emigrating from the bone marrow, suggesting that a low level of spontaneous, tonic signalling by the antigen receptor may be necessary for mature B cells to stay alive. Thus when signalling is depressed by the CD45 mutation, the low tonic signal only occurs when soluble HEL autoantigen binds with moderate avidity, skewing the recirculating repertoire that accumulates in favour of moderate avidity self-reactive clones and away from the non-self-reactive cells that would normally be favoured.

Cues deciding B cell fate in the periphery

Mature B cells use timing and co-stimuli as cues to determine wheth
immune or tolerant fate (Fig. 4B). Co-stimulatory cues—the body's r
infection—can come from: (1) microbes themselves, such as lipopolysaccha
double-stranded RNA; (2) from the innate immune system such as complement C
or (3) from other lymphocytes, such as CD40L and interleukin 4 from helper T cells. When these co-stimuli are coupled with an antigenic stimulus that appears as a sudden burst, this combination of timing and co-stimuli are both characteristic presentations for infectious antigens and act in concert as cues to drive B cells into clonal expansion and antibody production (Fig. 4B). In contrast, when antigen appears in a constant monotonous way in the absence of any co-stimulatory red flags, these are both characteristic presentations for self antigens. In this case, mature B cells take these as cues to adopt tolerant fates either by desensitizing and down-regulating their receptors (Goodnow et al 1989) or by dying as a result of follicular exclusion (Cyster & Goodnow 1995a).

Moderate avidity self-reactive B cells that are chronically binding self-ligands, and as a result have desensitized their receptors and become trapped in the T zones, must nevertheless face conflicting cues frequently. The most common of these circumstances arises when a B cell, by virtue of its self-reactivity, has captured a foreign antigen that is associated with self. Examples of such frequently occurring circumstances include the capture of virions whose envelopes are decorated with self proteins and carbohydrates, viral DNA binding proteins that are firmly attached to autologous chromatin, or bacterial or viral antigens that are complexed with autologous IgG. In each case, B cells that bind the self-component of these complexes will present the foreign peptide moieties to non-tolerant helper T cells in the T zones, eliciting CD40L expression on the T cells. At this stage, the B cell is faced with conflicting cues (Fig. 4C): on the one hand, chronic antigen-binding implies self-reactivity, but on the other hand reception of the mitogenic CD40L signal implies reactivity with infectious non-self. Desensitized self-reactive B cells that are trapped in the T zone in fact respond efficiently to CD40L (Cooke et al 1994, Eris et al 1994), exacerbating the dilemma.

The interacting pair of cell surface molecules, FasL and Fas (CD95), appear to function by resolving these type of conflicting cues to ensure that T cell help is not delivered inappropriately to self-reactive B cells. This role of Fas and FasL is indicated by the outcome of *in vivo* interactions (Rathmell et al 1995) between HEL-specific $CD4^+$ T cells and either naïve HEL-specific B cells that have grown up in animals where HEL antigen was not part of genetic self, or desensitized (anergic) HEL-specific B cells that have grown up in the face of chronic receptor engagement by soluble HEL self antigen.

When naïve B cells suddenly encounter HEL antigen just before interacting with T cells in an adoptive transfer, the B cells first become activated as judged by high expression of the T cell co-stimulatory molecule, CD86 (B7.2), and then are iggered into extensive clonal expansion and antibody production in a manner that is substantially dependent on CD40L (Cyster & Goodnow 1995a, Rathmell et al 1995, 1996). Fas comes to be expressed at high levels on the initially naïve B cells during their interactions with the T cells, but Fas does not trigger apoptosis in this context (Rathmell et al 1996). By contrast, HEL-specific B cells that have grown up chronically exposed to HEL do not become activated to express CD86 (B7.2) when they bind more HEL antigen, although they present HEL peptides efficiently (Cooke et al 1994, Ho et al 1994, Cyster & Goodnow 1995a). Upon interaction with HEL-specific T cells the self-reactive B cells begin to activate in a CD40L-dependent manner and up-regulate Fas, but FasL on the T cell then aborts the process by triggering their death. When the genes encoding Fas or FasL are defective in the B and T cell, respectively, CD40L-dependent activation of the self-reactive B cells cannot be aborted and the cells proceed into cell cycle (Rathmell et al 1995).

Fas is thus expressed both on naïve B cells that have been acutely activated by antigen and T cells, and on anergic self-reactive B cells that have been faced with the conflicting combination of chronic antigen exposure and T cell help. Fas only aborts clonal expansion of self-reactive cells, however, because their desensitized antigen receptors can no longer transmit signals that normally block the Fas-induced death signals. The real-life importance of resolving this conflict is driven home by the severe autoantibody disorders that occur in rare children with Fas deficiency (Rieux-Laucat et al 1995, Fisher et al 1995), and by the massive accumulation of autoantibody-producing B cells in the T zones of mice with Fas deficiency (Jacobson et al 1995).

Closing comments

The glimpses into tolerance and how it is balanced with immunity that I have described here illustrate how the different tools now available for genetic analysis in mice can illuminate complex physiological processes. The use of transgenic animals allows repertoire diversity to be fixed in many different ways, enabling specific cells to be tracked in their native context. Transgenesis also enables antigens to be added to the complement encoded as genetic self, and allows systematic changes in the way these antigens are presented to the immune system. The greatest opportunity for illuminating the balance between tolerance and immunity, however, comes from using these tools to pinpoint the effects of spontaneous and induced mutations in key molecules in the selection circuitry, illustrated by the PTP1C, CD45, FasL/Fas and CD40L mutations discussed here. The key limitation to this approach currently lies in the very small number of strong, single locus mutations known to affect tolerance and selection. To rectify that shortage, we are now embarking on a genome-wide screen for chemically induced mutations that disturb B and T cell selection and tolerance.

References

Bell SE, Goodnow CC 1994 A selective defect in IgM antigen receptor synthesis and transport causes loss of cell surface IgM expression on tolerant B lymphocytes. EMBO J 13:816–826

Burnet FM 1959 The clonal selection theory of acquired immunity. Vanderbilt University Press, Nashville, TN

Chan EY-T, MacLennan ICM 1993 Only a small proportion of splenic B cells in adults are short-lived virgin cells. Eur J Immunol 23:357–363

Cooke MP, Heath AW, Shokat KM et al 1994 Immunoglobulin signal transduction guides the specificity of B cell–T cell interactions and is blocked in tolerant self-reactive B cells. J Exp Med 179:425–438

Cyster JG, Goodnow CC 1995a Antigen-induced exclusion from follicles and anergy are separate and complementary processes that influence peripheral B cell fate. Immunity 3:691–701

Cyster JG, Goodnow CC 1995b Protein tyrosine phosphatase 1C negatively regulates antigen receptor signaling in B lymphocytes and determines thresholds for negative selection. Immunity 2:13–24

Cyster JG, Hartley SB, Goodnow CC 1994 Competition for follicular niches excludes self-reactive cells from the recirculating B-cell repertoire. Nature 371:389–395

Cyster JG, Healy JI, Kishihara K, Mak TW, Thomas ML, Goodnow CC 1996 Regulation of B lymphocyte negative and positive selection by tyrosine phosphatase CD45. Nature 381:325–328

Eris JM, Basten A, Brink RA, Doherty K, Kehry MR, Hodgkin PD 1994 Anergic self-reactive B cells present self antigen and respond normally to CD40-dependent T cell signals but are defective in antigen-receptor-mediated functions. Proc Natl Acad Sci USA 91:4392–4396

Fisher G, Rosenberg F, Straus S et al 1995 Dominant interfering Fas gene mutations impair apoptosis in a human autoimmune lymphoproliferative syndrome. Cell 81:935–946

Goodnow CC, Crosbie J, Adelstein S et al 1988 Altered immunoglobulin expression and functional silencing of self-reactive B lymphocytes in transgenic mice. Nature 334:676–682

Goodnow CC, Crosbie J, Jorgensen H, Brink RA, Basten A 1989 Induction of self-tolerance in mature peripheral B lymphocytes. Nature 342:385–391

Gu H, Tarlinton D, Muller W, Rajewsky K, Forster I 1991 Most peripheral B cells in mice are ligand selected. J Exp Med 173:1357–1371

Hartley SB, Goodnow CC 1994 Censoring of self-reactive B cells with a range of receptor affinities in transgenic mice expressing heavy chains for a lysozyme-specific antibody. Int Immunol 6:1417–1425

Hartley SB, Crosbie J, Brink R, Kantor AA, Basten A, Goodnow CC 1991 Elimination from peripheral lymphoid tissues of self-reactive B lymphocytes recognizing membrane-bound antigens. Nature 353:765–768

Hartley SB, Cooke MP, Fulcher DA et al 1993 Elimination of self-reactive B lymphocytes proceeds in two stages: arrested development and cell death. Cell 72:325–335

Ho WY, Cooke MP, Goodnow CC, Davis MM 1994 Resting and anergic B cells are defective in CD28-dependent costimulation of naive $CD4^+$ T cells. J Exp Med 179:1539–1549

Jacobson BA, Panka DJ, Nguyen K-A, Erikson J, Abbas A, Marshak-Rothstein A 1995 Anatomy of autoantibody production: dominant localization of antibody-producing cells to T cell zones in Fas-deficient mice. Immunity 3:509–519

Lederberg J 1959 Genes and antibodies. Science 129:1649–1653

Malynn BA, Yancopoulos GD, Barth JE, Bona CA, Alt FW 1990 Biased expression of J_H-proximal V_H genes occurs in the newly generated repertoire of neonatal and adult mice. J Exp Med 171:843–859

Nossal GJV 1983 Cellular mechanisms of immunologic tolerance. Annu Rev Immunol 1:33–62

Rathmell JC, Cooke MP, Grein J, Ho W, Davis MM, Goodnow CC 1995 Elimination of self-reactive B cells by CD4 T Cells is disrupted by the lpr mutation in CD95 (Fas/Apo-1). Nature 376:181–184

Rathmell JC, Townsend SE, Xu JC, Flavell RA, Goodnow CC 1996 Expansion or elimination of self-reactive B cells *in vivo*: dual roles for CD40 and Fas (CD95) ligands modulated by the B cell antigen receptor. Cell, in press

Rieux-Laucat F, LeDeist F, Hivroz C et al 1995 Mutations in Fas associated with human lymphoproliferative syndrome and autoimmunity. Science 268:1347–1349

Shokat KM, Goodnow CC 1995 Antigen-induced B cell death and elimination during germinal center immune responses. Nature 375:334–338

DISCUSSION

Nossal: In the real world, of course, self-antigen is always present. You mentioned that the immature cell is somehow wired to bias the system in favour of tolerance. The biochemical basis of this is not known, but there is one stream of work, for example that of John Monroe, which does offer at least the glimmer of a solution. It talks about a disassociation between the wiring for 'signal 1' coming from the antigen receptor, and 'signal 2' registering co-stimulatory activity. In the immature B cell, the antigen receptor is wired up properly but the machinery for co-signalling is not (Yellen et al 1991). What is the present status of this and how does such a notion fit into your scheme?

Goodnow: There are so many potentially important differences between an immature and mature B cell that it is hard to pinpoint which one is most significant. For example, immature B cells don't express the CD21 receptor for complement, which is an important co-stimulator for B cells, and they express lower levels of CD45, which is a positive player in the signalling process. In addition, unlike mature cells, they don't up-regulate B7.2 in response to antigen, so they cannot co-stimulate T cells. And, of course, mitogenic signalling from the antigen receptor is somehow uncoupled, perhaps because the NF-κB pathway doesn't function in the same way. I think the biochemical differences that Monroe has described are interesting, but they are going to be just a small subset of what is really a global change in the wiring of response machinery in immature versus mature cells.

Rajewsky: Are these anergic cells part of the normal physiology of the mouse? Do you have an estimate of the number of mature B cells in the mouse that are actually anergic? Also, in the moth-eaten mutant, according to your data many anergic cells would be expected to appear, but in reality there are only a few B cells present in the peripheral immune system of these mice, and these are of the BI subset. What can we extrapolate from the lysozyme system to the normal situation in the animal?

Goodnow: In the normal repertoire, the IgD^+ cells exhibit a wide range of IgM levels, apparently on a clone by clone basis. So far in the transgenic system, the only factor that causes IgM to vary from one clone to another is how much signal is coming chronically from outside the cell through the antigen receptor. We therefore believe that at one end of the normal continuum, IgM bright cells in the recirculating follicular pool are having almost no receptor stimulation by endogenous autoantigens. At the other end of that continuum are cells with very little IgM, getting close to the level on anergic cells in transgenic animals. The IgM^{lo} cells are probably binding continuously to some autoantigens, but not to the extent where they are fully desensitized or anergic. In response to constant signalling at different levels, the rheostat is essentially readjusting and it has a large zone of plasticity.

Rajewsky: But you wouldn't think that according to the mechanism you just described, these should be eliminated by T cells?

Goodnow: No, because I think the ones that actually react with sufficient avidity to do what we see with soluble lysozyme are going to be very low for IgM; one sees almost no cells of that phenotype in the normal repertoire. In the competition experiments we've done, they all get stuck in the T cell zones and are eliminated within two days of emerging from the bone marrow. A large fraction of the bone marrow B cell production does seem to be exported to the spleen (and to a lesser extent lymph nodes), it arrives in the T cell zones just as the anergic cells do, and then once the peripheral repertoire is full these new immigrants get stuck in the T cell zones and die within 1–2 d similarly to these anergic self-reactive cells. The big question concerns what is causing these new B cells to get stuck there. If you eradicate the mature peripheral B cell pool then these new immigrants can suddenly get in. Cells that have already passed through this checkpoint once (for example, from the thoracic duct lymph) win in this competition every time and get into a primary follicle. So there's some fundamental difference between the cells that have already successfully passed through one round of the musical chairs game of competing for follicular niches and cells that are just coming in from the bone marrow. We would suggest that the difference lies in the specificity of the recirculating cells, having already been filtered of self-reactive clones.

Nossal: Fulcher & Basten (1994) have a more operational and less religious answer to this question. They've done a study of the lifespan of anergic B cells in various situations. They find that a cell which is being mildly tickled all the time has a markedly reduced lifespan. There may be a virtually complete gradation, depending on the antigen concentration, between elimination in the bone marrow at one extreme to near normal lifespan with low concentrations of antigen at the other. So the repertoire may not be 'junked up' by lots of anti-self cells because the lifespan of the anergic B cell is so much shorter than that of normal B cells.

Shortman: I also have a question on the physiology of these desensitized cells. Will they withstand a cell division stimulus if you activate them into cell cycle? How quickly do they lose their desensitized state?

Goodnow: They are back to normal within 16 h in the presence of LPS or CD40 ligand.

Melchers: This is in contrast to T cells. Once T cells are anergic, don't they stay this way for a long time?

Goodnow: That's in a G0 state. As long as these cells are in G0 they will remain desensitized for a long time. Somehow, getting them out of G0 resets the rheostat to the current environment.

Melchers: If I remember correctly, Harald von Boehmer tried to stimulate his anergic T cells and failed.

von Boehmer: We did not use LPS. We used receptor antibodies and concanavalin A. We also gave massive amounts of IL-2, which didn't help at all. In the presence of antigen the lifespan of anergic cells was considerable (months).

Zinkernagel: If we look in the soluble VSV system, the anergy of the T helper (Th) cells is comparable to that of the B cells. If you take these Th cells out of the transgenic animal and then put them into a non-transgenic recipient, within two or three days they are fully functional.

In the HEL transgenic, what are the relevant cut-off points in terms of affinity/avidity of your receptor where you would think that things are being decided? If you infect mice with certain viruses you immediately start to get rheumatoid type of factor and all sorts of things.

Goodnow: That's why we think it's significant that these anergic cells under physiological states get stuck in the T zones. Those low affinity rheumatoid factors and anti-single-stranded DNAs that you see after immunization or infection are coming from cells that have moderate avidity self-reactivity for those antigens. In response to a very strong antigen receptor stimulus, cross-linking much more extensively by the virus particle or all the co-stimuli associated with Freund's adjuvant, you are getting those cells to make a transient T-dependent focus response, much as we can do with mHEL for these lysozyme-tolerant B cells. It is a focused response in the T zone that leads to a transient wave of short-lived plasma cells and doesn't generate a memory type of response because of the negative selection that occurs in the germinal centre. As you know, these low affinity rheumatoid factor responses have no anamnestic properties; the next time you immunize the same thing occurs without evidence of priming.

Zinkernagel: And where would you put the cut-off?

Goodnow: The cut-offs are going to be very tricky to place, because although in the lysozyme system we are in theory dealing with a monovalent antigen, it is actually more complicated than that, because we don't know how lysozyme is signalling. It may actually be acting similarly to fibroblast growth factor (FGF)-2; it is very basic and it does bind to things like chondroitin sulfate. It may be working much like the FGF receptor, hanging like crows on telegraph wires of the extracellular glycocalyx, weakly cross-linking the antigen receptors. Each antigen must be dealt with separately: the minute you introduce these variable degrees of *in vivo* cross-linking it is going to be tricky to come out with numbers.

Melchers: Is there an antigen presenting cell in the bone marrow for B cells? Not everything is membrane bound — DNA, for instance.

Goodnow: There are data showing that apoptotic cells display nucleosomal blobs on their surface; this might be the real *in vivo* display of DNA and other nuclear antigens.

Miller: What is the mechanism of death in the short-lived anergic B cells?

Goodnow: Death in the bone marrow due to the high avidity antigen versus death in the T zones due to the weak avidity antigen are biochemically very different pathways.

The former is a cell autonomous development arrest that prevents L-selectin and CD21, for example, from being expressed. On the other hand, for moderate avidity self-reactive clones in the periphery, some signalling pathway (our favourite candidate is the ERK pathway) is changing the migration characteristics of these B cells so that they are less able to go into the primary follicles. We think the latter case is similar to a bone marrow B cell unable to get next to a bone marrow stromal cell because of competing pro-B cells that are more strongly attracted, with the cell dying partly by growth factor withdrawal. Essentially, nothing is known about what the follicular dendritic cell mesh does or what the follicular niches do, so it is not possible to test this experimentally.

Paul: You hinted that although lysozyme is theoretically monovalent, by virtue of its charge it might multimerize. There are many proteins that don't fit that bill. We have always been concerned as to whether or not one does in fact achieve tolerance to authentically monovalent or monomeric epitopes. Does the repertoire for reactivity of B cells to authentically monomeric epitopes remain intact?

Goodnow: Which antigens in particular are you thinking of that are only seen in monomeric form?

Zinkernagel: Hormones.

Goodnow: But aren't they presented on the surface of cells bound to receptors?

Paul: If there are vastly more soluble molecules of a given type than there are cell-associated molecules of the same type, would you not achieve the same situation?

Goodnow: It may well be that for things that are purely monomeric you don't need B cell tolerance — that autoantibodies to these things wouldn't be a real problem.

Paul: The historical argument is that when you get T cell help you do get a response. Others would argue that this is self-limited; when the virus goes away so will the response. Is your view still that this part of the repertoire is present or do you think that there is in fact a way to sense the presence of monomers?

Goodnow: One of the things that we found in our germinal centre studies is that once you start making low affinity primary autoantibodies, suddenly this soluble lysozyme gets displayed in a form that looks much more like membrane lysozyme. We think that this fairly monovalent antigen is now being displayed as multivalent immune complexes. In the experimental situation this seems to slam the cells very hard.

Paul: One can take the opposite view that for a cell that had never undergone tolerance, the presence of small amounts of antibody would put it into a situation where the resultant immune complexes would be quite active in stimulating a response, resembling an immunogen such as VSV.

Goodnow: But once it gets into the germinal centre it gets negatively selected very stringently by the germinal centre death process. The only thing that you can draw in terms of a real experiment is the WHO program with the β-HCG vaccine, where it has proven hard to get a good sustained antibody response even when it is linked to a good foreign carrier such as tetanus toxoid. The only way I can explain this is that the ostensibly monomeric hormone is inducing very strong hapten-specific tolerance.

Tarlinton: One of the attributes of the moth-eaten viable mouse is its very high levels of sIg. Is it possible that the deletion you see is not actually a deletion but, for example, plasma cell production? Have you measured serum titres against HEL and sHEL?

Goodnow: As you say, there is a massively elevated spontaneous antibody production, and also more cells diverting to this Ly1 B cell phenotype, in the absence of lysozyme autoantigen. In mice expressing lysozyme, however, antibody secretion and plasma cells are undetectable.

Tarlinton: Serum titres are not elevated, for example?

Goodnow: In the bone marrow you see cells halted in development, just as in the membrane-bound lysozyme situation, with receptors massively down-regulated. I don't think those two processes are linked.

Hodgkin: You are emphasizing the difference between the immature and more mature B cells, but it seems to me there's a common principle, which is when either cell is activated by antigen recognition they're compelled to do something within a certain time: if they don't, they die. In the case of the mature B cells it's very clear that they're compelled to go to that T cell zone and find some other signal. In the case of the immature cell, they still seem to go to that same T cell zone. This seems an important parallel.

Can you relate this compulsion to do something within a certain time to the strength of the initial activation?

Goodnow: Until we have the genetic tools to dissect the things that are downstream from those strong or weak signals at different stages of development, in the same way that Fas provides for the very late step, I don't think we are going to be able to answer that question.

Hodgkin: The principal of 'get signal, do something or die' seems fairly clear to me. This was the system of B cell tolerance that Peter Bretscher and Mel Cohn suggested a long time ago (Bretscher & Cohn 1970).

Strasser: You said that the death of the B cells which see self-antigen in the membrane-bound form in the marrow is very different from the death of the anergic cells that see the antigen in soluble form. I'm not so sure about this: the lifespan of these two cells is pretty much the same.

Goodnow: No, that's incorrect. Under competition with other B cells their lifespan is similar, but cells that are continuously exposed to soluble lysozyme in the absence of competition for follicular niches live for an average of one week, and many of the cells live for two weeks, whereas the membrane-antigen exposed immature B cells live for 1–3 d regardless of competition.

Strasser: My point is that perhaps the major difference is that if a B cell sees the membrane-bound form of the antigen, it just can't leave the marrow; it is stuck to where the antibody is stuck. It is not really a difference in the cell death signalling pathways. Cells are retained and can't leave if their antigen receptors are bound to a stromal cell, whereas if the antigen is soluble then they can wander around and do what they want during their dying process.

Williams: As a haematologist, one of the most difficult sets of patients to take care of are children with Evan's syndrome, characterized by low level autoantibodies to red cells, platelets and white cells. After viral infections, they often develop high titre antibodies and get acutely ill. Currently the treatment of these children involves a 'sledge hammer' steroid therapy to reduce the size of the B cell clones. Knowing what you do about cell tolerances, does this help you in any way to design new therapies for this type of autoimmune disease?

Goodnow: Although we're getting more detail, the factors that promote tolerance that I was mentioning are the same ones that were defined empirically in the 1960s. You need to give antigen in a poorly aggregated form that can't be decorated with complement, free of endotoxin, and unable to elicit much T cell help.

References

Bretscher P, Cohn M 1970 A theory of self–nonself discrimination. Science 169:1047–1049

Fulcher DA, Basten A 1994 Reduced lifespan of anergic self-reactive B cells in a double-transgenic model. J Exp Med 179:125–134

Yellen AJ, Glenn W, Sutchadine VP, Cao X, Monroe JG 1991 Signaling through surface IgM in tolerance-susceptible murine B lymphocytes. J Immunol 146:1446–1454

Interleukin 4: signalling mechanisms and control of T cell differentiation

William E. Paul

Laboratory of Immunology, National Institute of Allergy and Infectious Diseases, National Institutes of Health, Bethesda, MD 20892, USA

Abstract. Interleukin 4 (IL-4) is a pleiotropic type I cytokine that controls both growth and differentiation among haemopoietic and non-haemopoietic cells. Its receptor is a heterodimer. One chain, the IL-4Rα chain, binds IL-4 with high affinity and determines the nature of the biochemical signals that are induced. The second chain, γc, is required for the induction of such signals. IL-4-mediated growth depends upon activation events that involve phosphorylation of Y497 of IL-4Rα, leading to the binding and phosphorylation of 4PS/IRS-2 in haemopoietic cells and of IRS-1 in non-haemopoietic cells. By contrast, IL-4-mediated differentiation events depend upon more distal regions of the IL-4Rα chain that include a series of STAT-6 binding sites. The distinctive roles of these receptor domains was verified by receptor-reconstruction experiments. The 'growth' and 'differentiation' domains of the IL-4Rα chain, independently expressed as chimeric structures with a truncated version of the IL-2Rβ chain, were shown to convey their functions to the hybrid receptor. The critical role of STAT-6 in IL-4-mediated gene activation and differentiation was made clear by the finding that lymphocytes from STAT-6 knockout mice are strikingly deficient in these functions but have retained the capacity to grow, at least partially, in response to IL-4. IL-4 plays a central role in determining the phenotype of naïve $CD4^+$ T cells. In the presence of IL-4, newly primed naïve T cells develop into IL-4 producers while in its absence they preferentially become γ-interferon (IFN-γ) producers. Recently, a specialized subpopulation of T cells, $CD4^+$/$NK1.1^+$ cells, has been shown to produce large amounts of IL-4 upon stimulation. Two examples of mice with deficiencies in these cells are described—β_2-microglobulin knockout mice and SJL mice. Both show defects in the development of IL-4-producing cells and in the increase in serum IgE in response to stimulation with the polyclonal stimulant anti-IgD. Both sets of mice have major diminutions in the number of $CD4^+$/ $NK1.1^+$ T cells, strongly indicating an important role of these cells in some but not all IgE responses to physiologic stimuli.

1997 The molecular basis of cellular defence mechanisms. Wiley, Chichester (Ciba Foundation Symposium 204) p 208–219

IL-4 is a type I cytokine produced by specialized subsets of T cells and by basophils and mast cells in response to receptor-mediated signalling events (Paul 1991). Receptors for IL-4 consist of two chains, one of which (the IL-4Rα chain) binds the cytokine with high affinity (Ohara & Paul 1987). The second chain, generally the γc chain (formerly the IL-2Rγ chain), contributes little to the binding energy but is essential for signalling (Russell et al 1993, Kondo et al 1993). Recently, a new type I cytokine receptor chain has been described that is essential for the binding of IL-13, a congener of IL-4 (Hilton et al 1996). It is possible that this represents a γ-like chain that has affinity for IL-13 but that may also substitute for γc in forming a transducing IL-4 receptor. IL-4 receptors are widely distributed, accounting for the broad functionality of IL-4. Most work has been done thus far on IL-4 action on haemopoietic cells, particularly B cells, T cells and macrophages.

IL-4 was first recognized as a B cell growth stimulant (Howard et al 1982), particularly acting with soluble anti-IgM antibodies and with sub-mitogenic concentrations of lipopolysaccharide. It also stimulates growth of T cells (Mosmann et al 1986, Hu-Li et al 1987) and can strikingly enhance IL-3-mediated growth of mast cells (Mosmann et al 1986). It is now recognized that many of its most important actions lie in the control of differentiation, particularly in B cells where it induces transcription of non-rearranged forms of the constant region of the γ1 (Stavnezer et al 1988) and εH chains (Rothman et al 1988) (Iγ1 and Iε) as a prelude to immunoglobulin class switching (Vitetta et al 1985, Coffman et al 1986). It also induces expression of CD23 (Defrance et al 1987) and class II MHC (Noelle et al 1986) molecules on B cells. IL-4 plays an essential role, as will be described subsequently, in the process through which naïve T cells become committed to the 'Th1' and 'Th2' phenotypes (Le Gros et al 1990, Swain et al 1990).

IL-4-mediated growth depends upon a receptor domain that includes the I4R motif

The huIL-4Rα chain consists of 825 residues, of which 569 make up its cytosolic portion. The cytosolic domain contains five tyrosines that are evolutionarily conserved in the sense that each tyrosine residue is centred on a set of conserved amino acids. These tyrosines are located at residues 496 (Y1), 575 (Y2), 603 (Y3), 631 (Y4) and 716 (Y5).

In order to examine the requirements for the IL-4Rα chain to transduce growth and differentiation responses to IL-4, we transfected truncation and point mutants of the human IL-4Rα chain into a subline of the mouse myeloid cell line 32D that overexpressed IRS-1 [32D.IRS-1], where growth could be studied (Keegan et al 1994), or into the B lymphoma line M12.4.1, where IL-4-mediated differentiation could be examined (Ryan et al 1996).

Our initial experiments involved transfection of cDNAs encoding the wild-type huIL-4Rα and a series of truncation mutants of the huIL-4Rα chain. These experiments had been preceded by the demonstration that mouse IL-4 did not stimulate DNA synthesis or cell growth by wild-type 32D cells. These cells fail to express either IRS-1 (insulin response substrate 1) or 4PS/IRS-2 (IL-4R phosphorylation substrate/insulin response substrate 2). 32D cells that had been stably transfected with a cDNA for IRS-1 (32D.IRS-1) showed both growth in response to IL-4 and prompt tyrosine phosphorylation of IRS-1 (Wang et al 1993). Indeed, it had been previously demonstrated that IL-4, insulin and insulin-like growth factor 1 (IGF-1) caused tyrosine phosphorylation of IRS-1 and/or IRS-2 in a variety of cell lines (Wang et al 1992).

HuIL-4Rα chains truncated at amino acids 657 and 557 made 32D.IRS-1 cells capable of responding to huIL-4 by growth and by the phosphorylation of IRS-1, whereas cells expressing a mutant truncated at amino acid 437 made neither response (Keegan et al 1994). This suggested that the region between 437 and 557 contained sequences critical to the transduction of a growth signal. The single Y in this region (Y1) is surrounded by a sequence that is quite homologous to a sequence surrounding a key Y (Y960) in the insulin receptor (White et al 1988). The consensus sequence has been designated the I4R motif. This motif contains an NPXY, now recognized as the core sequence of phosphotyrosine binding (PTB) domains (Kavanaugh & Williams 1994, Blaikie et al 1994). A series of stable transfectants of 32D.IRS-1 cells were prepared that expressed a full length huIL-4Rα chain in which Y1 had been replaced with F (Keegan et al 1994). These lines showed no tyrosine phosphorylation of IRS-1 in response to IL-4; most of these lines failed to grow in response to huIL-4 although they responded normally to mouse IL-4 and to IL-3.

IL-4-induced differentiation depends upon a receptor domain containing STAT-6 binding sites

To examine IL-4-mediated gene activation and differentiation, we prepared a series of stable transfectants of the B lymphoma line M12.4.1. M12.4.1 cells that overexpressed insulin or IGF-1 receptors showed no induction of CD23, class II MHC molecules, or germ-line Cε transcripts in response to insulin or IGF-1 (Ryan et al 1996), strongly suggesting that possession of an I4R motif and induction of 4PS/IRS-2 tyrosine phosphorylation was not sufficient to induce these gene activation/differentiation effects. When M12.4.1 cells expressing truncation mutants of huIL-4Rα chains were studied, it was found that truncation at 657 did not diminish gene activation while truncation at 437 ablated it. In contrast to growth induction in which the 557 truncation mutant was fully active, M12.4.1 cells expressing the 557 truncation mutant showed only meagre (~10%) induction of CD23, class II and Iε (Ryan et al 1996). This strongly suggests that the control of growth and differentiation are not identical.

The interval between 557 and 657 contains the conserved Y2, Y3 and Y4. M12.4.1 cells were prepared expressing point mutants of the IL-4Rα chain in which each of these Ys was independently changed to F, in which each combination of two Ys were mutated to Fs and in which all three Ys were mutated. All of the mutants except that in which all three Ys were mutated to F remained able to induce CD23 expression, although levels of receptor expression required for such induction were higher than were required with the wild-type receptor (Ryan et al 1996).

There was good correlation between induction of CD23 and the appearance in cell extracts of activated STAT-6, as judged by mobility shift analysis and anti-STAT-6-mediated 'supershifting'. Furthermore, the sequences surrounding Y3 and Y4 had previously been reported to be capable of binding STAT-6 (Hou et al 1994). These regions contain a GYKXF core sequence; Y2 has the sequence GYQXF, suggesting that it, too, may be a STAT-6 binding site.

Cells from STAT-6 knockout mice can grow in response to IL-4 but fail to undergo differentiation

To further test the differential control of growth and differentiation by the IL-4Rα chain, we used STAT-6 knockout mice prepared by Shimoda et al (1996). B cells, T cells, and mast cell lines from these STAT-6-deficient mice showed growth responses to IL-4, although at a somewhat lower level than the response of heterozygous littermate cells. However, B cells from STAT-6-deficient mice failed to respond to lipopolysaccharide (LPS) plus IL-4 with greater induction of CD23, class II MHC molecules or Thy1 than in response to LPS alone.

B cells purified from STAT-6-deficient mice failed to secrete IgE in response to LPS and IL-4. STAT-6-deficient mice had undetectable levels of serum IgE; their IgE levels in response to polyclonal *in vivo* activation with anti-IgD were less than 1% those of heterozygous littermates (Shimoda et al 1996). By contrast, serum IgG1 levels and IgG1 responses to anti-IgD were normal in STAT-6-deficient mice. These results indicate that STAT-6 is principally required for IL-4-mediated differentiation but has only a partial effect on growth. They also suggest that IgG1 can be induced either by an IL-4-independent mechanism or by an IL-4-dependent, STAT-6-independent mechanism.

SJT

Chimeric IL-2Rβ chains expressing either the IL-4Rα 'growth' or 'differentiation' domain mediate IL-2-dependent growth and differentiation, respectively

To further examine the independence of the control of growth and gene activation/ differentiation by IL-4Rα, we transfected cell lines with constructs coding for chimeras composed of a truncated form of the IL-2Rβ chain (Δ404) and either the 'growth' domain (residues 437–557) or the 'differentiation' domain (residues 558–657) of the IL-4Rα chain (Wang et al 1996). 32D.IRS-1 cells expressing the chimera containing the growth domain showed vigorous [^{3}H]thymidine uptake in response to IL-2, whereas cells expressing the chimera that contained the differentiation domain showed no detectable DNA synthesis in response to IL-2. In contrast, M12.4.1 cells expressing the chimera that contained the growth domain showed only modest induction of CD23 (similar to that induced by the Δ557 truncation mutant) whereas lines that expressed the chimera containing the differentiation domain showed full induction of CD23 in response to IL-2. These results provide strong evidence in favour of the hypothesis that growth and gene activation/differentiation mediated by the IL-4Rα chain are determined by separate domains of this receptor. In addition to the growth and differentiation domains, critical sequences in membrane-proximal regions of the receptor are certainly important. It appears that the IL-2Rβ chain together with the γc chain can provide these signals allowing the chimeric receptors to mediate their characteristic responses.

IL-4 exerts a critical role in the acquisition of the cytokine-producing phenotype of CD4$^+$ T cells

Naïve CD4$^+$ T cells secrete IL-2 upon *in vitro* challenge with polyclonal activators or with their cognate antigen and antigen-presenting cells (APCs). These cells can differentiate into cells capable of secreting distinct patterns of cytokines, of which the best known are those secreting a Th1-like pattern (IFN-γ and IL-2) and those secreting a Th2-like pattern (IL-4, IL-5, IL-6, IL-10 and IL-13) (Street et al 1990). We and others have demonstrated that IL-4 itself plays a key role in the process through which T cells make the decision to develop into Th1-like or Th2-like cells (Le Gros et al 1990, Swain et al 1990). When naïve CD4$^+$ T cells are stimulated with their cognate peptide and APC, they develop into IL-4-producing cells if IL-4 is present in the culture medium; these cells generally fail to secrete IFN-γ (Seder et al 1992, Hsieh et al 1992). In contrast, if IL-4 is excluded from the 'priming' culture, the cells that emerge are capable of secreting IFN-γ but not IL-4. Including IL-12 in the priming culture further enhances the capacity of cells cultured in the absence of IL-4 to become IFN-γ producers (Hsieh et al 1993) and allows some IFN-γ production even by cells that had been primed in the presence of IL-4 (Seder et al 1993).

This effect of IL-4 can now be shown to depend upon STAT-6. T cells from STAT-6 knockout mice cultured with immobilized anti-CD3 and IL-2 fail to develop into IL-4 producers even when IL-4 is added to the priming culture (Shimoda et al 1996). Indeed, these cells become IFN-γ producers to an extent comparable to that observed as a result of priming in the absence of IL-4.

CD4$^+$/NK1.1$^+$ T cells are a source of IL-4 that can act to cause naïve T cells to develop into cells capable of producing IL-4

The need for IL-4 at the time of initial priming of naïve T cells if these cells are to differentiate into IL-4-producing cells calls for the identification of potential sources of such IL-4. Recently, we showed that the great majority of the substantial amount of IL-4 produced within 30–90 min of the injection of monoclonal anti-CD3 antibody into mice is made by a unique population of CD4$^+$ T cells that express NK1.1 on their surface (Yoshimoto & Paul 1994). These cells are relatively rare in the spleen ($\sim$1% of spleen cells). They show a dominant expression of Vα14 and of Vβ8.2, 7 and 2. Many of these cells are specific for the class Ib molecule CD1 (Bendelac et al 1995).

Interestingly, although this cell population is rare among spleen cells and even less well represented among peripheral lymph node cells, they are a larger fraction of the $CD4^+$ T cells in the liver, Peyer's patch and portal blood (Watanabe et al 1995, Kenai et al 1995). In these populations, in excess of 40% of the $CD4^+$ T cells may express NK1.1. Indeed, liver T cells make large amounts of IL-4 in response to culture with anti-CD3.

The development of $CD4^+$/NK1.1$^+$ T cells is markedly impaired in β_2-microglobulin knockout mice (Ohteki & McDonald 1994). This presumably reflects the role of β_2-microglobulin in forming heterodimers with the CD-1 'heavy' chain and the lack (or diminished) expression of CD-1 in β_2-microglobulin knockout mice. Indeed, spleen cells from β_2-microglobulin knockout mice make little or no IL-4 in response to injection of anti-CD3 (Yoshimoto et al 1995a). β_2-microglobulin knockout mice show an impairment in the appearance of IL-4-producing cells 5 d after the injection of anti-IgD antibodies and the produce approximately 10-fold less IgE in response to this stimulation (Yoshimoto et al 1995a).

Furthermore, one can restore the capacity of sub-lethally irradiated β_2-microglobulin knockout mice to make IgE in response to anti-IgD by infusing both enriched populations of $CD4^+$/NK1.1$^+$ thymocytes and T-depleted spleen cells from normal (but not β_2-microglobulin knockout) mice (Yoshimoto et al 1995a). These results indicate that in response to a class of stimuli, of which anti-IgD is the prototype, the production of IgE depends upon $CD4^+$/NK1.1$^+$ T cells, presumably because these cells promptly produce IL-4 which in turn acts at the time of the priming of conventional T cells to induce them to develop into cells capable of secreting IL-4 in response to subsequent receptor-mediated stimulation.

However, β_2-microglobulin knockout mice can produce IgE in response to certain stimuli. Thus immunization of these mice with ovalbumin and alum leads to the secretion of IgE and to the appearance of IL-4-producing cells. This may be due either to the fact that the defect of β_2-microglobulin knockout mice in expression of $CD4^+$/NK1.1$^+$ T cells is only partial ($\sim$90%) or due to the production of IL-4 by other cell types. Indeed, it must also be considered that an IL-4-independent means of priming naïve cells to become Th2-like cells may exist. Efforts to evaluate these various possibilities are now in progress.

SJL mice have a deficiency in $CD4^+$/NK1.1$^+$ T cells

The poor IgE response of β_2-microglobulin knockout mice in response to anti-IgD was predicted on the basis of their deficiency in $CD4^+$/NK1.1$^+$ splenic T cells. SJL mice are known to have a defect in IgE in responses to certain stimuli (Yoshimoto et al 1995b). To examine the possibility that their defect might reflect a lack of $CD4^+$/NK1.1$^+$ T cells, we initially challenged SJL mice with anti-IgD. We observed that they made very poor IgE responses to this polyclonal stimulant and that, at 5 d after injection of anti-IgD, their spleen cells had essentially undetectable levels of mRNA for IL-4, in contrast to spleen cells of similarly treated BALB/c and C57BL/6 mice, both of which made good IgE responses and expressed substantial amounts of IL-4 mRNA

(Yoshimoto et al 1995a). Injection of anti-CD3 into SJL mice failed to elicit IL-4 mRNA or protein synthesis. SJL mice had few if any CD4$^+$/NK1.1$^+$ T cells in spleen or liver or thymus. These results strongly suggest that their defect in IL-4 and IgE production in response to anti-IgD stemmed from a deficiency in this cell population. Indeed, SJL mice are known to be particularly prone to the development of experimental allergic encephalomyelitis (Brown & McFarlin 1981) suggesting they tend to develop Th1 responses. This would be consistent with a defect in initiation of Th2 responses as a result of a deficit in CD4$^+$/NK1.1$^+$ T cells.

How do CD4$^+$/NK1.1$^+$ T cells participate in the commitment of naïve conventional T cells to the Th2-like phenotype?

How do CD4$^+$/NK1.1$^+$ T cells participate in the induction of IL-4-producing T cells? There are limited data to help answer this question. Several possibilities need to be considered. Firstly, it is possible that peptides derived from allergens and pathogens that induced Th2-like responses may bind to CD1 and either increase its density on the surface of APCs or form epitopes that are recognized by the receptors of CD4$^+$/NK1.1$^+$ T cells. This possibility requires that CD1 acts as a peptide-binding molecule in a manner akin to conventional class I MHC molecules or, as has been described for human isoforms of CD1, as a lipid-binding molecule (Beckman et al 1994).

An alternative possibility is that the level of expression of CD1 or of co-stimulatory molecules on relevant APCs is under the control of cells that directly or indirectly interact with allergens and Th2-inducing pathogens. This is a particularly attractive hypothesis and requires an extensive evaluation of the regulation of expression of CD1 and of the physiological requirements for activation of and cytokine secretion by CD4$^+$/NK1.1$^+$ T cells.

A final possibility is that CD4$^+$/NK1.1$^+$ T cells are under some type of tonic stimulation through interaction of their receptors with CD1 expressed in the gut and possibly the skin. Naïve, conventional T cells that encounter antigens in these locales are likely to be primed in the presence of IL-4 and thus to differentiate toward a Th2-like phenotype.

A substantial effort is warranted to determine how CD4$^+$/NK1.1$^+$ T cells act to aid in the development of Th2-like responses. Equally, it is essential to determine under what range of physiological conditions CD4$^+$/NK1.1$^+$ T cells are critical for priming conventional T cells to become IL-4 producers and under what circumstances other cell types, such as basophils or possibly other CD4$^+$ T cells that can produce IL-4, will prove to be the dominant source of the IL-4 that appears critical for this process.

References

Beckman EM, Porcelli SA, Morita CT, Behar SM, Furlong ST, Brenner MB 1994 Recognition of a lipid antigen by CD1-restricted $\alpha\beta^+$ T cells. Nature 372:691–694

Bendelac A, Lantz O, Quimby ME, Yewdell JW, Bennink JR, Brutkiewicz RR 1995 CD1 recognition by mouse NK1$^+$ T lymphocytes. Science 268:863–865

Blaikie P, Immanuel D, Wu J, Li N, Yajnik V, Margolis B 1994 A region in Shc distinct from the SH2 domain can bind tyrosine-phosphorylated growth factor receptors. J Biol Chem 269:32031–32034

Brown AM, McFarlin DE 1981 Relapsing experimental allergic encephalomyelitis in the SJL/J mouse. Lab Invest 45:278–284

Coffman RL, Ohara J, Bond MW, Carty J, Zlotnik A, Paul WE 1986 B cell stimulatory factor-1 enhances the IgE response of lipopolysaccharide-activated B cells. J Immunol 136:4538–4541

Defrance T, Aubry JP, Rousset F et al 1987 Human recombinant interleukin 4 induces Fcε receptors (CD23) on normal human B lymphocytes. J Exp Med 165:459–467

Hilton DJ, Zhang JG, Metcalf D, Alexander WS, Nicola N, Willson TA 1996 Cloning and characterization of a binding subunit of the interleukin 13 receptor that is also a component of the interleukin 4 receptor. Proc Natl Acad Sci USA 93:497–501

Hou J, Schindler U, Henzel WJ, Ho TC, Brasseur M, McKnight SL 1994 An interleukin-4-induced transcription factor: IL-4 Stat. Science 265:1701–1706

Howard M, Farrar J, Hilfiker M et al 1982 Identification of a T cell-derived B cell growth factor distinct from interleukin 2. J Exp Med 155:914–923

Hsieh CS, Heimberger AB, Gold JS, O'Garra A, Murphy KM 1992 Differential regulation of T helper phenotype development by interleukins 4 and 10 in an $\alpha\beta$ T-cell receptor transgenic system. Proc Natl Acad Sci USA 89:6065–6069

Hsieh CS, Macatonia SE, Tripp CS, Wolf SF, O'Garra A, Murphy KM 1993 Development of T_H1 CD4$^+$ T cells through IL-12 produced by *Listeria*-induced macrophages. Science 260:547–549

Hu-Li J, Shevach EM, Mizuguchi J, Ohara J, Mosmann T, Paul WE 1987 B cell stimulatory factor 1 (interleukin 4) is a potent costimulant for normal resting T lymphocytes. J Exp Med 165:157–172

Kavanaugh WM, Williams LT 1994 An alternative to SH2 domains for binding tyrosine-phosphorylated proteins. Science 266:1862–1865

Keegan AD, Nelms K, White M, Wang LM, Pierce JH, Paul WE 1994 An IL-4 receptor region containing an insulin receptor motif is important for IL-4-mediated IRS-1 phosphorylation and cell growth. Cell 76:811–820

Kenai H, Matsuzaki G, Lin T, Yoshida H, Nomoto K 1995 Precursor cells to CD3-intermediate (CD3int) liver mononuclear cells in the adult liver: further evidence for the extrathymic development of CD3int liver mononuclear cells. Eur J Immunol 25:3365–3369

Kondo M, Takeshita T, Ishii N et al 1993 Sharing of the interleukin-2 (IL-2) receptor γ chain between receptors for IL-2 and IL-4. Science 262:1874–1877

Le Gros G, Ben-Sasson SZ, Seder R, Finkelman FD, Paul WE 1990 Generation of interleukin 4 (IL-4)-producing cells *in vivo* and *in vitro*: IL-2 and IL-4 are required for *in vitro* generation of IL-4-producing cells. J Exp Med 172:921–929

Mosmann TR, Bond MW, Coffman RL, Ohara J, Paul WE 1986 T-cell and mast cell lines respond to B-cell stimulatory factor 1. Proc Natl Acad Sci USA 83:5654–5658

Noelle RJ, Kuziel WA, Maliszewski CR, McAdams E, Vitetta ES, Tucker PW 1986 Regulation of the expression of multiple class II genes in murine B cells by B cell stimulatory factor-1 (BSF-1). J Immunol 137:1718–1723

Ohara J, Paul WE 1987 Receptors for B-cell stimulatory factor-1 expressed on cells of haematopoietic lineage. Nature 325:537–540

Ohteki T, MacDonald HR 1994 Major histocompatibility complex class I related molecules control the development of CD4$^+$8$^-$ and CD4$^-$8$^-$ subsets of natural killer 1.1$^+$ T cell receptor-α/β^+ cells in the liver of mice. J Exp Med 180:699–704

Paul WE 1991 Interleukin-4: a prototypic immunoregulatory lymphokine. Blood 77:1859–1870

Rothman P, Lutzker S, Cook W, Coffman R, Alt FW 1988 Mitogen plus interleukin 4 induction of C epsilon transcripts in B lymphoid cells. J Exp Med 168:2385–2389
Russell SM, Keegan AD, Harada N et al 1993 Interleukin-2 receptor γ chain: a functional component of the interleukin-4 receptor. Science 262:1880–1883
Ryan JJ, McReynolds LJ, Keegan A et al 1996 Growth and gene expression are predominantly controlled by distinct regions of the human IL-4 receptor. Immunity 4:123–132
Seder RA, Paul WE, Davis MM, Fazekas de St Groth BF 1992 The presence of interleukin 4 during *in vitro* priming determines the lymphokine-producing potential of $CD4^+$ T cells from T cell receptor transgenic mice. J Exp Med 176:1091–1098
Seder RA, Gazzinelli R, Sher A, Paul WE 1993 Interleukin 12 acts directly on $CD4^+$ T cells to enhance priming for interferon γ production and diminishes interleukin 4 inhibition of such priming. Proc Natl Acad Sci USA 90:10188–10192
Shimoda K, van Deursen J, Sangster MY et al 1996 Lack of IL-4-induced Th2 response and IgE class switching in mice with disrupted Stat6 gene. Nature 380:630–633
Stavnezer J, Radcliffe G, Lin YC et al 1988 Immunoglobulin heavy-chain switching may be directed by prior induction of transcripts from constant-region genes. Proc Natl Acad Sci USA 85:7704–7708
Street NE, Schumacher JH, Fong TA et al 1990 Heterogeneity of mouse helper T cells. Evidence from bulk cultures and limiting dilution cloning for precursors of Th1 and Th2 cells. J Immunol 144:1629–1639
Swain SL, Weinberg AD, English M, Huston G 1990 IL-4 directs the development of Th2-like helper effectors. J Immunol 145:3796–3806
Vitetta ES, Ohara J, Myers CD, Layton JE, Krammer PH, Paul WE 1985 Serological, biochemical, and functional identity of B cell-stimulatory factor 1 and B cell differentiation factor for IgG1. J Exp Med 162:1726–1731
Wang LM, Keegan AD, Paul WE, Heidaran MA, Gutkind JS, Pierce JH 1992 IL-4 activates a distinct signal transduction cascade from IL-3 in factor-dependent myeloid cells. EMBO J 11:4899–4908
Wang LM, Myers MJ, Sun XJ, Aaronson SA, White M, Pierce JH 1993 IRS-1: essential for insulin- and IL-4-stimulated mitogenesis in hematopoietic cells. Science 261:1591–1594
Wang HY, Paul WE, Keegan AD 1996 IL-4 function can be transferred to the IL-2 receptor by tyrosine containing sequences found in the IL-4 receptor α chain. Immunity 4:113–121
Watanabe H, Miyaji C, Kawachi Y et al 1995 Relationships between intermediate TCR cells and $NK1.1^+$ T cells in various immune organs. $NK1.1^+$ T cells are present within a population of intermediate TCR cells. J Immunol 155:1972–1983
White MF, Livingston JN, Backer JM et al 1988 Mutation of the insulin receptor at tyrosine 960 inhibits signal transmission but does not affect its tyrosine kinase activity. Cell 54:641–649
Yoshimoto T, Paul WE 1994 CD4pos, NK1.1pos T cells promptly produce interleukin 4 in response to *in vivo* challenge with anti-CD3. J Exp Med 179:1285–1295
Yoshimoto T, Bendelac A, Watson C, Hu-Li J, Paul WE 1995a Role of $NK1.1^+$ T cells in a T_H2 response and in immunoglobulin E production. Science 270:1845–1847
Yoshimoto T, Bendelac A, Hu-Li J, Paul WE 1995b Defective IgE production by SJL mice is linked to the absence of CD4+, NK1.1+ T cells that promptly produce interleukin 4. Proc Natl Acad Sci USA 92:11931–11934

DISCUSSION

Melchers: In the STAT-6 knockout mouse, do you ever see with these different receptor tails a differentiation into signals that affect $\gamma 1$ versus ε, transcription versus switching, and is there a differential signalling towards either transcription or switching?

Paul: If we do the *in vitro* study with LPS plus IL-4, we don't see any IgE secreted. We do see some IgG1.

Melchers: How about on the molecular level of transcription and switching?

Paul: We haven't yet looked at germline transcripts for that. IL-4 *in vitro* definitely causes a major degree of switching to IgG1. On the other hand we have never succeeded in blocking the production of IgG1 in response to anti-IgD or to *Nippostrongylus brasiliensis* with neutralizing amounts of anti-IL-4 antibody *in vivo*, although we can inhibit IgE production. IL-4 knockouts do have an IgG1 defect, but it is not complete, whereas the STAT-6 knockouts we have examined do not display an IgG1 defect.

Metcalf: The data on the IL-4 receptor very nicely complement the data from the analysis of granulocyte/macrophage colony-stimulating factor receptor deletions. You've gone one step further in that you have reassembled them and done the positive side of the experiment rather than the deletion side. You assembled them in the correct sequence: have you considered putting the differentiation segment membrane-proximal to the growth signal? Is there a physical reason why the sequence regulating mitosis is always membrane-proximal to the sequence controlling differentiation?

Paul: We have not studied that. Instead, we have replaced the sequence surrounding the tyrosine in the I4R motif with what we believe is an authentic STAT-6 site in the hope that we would induce efficient differentiation. The experiment has so far been rather a nice failure.

Metcalf: In his presentation, Nick Nicola said that the deletion experiments do not actually exclude the possibility that you might need a Box1/Box2 region for differentiation.

Paul: That was why the IL-2 receptor was such a good choice, although at the time we chose it simply for convenience. We have Box1/Box2 and both JAK1 and JAK3 activation. I wouldn't want to say that without those we would get the induction of IL-4-type differentiation events.

The question arises as to whether the exact location of the 'growth' and 'differentiation' domains in the receptor is significant. One interpretation would be that in the non-ligated receptor the two domains are folded against each other and act basically to restrict access of potential binding molecules. The first phosphorylation event may open the receptor and make it available for docking by such molecules. That's why we think that the Y497F mutation prevents differentiation, because it may prevent the receptor from achieving an open conformation. It is an attractive idea.

Burgess: Sandra Nicholson and Julie Layton have very similar observations for the G-CSF receptor in separating differentiation and proliferative effects. In the IL-4 receptor there's a region you described between positions 437 and 555 which is sufficient for growth but doesn't seem to be involved in any of the systems that you talked about. How do you explain this?

Paul: That was the argument that this sequence surrounding Tyr1 was principally involved with growth and that it wasn't a good differentiation signal. It turns out that the sequence surrounding this tyrosine in the IL-4 receptor has a limited homology to the STAT-6 binding sites, whereas the sequence surrounding in the comparable tyrosine in the insulin receptor has none at all. I think that a modest differentiation effect is mediated by the 'growth' domain, but I would argue that that region is *principally* involved in growth determination.

Burgess: Which amino acid would be phosphorylated at 557? Would that induce the growth signalling?

Paul: We haven't ablated the JAK1/JAK3 sites.

Burgess: Do you think JAK1/JAK3 are phosphorylating that tyrosine?

Paul: Yes. I don't have direct evidence, but I believe that JAK1 and JAK3 are probably phosphorylating tyrosine 497.

Burgess: That would be one of the first implications that the JAKs are involved in a proliferative response.

Paul: We have examined a JAK1 mutant; it doesn't phosphorylate Y497 in response to IL-4.

Burgess: Doesn't that worry you?

Paul: No, because the kinase is gone. We argue that JAK1 and JAK3 are important in the growth effect.

Burgess: So those mutants don't grow?

Paul: We don't have good data, but we can show there is no phosphorylation of IRS-1.

Strasser: One of the well known effects of IL-4-mediated signalling is cell survival, as opposed to proliferation. Do you think in your 32D cell system you really measured the mitogenic signal or might you have measured a survival signal? 32D cells are certainly somewhat transformed: perhaps there is an inherent mutation towards growth and as long as you give the survival signal (which IL-4 can do) the cells can grow, at least for a while. You mentioned that the growth was actually poor. What were the survival data on these cells?

Paul: Naïve T cells die rapidly. IL-4 prevents them from dying. With the STAT-6 knockouts we can at least test whether STAT-6 is required to mediate this survival effect. Preliminary data suggest it is not. We would like to have an IRS-2 knockout; that would be another way to tackle this problem.

You are asking whether we can distinguish growth and survival in this system. So far we can't, since both processes appear to depend on the same region.

Dexter: Coming back to the previous question, because these 32D cells are differentiation-inducible one of the things that you could have looked for is the degree of differentiation in the presence of IL-4. They have been reported to go to granulocytes and other cell lineages. Did you do that?

Paul: No, but we would like to.

Dexter: One question you could address concerns what happens in the presence of granulocyte colony-stimulating factor when you add IL-4.

I was a bit confused, because in the early part of your paper you indicated a role for IL-4 in B cell differentiation, and yet your STAT-6 knockouts show that STAT-6 is not essential for that.

Paul: In the STAT-6 knockout mice we don't get CD23 induction, development of Th2 cells nor switching to IgE in response to IL-4. We would argue that it is consistent.

Nicola: The IL-4 receptor differs from other cytokine receptors in that the NPXY site, which is an IRS-2 binding site in the IL-4 receptor, is a Shc binding site for most other cytokine receptors. Does IL-4 activate Shc?

Paul: Shc does bind to that site; we have also picked up another substrate that binds there. In addition to the NPXY motif, upstream residues are also important; different substrates rely differently on these residues.

Nicola: The other general difference is that when you eliminate your STAT-6 binding sites you lose STAT-6 activation, whereas in some of the other cytokine receptors there are alternate mechanisms of STAT activation that don't involve STAT binding to the receptors.

B lymphocyte physiology: the beginning and the end

G. J. V. Nossal

The Walter and Eliza Hall Institute of Medical Research, Post Office, Royal Melbourne Hospital, Victoria 3050, Australia

Abstract. Whereas lymphatic tissues were implicated as the chief sites for antibody production almost 100 years ago, three findings cemented the role of the plasma cell as the actual producer, namely the histological and tissue culture studies of Fagraeus, the beautiful immunofluorescent approach of Coons and his group, and the micromanipulation approach to the study of antibody formation by single cells introduced by Lederberg and myself. Proof that antibody-forming cells derived from B, rather than T, cells had to await the studies of Miller and Mitchell, with some minor technical contributions from myself. The physiology of the germinal centre also has a long history, and recently much interest has surrounded the relative roles of germinal centres as sites of somatic immunoglobulin V gene hypermutation and selection of high affinity B cells, versus the roles of extra-follicular proliferative foci as sources of the primary antibody response. Tolerance and self-antigens within the primary B lymphocyte repertoire is secured by clonal deletion within the bone marrow of those cells which are operationally the most threatening to the body, and a second level of functional impairment of B cell activity and life-span, capable of being induced centrally or peripherally, termed clonal anergy. As these concepts became progressively more refined through transgenic models of immunological tolerance, we turned our attention more towards tolerance within the secondary B lymphocyte repertoire. This is generated primarily inside germinal centres, where it appears that two quite separate mechanisms act as a bulwark against the possible creation of hypermutated anti-self cells. The first is that germinal centre activity and memory cell generation are dependent on antigen-specific, germinal centre-seeking $CD4^+$ T cells, and if a putative anti-self mutant B cell gets no help because of T cell tolerance, it will not expand further. The second is a very specific mechanism confined to the germinal centre whereby antigen-specific B cells are especially sensitive to antigen-induced apoptosis if soluble, deaggregated antigen is presented to them before they reach the 'rescue' signal of follicular dendritic-cell-bound antigen. While some of the death in germinal centres is clearly apoptotic in nature, a further phenomenon observed electron microscopically relates to the formation of type B dark cells. It is not yet clear whether the DNA in this type of dying cell is cleaved. Transgenic expression of *bcl-2* in germinal centre B lymphocytes confers incomplete protection from apoptosis caused by soluble antigen. The suggestion that at least some of this apoptosis is mediated via Fas–Fas ligand interactions is prompted by the observation that the apoptotic phenomenon is markedly reduced in *lpr* mice.

1997 The molecular basis of cellular defence mechanisms. Wiley, Chichester (Ciba Foundation Symposium 204) p 220–231

1957 was the year that Talmage (1957) first presented the clonal selection theory, as did Burnet (1957) in a more elaborate form. This was also the year I entered cellular immunology, so it seemed natural to address the problem which was engaging my mentor's mind. Moreover, a critical feature of clonal selection was the proposal of clonal deletion as the basis of immunological tolerance, so the development of robust tolerance models using antibody formation as a read-out was a natural extension of my work. The first hint that clonal selection might be correct came from the finding that each cell always made just one antibody (Nossal & Lederberg 1958). Little did I realize that formal proof of clonal selection would only come 18 years later (Nossal & Pike 1976)! The methodological constraints to the preparation of antigen-specific B cells from unimmunized mice were formidable, and the microculture techniques for cloning single lymphocytes were quite demanding. The years in between, however, were not wasted as they saw our single-cell techniques applied to diverse problems, such as the disproof of the direct template hypothesis (Nossal et al 1965); the switch of antibody synthesis from IgM to IgG without change of specificity (Nossal et al 1964a); and, as a footnote to the major thrust of Graham Mitchell and Jacques Miller, the demonstration that B cells, not T cells, make antibody (Nossal et al 1968a). When an exact method for the enumeration of antibody-forming cell precursors became available through B cell cloning, it was possible to address the issue of immunological tolerance within the primary B cell repertoire. This became our major preoccupation for about a decade (reviewed in Nossal 1983) and led to a surprising result. High doses of toleragens could indeed cause clonal deletion of antigen-specific B cells, but very much lower doses exhibited in fetal life or shortly after birth caused a non-deletional form of tolerance which we termed *clonal anergy* (Nossal & Pike 1980). It has been satisfying to see the concept of clonal anergy also applied to the T cell (Lamb et al 1983, Jenkins et al 1987) and verified as a major mechanism for the B cell by the application of transgene technology (reviewed by Goodnow 1992).

'Real' antibody formation occurs not in single cell microcultures but in lymphoid organs with an elaborate and highly interesting microarchitecture. Since about 1962, the structure of lymph nodes and spleen, and particularly of the antigen-trapping cells, has been a major preoccupation of our laboratory. We were fortunate enough to discover the antigen-capturing power of primary lymphoid follicles (Nossal et al 1964b) based on follicular dendritic cells (FDCs) which hold antigen complexed to antibody on their surface for very long periods (Nossal et al 1968b). With the strong bacterial flagellar antigens that we used, the details of the rapid emergence of germinal centres adjacent to these antigen depots could be charted. The critical involvement of T cells in the process became apparent later, there being a virtual absence of germinal centre formation in young nude mice despite the presence of normal antigen-trapping mechanisms. Our more recent interest in germinal centres came about in a round-about way.

High affinity antibodies usually result from hypermutated immunoglobulin V genes and germinal centres are critical to hypermutation

Having devoted considerable energy to establishing B cell tolerance models using exogenously introduced antigens, we sought to use the clonal approach to distinguish between clonal deletion and clonal anergy for cells reactive with authentic self antigens. How does the number of anti-human serum albumin B cells in a mouse compare with the number of anti-murine serum albumin B cells? What this analysis soon revealed was that the proportion of high affinity anti-protein B cells within the primary repertoire was vanishingly small even for a foreign protein—too small, in fact, to work with. Clearly most anti-protein antibody responses begin with B cells of relatively low affinity, the response then being shaped by somatic Ig V gene hypermutation and selection of high affinity variants. It was becoming clear (reviewed in Nossal 1994) that most, if not all, of this hypermutation was taking place in germinal centres, with antigen on FDCs playing a critical role in the selection process. 'Unfit' B cells within the centre died rapidly through apoptosis and were removed by 'tingible body' macrophages. This raised the question of whether there were special mechanisms preventing the emergence of cells bearing mutations which conferred anti-self potential on the cell. Would such cells be stimulated?

Soluble antigen stops emergence of high affinity B cells

Having failed to establish a system dependent on authentic self antigens, and wishing to explore physiological rather than transgenic immune responses, we developed a model where the toleragen was a freshly deaggregated low-substitution hapten–protein conjugate. Introduced before or after immunization with a high substitution conjugate of the same antigen, the toleragen substantially lowered resultant antibody titres and moreover lowered or abrogated subsequent secondary responses. With the hapten (4-hydroxy-3-nitrophenyl acetyl; NP) and the C57BL/6 mouse strain, it became possible to explore many facets of this tolerance. Single cell cloning studies indeed showed a failure to generate high affinity, isotype-switched antibody-forming cell precursors (Karvelas & Nossal 1992) and, interestingly, this could be achieved after the immune response had started (Nossal et al 1993). Flow cytometric analysis confirmed a deficit of isotype-switched antigen-binding cells (Pulendran et al 1994). T cell tolerance contributed to this failure of emergence of high affinity B cells (Karvelas & Nossal 1992) and a major direct effect on the germinal centre B cell population was also noted. This was demonstrated most directly by looking for apoptosis of B cells in germinal centres by the TUNEL technique (Pulendran et al 1995a). As cells in germinal centres display a high rate of death in any case, a background level of apoptosis was established. Then, deaggregated antigen was injected and within 1 h there was a significant rise in the number of TUNEL-positive B cells from a mean of about five per germinal centre section to about 15. The peak was over 30 apoptotic cells per germinal centre section at 4 h, following which the number gradually fell, subsiding to background levels by 24 h. This did not happen with carrier alone or with irrelevant protein. In other words, the soluble antigen caused a wave of massive cell death in germinal centre B cells with specificity for NP, which subsided within a day. Early after the pulse of soluble antigen, the TUNEL-positive fragmented nuclear material was distributed in a characteristic fashion, mainly as whole nuclei clustered inside macrophages. It is clear that macrophages are capable of engulfing apoptotic cells quite quickly and certainly within less than 1 h. However, some scattered single apoptotic cells could also be seen, presumably representing cells that had died most recently.

Apoptosis induced rapidly after introduction of deaggregated antigen was confined to the germinal centre. Of course, there were occasional apoptotic cells in the white and red pulp, but their number was not increased soon after soluble antigen. Specifically, foci of extrafollicular antibody-forming cells (AFCs) usually found near the junction of white and red pulp, did not show the same susceptibility to antigen-induced death. While the cells within them do spontaneously die by apoptosis (Smith et al 1996), this is not increased by soluble antigen. It is true that 12 h or so after soluble antigen, macrophages filled with fine, granular TUNEL-positive material could be found throughout the spleen and particularly in the marginal zone and the red pulp. The most probable interpretation is that these represented 'tingible body' macrophages which had migrated out of germinal centres and had degraded the nuclear material into fine debris within phagolysosomes.

Further insights into the cell death process within germinal centres was gained via electron microscopy (B. Pulendran, R. van Driel & G. J. V. Nossal, unpublished observations) (Fig. 1). This confirmed the TUNEL results, in that the dead cells in germinal centres showed the features typical of apoptosis, including loss of nuclear volume, chromatin changes lending a dense granular appearance to the nucleus, dilatation of the endoplasmic reticulum, blebbing of the cell membrane where this was still visible, but relatively good preservation of cytoplasmic organelles. Again, the majority of apoptotic cells and/or nuclei were inside macrophages, some could be seen adjacent to macrophages but a few were scattered as single cells. In a minority of cases, a different type of electron microscopic picture was seen. Cells typical of type B dark cells occurred singly or in clusters. These showed abnormally electron-dense nuclei but with relatively normal chromatin distribution. Membrane blebbing was not apparent, but the mitochondria were grossly swollen and the cytoplasm, also abnormally electron-dense, formed thin processes extending some distance between neighbouring cells. Type B dark cells do not fit neatly into either apoptotic or necrotic pathways. Interestingly, apoptotic cells and dark cells were sometimes seen in close apposition to each other. Given the huge interest in apoptosis at the present time, the relationship between these two types of dying cells deserves much closer examination.

It might have been tidier if we could have argued that soluble antigen had the capacity to stop the germinal centre reaction dead in its tracks. Such was not the case, however. In fact, germinal centres in general were *larger* in tolerant than in control

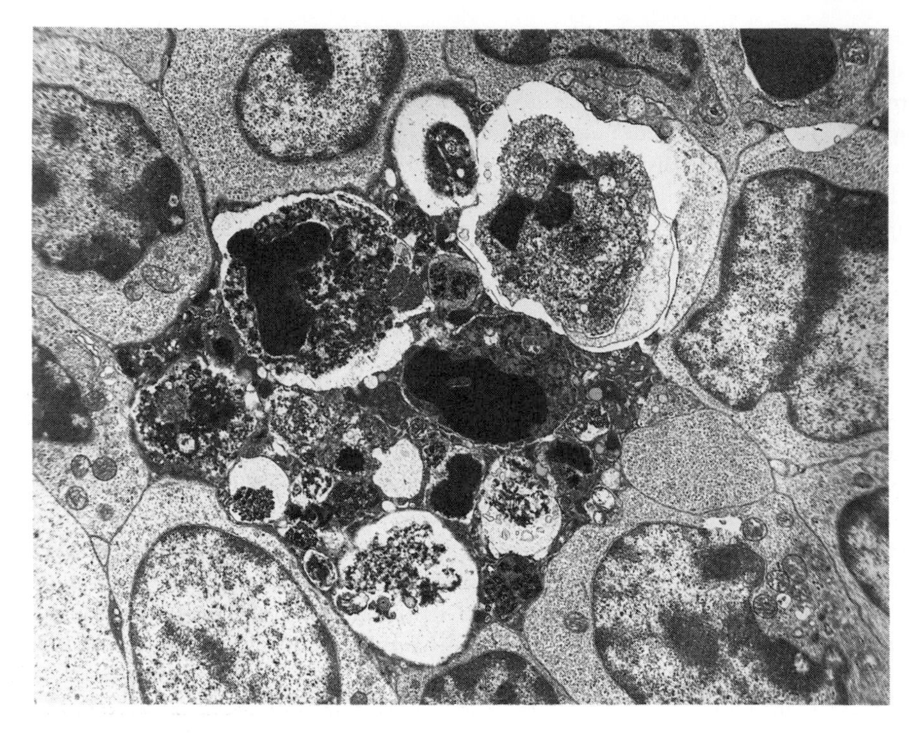

FIG. 1. Transmission electron micrograph of a cluster of germinal centre B cells showing varying degrees of apoptosis, almost certainly recently engulfed by a macrophage. Note the electron-dense, homogeneous appearance of the nuclei and nuclear fragments. The phagolysosomes at the lower end show more advanced degradation. Several cells show blebbing of the membrane. At the top right hand corner, an isolated cell shows very early apoptosis, with a characteristic, somewhat granular homogeneity of the chromatin.

mice. The majority of these were not λ-positive, i.e. were not founded by the V_H 186.2$^+$ $\lambda 1^+$ B cells characteristic of the primary T-dependent response of C57BL/6 mice to NP conjugates. It is possible that these large, apparently antigen non-specific germinal centres could have represented a reaction to immune complexes and/or to the carrier. They certainly are very active as shown by a greater number of cells in S phase. Moreover, the toleragen injected at a stage where well-developed extrafollicular foci of AFCs are present, enhanced both specific anti-NP and general IgG1 secretion. These studies further document that NP-specific germinal centres and extrafollicular foci are differentially regulated.

The special susceptibility of germinal centre B cells to apoptosis following contact with soluble antigen has also been studied by Shokat et al (1995) and Han et al (1995). While there are some subtle differences (discussed below), both these groups independently reached a similar conclusion to ours.

The influence of genes capable of regulating cell death on various phases of the germinal centre reaction

Bcl-2

As mentioned above, there has been a veritable explosion of work on cell death-related genes over the last few years, and particularly on the genes *bcl-2* and *fas*. Regretfully, our exploration of the effects of these genes on soluble antigen-induced or spontaneous apoptosis in germinal centres is not yet as definitive as we would like, in part because the numbers of relevant mice on a C57BL/6 background is still a limitation, and in part because most of the mice available to us are not yet specific-pathogen free and thus do not have the commendably low spontaneous (or pre-antigenic) germinal centre formation which characterizes the spleens of our standard C57BL/6 SPF mice. With these limitations, we can make the following statements about *bcl-2* and *fas* in respect of germinal centres (Smith et al 1994, 1995, Pulendran et al 1995a,b, B. Pulendran, M. Karvelas, G. Kannourakis & G. V. J. Nossal, unpublished observations). Mice transgenic for *bcl-2* (an anti-apoptosis gene) express *bcl-2* in germinal centres and display markedly increased life-span not only of IgM and IgG antibody-forming cells but also of NP^+ $IgG1^+$ $B220^+$ IgM^- IgD^- memory cells. At the peak of the germinal centre reaction, the total number of such cells was threefold higher in transgenics than in control, and from day 42 on, the number of antigen-specific memory B cells persisting in the spleens was 20-fold raised. This difference lasted till at least 215 d after immunization. This might have suggested that the piling up of possibly poorly selected and unnecessary B cells within germinal centres could have interfered with affinity maturation of the immune response. Examination of this question by a comparison of relative binding of serum IgG anti-NP antibody over the first 21 d of the response to NP_2 versus NP_{15}-bovine serum albumin (BSA) ELISA plate coats lent no support to this view. Moreover, at day 10 of the response, of the V_H genes used that were found to be mutated, the mutation frequency was not significantly different (1 in 88 base pairs in the controls, and 1 in 106 in the *bcl-2* transgenic mice). Furthermore, the critical tryptophan to leucine exchange frequency at position 33 in the first complementarity-determining region, which alone enhances antibody affinity 10-fold, was 42% and 33%, respectively, in the above groups. Thus effective selection of high affinity mutants was occurring. There was one striking difference, however, in that only 35% of the *bcl-2* transgenic sequences were multiply mutated, versus 86% in a small sample (14 sequences) of controls. This suggested a prolonged retention of non-mutated cells by the transgenic mice.

We have now had a chance to perform a much more extensive series of sequence analyses of V_H genes from single cells sorted from spleens 49 d after immunization. These cells, harvested long after the peak of the germinal centre reaction, represented authentic isotype-switched memory B cells. Here the *bcl-2* transgenic cells exhibited far fewer affinity-enhancing position 33 mutations and only half the overall mutation frequency. In other words, the iterative process of mutation and selection had been significantly interfered with. It is of interest that the effect was only partial, and the end result perhaps not of great functional significance.

A further point of interest was to examine the extrafollicular foci of AFCs which arise over the first 7–8 d of the primary response and then rapidly decline (Smith et al 1996). Examination of a likely site of AFC migration, the bone marrow, suggested that the decline of AFCs within the spleen was unlikely to be due to migration. TUNEL staining showed that apoptosis *in situ* was likely to be the main process at work, as 3.1% of cytoplasmic IgG1$^+$ cells at day 7 after immunization but 9% by day 11 after immunization were TUNEL positive. Moreover, foci persisted much longer and were of greater size in *bcl-2* transgenic mice.

We also examined germinal centre-specific soluble-antigen-induced apoptosis in *bcl-2* transgenic mice. First, prior to the soluble antigen pulse, there appeared to be fewer TUNEL-positive cells in *bcl-2* transgenic mouse germinal centres than in littermate controls. As in the above studies, there was the suggestion that unwanted cells persisted to a greater degree. The soluble antigen did cause extra cell death in the transgenic mice though there were fewer TUNEL-positive cells than in controls. In other words, *bcl-2* only partially protected against this form of cell death. This finding is in contrast to that of Shokat et al (1995) for reasons that are not clear. Our results raise the possibility that the antigen-induced cell death is different from the 'death by neglect' of cells not chosen within the germinal centre because of raised affinity.

Fas

Another important cell death-related gene is the *fas* gene, where ligation of the gene product by the Fas ligand leads to cell death. It might be expected, therefore, that a functional deletion of the *fas* gene might impede some cell death events. The Fas ligand is expressed on activated T cells and has recently been shown to be present on activated B cells as well (Hahne et al 1996). The spontaneous *lpr* mutation inactivates the *fas* gene, thus *lpr* mice permit study of the question of how involved this gene is in the germinal centre reaction.

We have established (Smith et al 1995) that Fas is highly expressed on normal murine germinal centre B cells but the absence of Fas did not influence the kinetics of appearance either of AFC or of NP$^+$ IgG1$^+$ B220$^+$ IgD$^-$ IgM$^-$ memory B cells. Furthermore, affinity maturation of anti-NP IgG1 antibody was not influenced by the *lpr* mutation. The hypermutation and selection process within the germinal centre appeared to be normal in that single NP-specific memory cells sorted 41 d after immunization showed a mean of 9.0 V_H gene mutations versus 8.3 in controls. Also, the percentage of tryptophan to leucine mutations at position 33 in CDR-1 was similar, 63% versus 55%. Thus Fas does not seem to be required for correct germinal centre functioning or for related aspects of B cell regulation.

For this reason, we have been very interested in the results of TUNEL experiments on the soluble antigen-induced cell death in germinal centres of NP-immunized *lpr* mice. One small study already published suggested that this would be unaffected by the *fas* gene mutation (Han et al 1995). In contrast, we made two observations. First, the 'background' number of TUNEL-positive cells seen in germinal centres in the absence of the soluble antigen pulse was no different between wild type and *lpr* mice. Thus the Bcl-2 preventable 'death by neglect' does not require Fas–Fas ligand interactions. However, the soluble antigen-induced apoptosis was markedly inhibited in *lpr* mice. Thus there was an element of Bcl-2-unprotectable, Fas-dependent cell death in the process. This is puzzling, but it should be remembered that activated B cells do express Fas ligand (Hahne et al 1996), so both suicidal and fratricidal processes could be considered. Furthermore, there are activated T cells in germinal centres, which are Fas ligand positive. We do have a possible explanation for the differences between our results and those of Han et al (1995). We have used very low substitution NP conjugates (NP_1-HSA) in order to mimic soluble self antigens, whereas Han et al (1995) used NIP_{15}-HSA, NIP being a hapten having a higher affinity than NP for the B cells in question. As a result, they saw more dramatic cell death in normal C57BL/6 immunized mice. When we went to a high conjugation ratio, we also saw more apoptosis, and now there was a definite (though still reduced) effect in *lpr* mice. Under these extremely cross-linking conditions, different mechanisms may have been at work. The results of Han et al (1995) and ours are in agreement about the irrelevance of Fas in the regular, spontaneous apoptosis within germinal centres.

The above results do not fit neatly into the two accepted methods of lymphocyte death, namely 'death by neglect', which is inhibited by Bcl-2, and activation-induced death, which frequently involves Fas (Strasser 1995). The partial nature of the effects of the gene mutations suggests a complex situation. Complement fixation appears to have been excluded as a cause of antigen-induced B cell death (Han et al 1995). Up-regulation of Fas or Fas ligand on the B cells is not excluded as an effect of antigen, though the death is rather fast for that. Certainly, germinal centre B cells are unduly susceptible to apoptosis after receptor cross-linking, and, in an *in vivo* situation, the Fas-dependent mechanism may be only one of several pathways at work.

Effects of mutation in the tyrosine kinase Btk on T cell-dependent B cell responses

We have also embarked on a study of the primary and secondary anti-NP responses of CBA/N mice, which suffer from an X-linked immune deficiency (*xid*) because of a point mutation in the tyrosine kinase Btk (Ridderstad et al 1996). The *xid* mutation causes male CBA/N mice to be deficient in T-independent responses to some antigens, and to have a reduced primary T-dependent response. To retain the advantages of the anti-NP response using chiefly V_H186.2 (which requires a C57BL/6 background), we investigated (CBA/N × C57BL/6) F1 mice. Germinal centre formation, generation of AFCs and appearance of high affinity AFCs were markedly deficient in male (*xid*) mice compared to females (heterozygous for the mutation and thus phenotypically normal). The frequency of NP-specific memory B cells generated in the primary response was reduced 10-fold. Despite this, and somewhat remarkably, the secondary AFC responses were normal in *xid* mice, as were the appearance of clusters of extrafollicular AFCs in the secondary response, and the number and size of germinal centres after rechallenge. Moreover, single NP-specific AFCs from day 5 of the secondary response, i.e. IgM^- IgD^- $Syndecan^+$ $IgG1^+$ NP^+ cells, were examined for mutations in the case of those cells using the V_H 186.2 gene. All of 29 cells were extensively hypermutated and 55% of them had the tryptophan to leucine exchange, features not expected from the poor germinal centre development in the primary response and the lowered memory B cell generation. The results are a sobering reminder that the limiting factors during *in vivo* immune responses are poorly understood. It is clear that even a very small number of suitable memory B cells can mount an excellent secondary response, particularly (as in this case) where a primed helper T cell population is available.

Conclusions

The remarkable adventure of the last 40 years has produced the answer to the antibody puzzle which was so impenetrable during the early days of the 'instruction versus selection' debate (Lederberg 1959). Who could have foreseen the intricacies of multiple sets of Ig minigenes, their random somatic assembly on a case by case basis during B cell genesis, the huge regulatory role of T cells on B cell physiology or the multiple pathways to tolerance? Who could have foretold the cleverness of a system which has T cells seeing linear peptides of inaccessible proteins from viral and other pathogens while B cells looked after the conformational determinants of proteins from their outside? The grand adventure of cellular and molecular immunology which has brought us to this point must surely stand as one of the finest chapters in the history of biology. To have been an eyewitness to the unfolding drama has been a rare privilege.

As the broad outlines of the immune system's strategic design, the cellular and molecular anatomy, as it were, now stand revealed, the ineffable challenge of understanding its physiology remains. The 21st century will belong to immunoregulation. We must increasingly come to grips with the myriad of cellular and molecular signalling mechanisms and feedback loops, not only for intellectual satisfaction but also for major tools of disease control. At the heart of the problem still lies self/non-self discrimination. This brief paper has sought to highlight some of the mechanisms that guard against the entry of somatically mutated anti-self B cells into the secondary repertoire. In revealing some new mechanisms, it has also shown how much more we need to know about the control of B cell death. The refined methods now available for the enumeration, sorting, characterization, culture and genetic analysis of single B cells offer some great opportunities for progressing the story. The end is nowhere in sight!

Acknowledgements

Special thanks are due to the members of my laboratory down the years who have done this work, and particularly to David Tarlinton, Maria Karvelas, Bali Pulendran, Ken Smith, Anna Ridderstad and Amanda Light.

References

Burnet FM 1957 A modification of Jerne's theory of antibody production using the concept of clonal selection. Aust J Sci 20:67–69

Goodnow CC 1992 Transgenic mice and analysis of B cell tolerance. Annu Rev Immunol 10:489–518

Hahne M, Renno T, Schroeter M et al 1996 Activated B cells express functional Fas ligand. Eur J Immunol 26:721–724

Han S, Zheng B, Dal Porto J, Kelsoe G 1995 *In situ* studies of the primary immune response to (4-hydroxy-3-nitrophenyl)acetyl IV. Affinity-dependent, antigen-driven B cell apoptosis in germinal centers as a mechanism for maintaining self-tolerance. J Exp Med 182:1653–1644

Jenkins MR, Pardoll DM, Mizuguchi J, Chused TM, Schwartz RH 1987 Molecular events in the induction of a nonresponsive state of interleukin-2 producing helper T lymphocyte clones. Proc Natl Acad Sci USA 84:5409–5414

Karvelas M, Nossal GJV 1992 Memory cell generation ablated by soluble protein antigen by means of effects on T- and B-lymphocyte compartments. Proc Natl Acad Sci USA 89:3150–3154

Lamb JR, Skidmore BJ, Green N, Chiller JM, Feldmann M 1983 Induction of tolerance in influenza virus-immune T lymphocyte clones with synthetic peptides of influenza hemaglutinin. J Exp Med 157:1434–1447

Lederberg J 1959 Genes and antibodies. Science 129:1649–1653

Nossal GJV 1983 Cellular mechanisms of immunologic tolerance. Annu Rev Immunol 1:33–62

Nossal GJV 1994 Twenty-five years of germinal centre physiology: implications for tolerance in the secondary B cell repertoire. Scand J Immunol 40:575–578

Nossal GJV, Lederberg J 1958 Antibody production by single cells. Nature 181:1419–1420

Nossal GJV, Pike BL 1976 Single cell studies on the antibody-forming potential of fractionated, hapten-specific B lymphocytes. Immunology 30:189–202

Nossal GJV, Pike BL 1980 Clonal anergy: persistence in tolerant mice of antigen-binding B lymphocytes incapable of responding to antigen or mitogen. Proc Natl Acad Sci USA 77:1602–1606

Nossal GJV, Szenberg A, Ada GL, Austin CM 1964a Single cell studies on 19S antibody formation. J Exp Med 119:485–502

Nossal GJV, Ada GL, Austin CM 1964b Antigens in immunity. IV. Cellular localization of ^{125}I- and ^{131}I-labelled flagella in lymph nodes. Aust J Exp Biol 42:311–330

Nossal GJV, Ada GL, Austin CM 1965 Antigens in immunity. IX. The antigen content of single antibody-forming cells. J Exp Med 121:945–954

Nossal GJV, Cunningham A, Mitchell GF and Miller JFAP 1968a Cell to cell interaction in the immune response. III. Chromosomal marker analysis of single antibody-forming cells in reconstituted, irradiated or thymectomized mice. J Exp Med 128:839–853

Nossal GJV, Abbot A, Mitchell J, Lummus Z 1968b Antigens in immunity. XV. Ultra-structural features of antigen capture in primary and secondary lymphoid follicles. J Exp Med 127:277–290

Nossal GJV, Karvelas M, Pulendran B 1993 Soluble antigen profoundly reduces memory B-cell numbers even when given after challenge immunization. Proc Natl Acad Sci USA 90:3088–3092

Pulendran B, Karvelas M, Nossal GJV 1994 A form of immunologic tolerance through impairment of germinal center development. Proc Natl Acad Sci USA 91:2639–2643

Pulendran B, Kannourakis G, Nouri S, Smith KCG, Nossal GJV 1995a Soluble antigen can cause enhanced apoptosis of germinal-centre B cells. Nature 375:331–334

Pulendran B, Smith KCG, Nossal GJV 1995b Soluble antigen can impede affinity maturation and the germinal center reaction but enhance extrafollicular immunoglobulin production. J Immunol 155:1141–1150
Ridderstad A, Nossal GJV, Tarlinton DM 1996 The *xid* mutation diminishes memory B cell generation, but does not affect somatic hypermutation and selection. J Immunol 157:3357–3365
Shokat KM, Harris AW, Goodnow CC 1995 Antigen-induced B cell death and elimination during germinal center immune responses. Nature 375:334–338
Smith KCG, Weiss U, Rajewsky K, Nossal GJV, Tarlinton DM 1994 Bcl-2 increases memory B cell recruitment but does not perturb selection in germinal centers. Immunity 1:803–818
Smith KCG, Nossal GJV, Tarlinton DM 1995 FAS is highly expressed in the germinal center but is not required for regulation of the B-cell response to antigen. Proc Natl Acad Sci USA 92:11628–11632
Smith KCG, Hewitson TD, Nossal GJV, Tarlinton DM 1996 The phenotype and fate of the antibody-forming cells of the splenic foci. Eur J Immunol 26:44–448
Strasser A 1995 Death of a T cell. Nature 373:385–386
Talmage DW 1957 Allergy and immunology. Annu Rev Med 8:239–256

DISCUSSION

Zinkernagel: What type of adjuvant did you use for these responses where you distinguished between affinity maturation versus non-affinity maturation?

Nossal: In these experiments we have used just antigen adsorbed to alum. The germinal centre architecture is far better if you don't use strong adjuvants such as *Bordetella pertussis*.

Zinkernagel: Do you have evidence suggesting that the affinity maturation of that change in usage of particular amino acids in the variable regions is dependent or independent of whether the antigen is stored solely on FDCs versus in other forms of antigen?

Nossal: We can answer that reasonably precisely. From earlier work we have examples of where there is virtually no antigen except on FDCs (Nossal & Ada 1971). We're quite convinced that the driver of these germinal centre events is the FDC. However, a cell has to enter the follicle first. For the cell to enter the follicle it must interact with a T cell, and that T cell would normally be stimulated by the interdigitating dendritic cells in the spleen. These are fairly early events, and I would suggest that the key selective cell in the germinal centre is the FDC.

If you add a lot of soluble antigen, as far as affinity maturation is concerned this makes things worse. This does not give the FDC-bound antigen the chance to test the very good cells from the only fairly good cells.

Melchers: When you give antigen in the middle of a response and you inhibit or somehow reduce the ratio of mutated receptors, are you simply saying that you have now increased the concentration of antigen so that the requirement for finding a good mutation is less stringent?

Nossal: I think that is part of it. The other part is that your high affinity cells are going to suffer selectively by this rapid apoptotic cell death encountered when germinal centre B cells meet soluble antigen. When we used highly deaggregated NP_1-HSA to do the killing, it only killed high affinity cells. When we went to NIP_{15}-HSA as a toleragen, we started to kill the low affinity cells as well, thus many more. There are two things going on. First, as you say, there is a lowering of the threshold for affinity maturation. Secondly, there is a selective killing by the action on the best-affinity cells — in this case the mutated cells.

Hodgkin: I have a question concerning the role of Fas in the antigen-induced killing. Have you looked for antigen-induced Fas expression in the germinal centre in the short term?

Nossal: We haven't done a lot of work on this, but it is clear that both the activated T cells and the activated B cells are Fas^+.

Tarlinton: Day 14 germinal centre B cells are certainly Fas^+ in the antigen-specific system.

Reference

Nossal GJV, Ada GL 1971 Antigens, lymphoid cells and the immune response. Academic Press, New York

Final discussion

Metcalf: I would like to consider real-life infections. Why does a deficiency in granulocytes have such a major impact on susceptibility to a range of bacterial and fungal infections? Does this tell us that immune responses — elegant as they might be — are relatively impotent?

Nossal: The adaptive immune system has two functions, the more important of which is to prevent reinfection. Its crucial function is to stop the species from being decimated by individuals catching a disease several times rather than only once. However, the immune system clearly also has a role in overcoming first-time infections. At the beginning of this century, Metchnikov showed that lobar pneumonia resolves when macrophages are helped by antibodies to become efficient at phagocytosis. The adaptive immune system isn't tremendously effective at squelching a primary infection from day one, but it has weapons such as interferon, IgM antibody and opsonization capability. I would argue, though, that this is a side benefit: its real function is to prevent reinfection.

Zinkernagel: In general I agree, but there are different levels at which this question can be addressed. For instance, knockout mice have shown that immunology is completely irrelevant or inefficient against most viral infections in the absence of interferon $\alpha\beta$. The same is true for interferon γ: many bacterial infections are so much worse without it that immunology cannot reverse the situation. I would argue that the vital function of the adaptive immune system is not to protect the individual from reinfection, but rather to give the mother the chance to build up an 'account' of relevant protection to transmit to the offspring.

Nossal: I agree that it is profoundly important to keep offspring alive while their immune system is so frail. Through the antibodies transmitted via the placenta, and to a lesser extent the milk, there is a six month period where the full experience of the mother in terms of all of the antigens that have been circulating in that community for the past 30 years are transmitted to the offspring, allowing it some time to mature its own immune system.

Mathis: When you say that the lack of granulocytes is so detrimental, is this a direct effect of the lack of granulocytes or might it be that their absence changes the quality of the immune response?

Metcalf: I don't think anyone has looked at the consequences for lymphocytes of a simple granulocyte or macrophage deficit. There are suitable models available for study, but no one has examined the behaviour of T or B cells in these models.

Nossal: What happens in children with Kostmann's syndrome?

Williams: They're pretty normal with respect to antibody responses and T and B cell numbers.

von Boehmer: You have only posed half the question: you say that if you have an immune system but no granulocytes the defence is poor: what happens if you have granulocytes but no immune system?

Nossal: Then you have, for example, agammaglobulinaemic children who are susceptible to every bacterial infection. One could make a good case for all of the eight white cells of the blood having a pretty important role. This takes us back to the argument about redundancy: there appear to be some redundancies in the immune system, but every player has got a role, and the system works best as an ensemble.

Shortman: There's an obvious link between the more primitive non-specific aspects of the defence system and the sophisticated response of, for example, T cells. It seems you can't initiate a primary T cell response without having a dendritic cell response. The dendritic cells appear to respond to fairly crude signals. Polly Matzinger is pushing this argument strongly. To get a dendritic cell to pick up antigen, go to the lymph node and start talking to T cells requires a set of signals, and this is an obvious point of interaction between the two systems.

The importance of quantitative studies on growth factor–receptor interactions: EGF and IL-4

Burgess: I wanted to show some data we have obtained using the biosensor. We have looked at a peptide interacting with an SH2 domain, a critical reaction in biology which occurs with reasonably high affinity. We load up the biosensor with unmodified or phosphorylated peptide, pass SH2 peptide domains over the chip and then watch the off-rate. In the absence of competing peptide, the off-rate is about $10^{-3} s^{-1}$, which is well within the capability of the instrument to measure. However, that off-rate is actually biased by reassociation; to measure the actual off rate you have to extrapolate almost to zero time, which is subject to considerable error. If we add competing peptide the off-rate is much faster, about $10^{-2} s^{-1}$. This is at the limit of our ability to measure. I'll illustrate that first concept from our studies in which we have looked at soluble epidermal growth factor (EGF) receptor passing over a blank on an EGF chip. The blank chip shows a refractive index problem. In the presence of a competing peptide, you can compete most of the specific binding to the EGF on the chip and the off-rate is not very different from the rate of the refractive index change. Again, we are limited by sensitivity.

For his PhD, Ian Davis studied a mutant of interleukin (IL)-4 with one amino acid change at position 128 (tyrosine to aspartic acid), which seemed to be almost identical in structure to wild-type IL-4. The affinity of binding of that molecule to its human receptor is identical to normal IL-4, but this an antagonist and not an activator of the murine IL-4 receptor. Interestingly, if you take the human IL-4 receptor and put it into mouse cells, this mutant IL-4 is a perfectly good agonist. This does not have to do with the way this triggers the IL-4 receptor, but rather a difference in the way the mutant IL-4 interacts with the other chains in the receptor complex.

IL-4 receptors are found on other cell types, including colonic epithelial cells. Ian Davis spent three years of his PhD wondering why these receptors are there—he couldn't find any biological effect of IL-4 on colonic cells. Eventually he and Normand Pouliot found that IL-4 is probably the most potent spreading factor for colonic cells.

Finally, I want to address this issue of redundancy versus complexity. We are dealing with extraordinarily complex systems. EGF represents a good example of this complexity. The paradigm built up over about 12 years was that EGF bound to its receptor, caused dimerization and then there was activation of the receptor kinase. In fact it is not quite as simple as that. We no longer believe that the EGF receptor dimer is symmetrical; only one of the kinase domains appears to be activated. Towards the end of the 1980s it was discovered that there are actually four fairly closely related EGF receptors. We knew that EGF bound to ErbB1 but it didn't seem to react with the other EGF receptor family members, until it was realized that all four receptors form heterodimers, and each of the dimers has different binding properties not only for EGF but also for the other EGF-like ligands. We don't know the biological consequences of ligand binding to heterodimers, although it is known that the EGF receptor (ErbB1) can signal as a heterodimer with some of the other receptors. There's now a lot of evidence building up in the cytokine systems that dimers are probably not the active form of most receptors; instead higher order oligomer forms deliver the signals.

Therefore, with the range of receptors and regulators we are dealing with we are not talking about redundancy, but complexity.

Davis: We've shown convincingly that with the biosensor we can measure an off-rate in the order of 1–3 sec^{-1} and van der Merwe et al (1994) have published a variant of the standard techniques where they claim to measure an off rate of 4 sec^{-1}. I therefore think it is possible to get a bit more sensitivity than you have shown; part of the problem might have been to do with the concentrations of SH2 you used. What were these?

Burgess: In the order of 10–50 nM. Major problems are caused by ion exchange effects on the biochip matrix and there is reassociation at the concentrations we're dealing with here. At high concentrations there are problems with diffusion-control re-binding and interpretation of the off-rate then becomes even more difficult. Consequently, you may calculate a number, but whether that's the real off-rate becomes difficult to tell.

Davis: We may also be helped by having an abysmally slow on-rate so we can control the re-binding simply by adjusting the flow. However, with a robust affinity such as yours, that may not be enough.

Nossal: I was reminded as you spoke of the time that Howard and Renée Dintzis were with me, and we were struggling with their concept of the 'immunon' (Dintzis et al 1976). This predicted that immune triggering required an assemblage of about 20 receptors clustering together. Does anyone else think that more complex assemblies than just dimerization are involved in triggering?

Nicola: One of the things we have learned for almost all of the receptors is that even the ones that traditionally have been thought to act as monomers don't. The logical reason for this is that it is difficult to transmit a conformational change with a binding event that occurs on one side of the membrane to some sort of conformational change inside the membrane, through what's usually a single rigid helix in the membrane. The easiest way to solve that problem is to translate a conformational change on the outside of the cell which leads to an aggregation inside the cell. Looking at something like the growth hormone X-ray structure it appears that two receptor chains may be enough, but other chains may become associated with that. I suspect that things will end up pretty much like the T cell receptor with about 15–16 associated chains that are probably required for a full biological response. This also tallies with the sorts of arguments we have been hearing about the off-rate, which for a long time I have thought to be the most important component of a binding interaction leading to a biological response, because it determines how long you can hold the complex together and therefore maintain the signalling machinery. What we are learning about cytokine receptors and what I've been hearing about the ways of stimulating different immune cells may have many parallels.

Metcalf: On an average haemopoietic cell there might only be 200–300 haemopoietic regulator receptors. We have heard in a number of the systems how there is promiscuity with regard to subunit sharing, and also how rapid the on-rates are for complex formation. For this to be able to happen demands some spatial clustering of receptor chains of different types. We know that these receptor chains are not pre-associated, but they're available for rapid association. I wonder what mechanism achieves such spatial clustering. Is there a gene coding for a membrane region that allows this to happen? We tend to think of receptors as drifting randomly in a sea of lipoprotein but I don't think this can be correct.

Mosmann: The NK 1.1^+ T cells are clearly a potentially potent source of IL-4. What is the current knowledge of their specificity? If they have a restricted T cell receptor repertoire, this suggests they have some very particular specificity.

Paul: The current view is that these cells may use CD1 as their target. The evidence for this is that hybridomas derived from these cells can be stimulated to secrete cytokines when they're confronted with CD1-transfected cells. There are many complexities, though. Firstly, I understand from the modelling of CD1 that it is a possibility that CD1 might be stable without a peptide within it. On the other hand, it is clear that there are cells that express CD1 which don't stimulate T cells that appear to be CD1 specific, whereas other cells that express CD1 are good stimulants. This could be due to loading of CD1 with different ligands in different cell types. I would say we are far from understanding the details of this specificity; translating those details into physiological significance is hard. If our thesis is correct that the activation of these cells early in immune responses provides a bolus of cytokine that acts to determine the decision that the cell will make, the question arises as to how that production is mediated. Is the pathogen causing CD1 to be expressed? Is a peptide from the pathogen binding to the CD1 site?

Mathis: I never really understood how you see the whole thing getting started, because taking into account the different ways one can make a Th2 response, do you think all these factors are able to stimulate that very restricted T cell population?

Paul: Presumably, that question concerns the extent to which I believe that the $NK1.1^+$ $CD4^+$ cell is responsible for biasing responses to be Th2-like. The answer is that I don't think it is the entire story. For example, in β_2-microglobulin knockouts it is clear that certain responses are deficient but others are not. If you look carefully, the β_2-microglobulin knockouts have about 10% of the normal level of $NK1.1^+$ $CD4^+$ cells. On the other hand, basophils are a terrific source of IL-4. I've been struggling for years to understand what would stimulate them physiologically to make IL-4 at the outset of responses, since the best stimulus is cross-linking of $Fc_{\varepsilon}RI$. That seems an unlikely stimulant since it would require pre-existing IgE.

Mosmann: It is also possible that if there is a peptide involved, what you're looking at is a self-distress signal, rather than a common pathogen antigen. This could be triggered by a variety of non-antigen-specific stimuli.

Paul: Although you could argue that pathogens would have some peptide that would bind and allow expression, the alternative is that you may simply have an interaction with some T cells, causing the production of a factor that up-regulates CD1 expression.

References

Dinitzis HM, Dinitzis RZ, Vogelstein B 1976 Molecular determinants of immunogenicity: the immunon model of immune response. Proc Natl Acad Sci USA 73:3671–3675

van der Merwe A, Barclay AN, Mason DW et al 1994 Human cell adhesion molecule CD2 binds CD58 (LFA-3) with a very low affinity and an extremely fast dissociation rate, but does not bind CD48 or CD59. Biochemistry 33:10149–10160

Summing-up

Sir Gustav Nossal

The Walter and Eliza Hall Institute of Medical Research, Post Office, Royal Melbourne Hospital, Victoria 3050, Australia

Have we succeeded in finding common ground between the three elements of nature's defence system we have been considering, namely macrophages and granulocytes, cell-mediated immunity and antibody formation?

I think we have been partially successful. There are four areas in which we have found a good deal of common ground. We are all absolutely committed to understanding ligands reaching a cell, cell surface receptors for them, and the downstream signalling system. This is clearly one of the huge unifying thrusts of biology. We have within each of the three areas a parallel commitment to understanding cellular lineages, subsets of lineages and differentiation branch points. Again, this sets us in the mainstream of one of the large puzzles of biology: we are occupying one subset of molecular embryology in seeking to understand this. All teams have been puzzled by conservatism—on the one hand the pleiotropism of action of growth factors discovered in one system and found to have profound effects in a completely different bodily system, and on the other hand the apparent similarity of different factors which in a variety of biological assays sometimes can hardly be distinguished from one another. We have also demonstrated how avid we are to transfer these reductionistic insights into real-life situations and to therapeutic possibilities.

Of course, we have also found dissimilarities between the three fields. The essence of science is such that there will always be topics which only interest a tiny in-group. In this respect we have been three in-groups, although we have done reasonably well in at least transferring the fascination of some of these particularities among one another.

Nevertheless, I think we have also partially failed. There is something random and chaotic about the generator of diversity in the immune system. That the receptor in this particular case doesn't know what it is going to have to recognize next creates complexities and possibilities for contradictory results from different models which the non-immunologists find particularly and peculiarly taxing. I wish it were not so. I don't know of any other bodily system where the recognition unit has an affinity differential of a million-fold. This means almost by definition that there are ever so many shades of grey in the study of immune induction or immunological tolerance. That must seem foolishly imprecise to someone dealing with receptors that the aeons of evolution have fashioned to have an affinity at the nanomolar level. Because of the fact that it is not possible for non-immunologists to worm their way into the particular models which individual scientists and small groups have developed to huge degrees of sophistication, the non-immunologist is perturbed by the apparent contradictory conclusions, for instance in tolerance. When such a discrepancy is looked at closely, it turns out that each investigator has tweaked the system in a slightly different way and they're almost always both right. Doubtless, if I were a haematologist I would now tell you about the fascinations and particularities of that system.

We've succeeded beyond my hopes, but we've not succeeded perfectly. We've addressed with conviction the task set for us in debating the molecular basis of cellular defence mechanisms and I think we will find the book resulting from the symposium to be of lasting value and fascination.

Index of contributors

Non-participating co-authors are indicated by asterisks. Entries in bold type indicate papers; other entries refer to discussion contributions.

Indexes compiled by Liza Weinkove

Subject index